DATE DUE			

Interpretation of X-Ray Powder Diffraction Patterns

Interpretation of X-Ray Powder Diffraction Patterns

H. Lipson D.SC. F.R.S.
Professor of Physics

H. Steeple PH.D.
Lecturer in Physics
University of Manchester Institute of Science and Technology

Macmillan · London
St Martin's Press · New York

© *H. Lipson and H. Steeple* 1970

First Published in 1970

Library of Congress Catalog Card Number
75–107906

Published by
MACMILLAN AND CO LTD
London and Basingstoke
Associated Companies in New York, Toronto
Melbourne, Johannesburg and Madras

Printed in Great Britain by
WILLIAM CLOWES AND SONS LTD
London and Beccles

Preface

The decision to separate the book by Henry, Lipson and Wooster into three parts has been taken in the light of the considerable growth in the subject of X-ray diffraction in recent years. It is now hardly practicable to produce an undergraduate textbook covering all branches of the subject, and this present book aims at dealing only with powder methods. Opportunity has been taken to bring certain parts up to date and to expand other sections—which in the earlier work were purely descriptive—into more logical and satisfying sequences.

It is true that powder methods are of less fundamental significance than single-crystal methods. Nevertheless they are of much more general importance, and we believe that no X-ray crystallographer should feel that his education is complete if he is not aware of the information that powder methods can give. Many results—the structures of many of the metallic elements, for example—were first obtained from powder photographs, and the method still provides a first attack on a new problem, even if only to decide whether the subject of interest is crystalline or not.

We have aimed to make the book complete in itself. It would have been tempting to rely on the other volumes for introductions to topics such as symmetry and the reciprocal lattice. But we felt that the impact would be greater if the student could refer to earlier chapters in the book for the basis of the later material and we have therefore included an outline of some of the topics that our other colleagues hope to write about in companion books.

The problems at the end of the book form a new feature. We hope that these will provide a means for students to test the efficiency with which the various procedures have been presented.

We wish to thank various people who have helped with the production of this book. Miss M. W. Allen, Miss V. A. Flinn and Mrs. E. B. Midgley have typed the script, and Mrs. Judith Cohen (née Lipson) has given considerable assistance with the production of suitable problems. Mr. M. Rechowicz read through the final draft and, with his aid, we hope that any inconsistencies will have been eliminated.

H. LIPSON
H. STEEPLE

August, 1968

Contents

CHAPTER 1

Crystal Lattices and Crystal Symmetry

1.1 Nature of Crystals

1.1.1 The crystalline state

This book is concerned with the study of polycrystalline materials, but to understand these we must begin by considering the nature of single crystals.

A crystalline solid can be formed from the liquid phase in such a way that it is bounded entirely by naturally occurring plane faces. The facility with which such a crystal can be grown varies from substance to substance, but external perfection of shape is rarely achieved even with tractable substances and stringent conditions. Crystals grown from the same solution usually show considerable variation in shape and sometimes the differences are so great that it is not certain, from a study of the shapes alone, that only one type of crystal is being deposited.

Characteristics other than crystal shape must therefore be considered. One of the first discoveries in crystallography was that the angles between the faces are invariant for a given crystalline material so that the measurement of these angles is sufficient to characterize the material. This constancy of angle is now known to arise from the orderly arrangement of atoms or groups of atoms within the crystal. These groups are repeated by **translation** in the same orientation at regular intervals in three dimensions. A two-dimensional example is shown in fig. 1.1. The effect of the translations is to produce, within the crystal, **lattice planes** which are more or less densely populated with these groups of atoms. Crystal faces grow parallel to densely populated planes and the greater the density the more important morphologically is the associated face; hence the external shape of the perfect crystal is closely related to that of the internal unit of pattern. Because the angles between adjacent planes of atoms do not change, the angles

1*

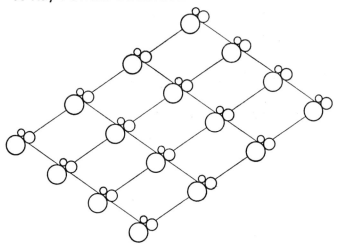

Fig. 1.1   Unit of pattern repeated by regular translations in two dimensions

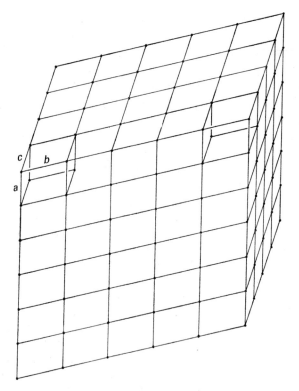

Fig. 1.2   Construction of a three-dimensional lattice by repeated translations
of a unit cell

between the corresponding faces will be constant, irrespective of the extent of development of the faces.

We can consider the crystal as a whole in terms of the normals drawn from a point to each of the faces. The normals will always have the same orientation relative to one another and, irrespective of the shapes of the faces, this group of lines will display the full characteristics of the perfect crystal.

If the unit of pattern is represented by a point, then the regular repetition by translation generates a three-dimensional array of points called a **crystal lattice** which can be imagined to extend indefinitely in all directions in space. The distances between the points in any direction are proportional to the repeat distances for the units of pattern in the same direction in the crystal itself. A particular lattice can be specified by drawing from one of the lattice points as origin non-coplanar vectors to three neighbouring points. The lattice is then specified by the lengths a, b and c of these vectors and by the angles α, β and γ between b and c, c and a, and a and b respectively. The parallelepiped defined by the vectors a, b and c is called the **unit cell** of the lattice and the repetition of this cell by repeated translations a, b and c, in the directions of a, b and c comprises the lattice (fig. 1.2).

1.1.2 Law of Rational indices

If straight lines are drawn through regular two-dimensional point lattice the points on the line will, by definition of a lattice, be evenly spaced. Each of these lines will be one of a family of parallel equidistant lines and the linear density of the points on each will be related to the line spacing; the wider the spacing the denser are the points (fig. 1.3). Further, if any family is extended throughout the lattice every lattice point will lie on one or other of the lines comprising the family. Thus the points at the corners of a unit cell will lie on two or more of the lines of any family which we might choose to construct, and if the spacing is sufficiently close there will be other lines between these which will necessarily cut the edges of the unit cell in an integral number of equal intercepts; the number need not, of course, be the same for each edge (fig. 1.3).

The same considerations apply to a three-dimensional lattice, but now parallel equidistant planes will be drawn through the lattice points. Again, provided the interplanar spacing is sufficiently close, any one set of planes will make integral numbers of intercepts, say h, k and l, on the unit-cell edges a, b and c respectively. The indices (hkl), enclosed in brackets, define the family. The intercepts on the edges a, b and c will be a/h, b/k and c/l respectively; if the intercept is of length a (or b or c) then the value of h (or k or l) will be 1, if it is one half of the unit-cell edge then the corresponding index will be 2, and so on. Fig. 1.4 shows a few of the family of (342) planes.

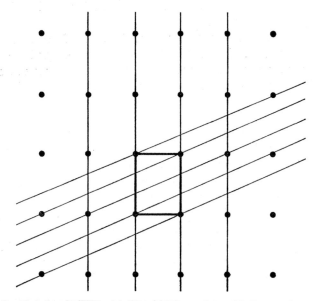

Fig. 1.3　Variation in linear density of lattice points with the spacing of lines drawn in the lattice. The intercepts on the cell edges are also shown

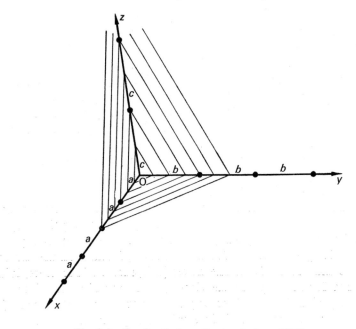

Fig. 1.4　Family of planes having indices (342)

If a family of planes happens to be parallel to one axis then the intercept on that axis will be infinite and the corresponding index will be zero. Intercepts on the negative direction of a unit-cell vector give rise to negative indices, written as $\bar{h}$, $\bar{k}$ or $\bar{l}$.

Since crystal faces grow parallel to planes which are associated with large numbers of atoms it follows that the faces will be parallel to the widely spaced planes which, because they are widely spaced, will have low (hkl) values. The indices of crystal faces, as opposed to those of planes within the crystal, are defined from morphological considerations and with reference to axes which may or may not be parallel to the unit-cell vectors chosen for the internal structure, but nevertheless, these indices are also always integral and low in value (Phillips 1963). These results are embodied in a rule known as the **law of rational indices** and the indices (hkl), whether applied to planes or faces, are called **Miller indices**. Often in the hexagonal system (section 1.2.3) the indices are referred to four axes x, y, u and z (1.2.4) and are expressed in the form $(hkil)$ where $h + k + i = 0$. This modification was introduced by Bravais and consequently indices of this kind are named **Miller–Bravais** indices.

1.1.3 Symmetry

It is often found that the face normals of a crystal do not merely satisfy the law of rational indices but have extra relationships as well. For example, they may occur generally in pairs pointing in opposite directions, indicating that the crystal has opposite faces parallel to each other. They may be inclined at specific angles, such as 90° or 120° to one other. These relationships indicate crystal symmetry.

A body is said to be symmetrical if it can be divided into exactly similar parts that can be brought into coincidence by some sort of movement, other than mere translation. Continued repetition of this movement will lead eventually, in a small number of steps, to the return of the body to its original disposition. For example, a cube rotated about an axis through its centre parallel to an edge will appear to be the same after rotation through 90°, and will be in its original disposition after four such operations.

The cube is therefore said to have a four-fold axis, and this is one example of a **symmetry element**. The cube has, however, many other symmetry elements and we shall give a more general discussion of the possibilities in section 1.2.1.

1.1.4 Zones and zone axes

A set of external crystal faces, or a set of internal lattice planes, which intersect in parallel edges is called a **zone**, and the common direction of the edges in space is called the **zone axis**. With reference to the internal structure, the planes of a zone need not have the same spacing and they need not all belong to the same form (1.2.5). Indeed, the only feature which is common to all the planes of a zone is the zone axis, and it is therefore the indices of this axis which are used to specify the zone.

In terms of the crystal lattice, these indices are defined as follows. The zone axis is a direction in the lattice and is therefore parallel to some row of points in the lattice which for the purposes of the definition may be taken as that which passes through the origin in the direction of the axis. The first point out from the origin along this row will have coordinates, referred to the crystallographic axes, which will be integral multiples, say ua, vb, wc, of the three vectors defining the unit cell (fig. 1.5), and the zone-

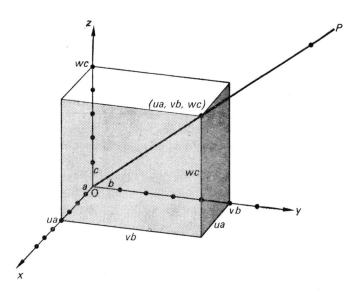

Fig. 1.5   Coordinates (ua, vb, wc) of the first lattice point from the origin along the zone-axis row denoted by OP

axis indices are then defined as $[uvw]$. When a set of zone axes is related by symmetry, the resulting form of similar directions is written $\langle uvw \rangle$.

The relationship between h, k, l and u, v, w can be determined by considering the general equation

$$Ax + By + Cz + D = 0$$

for a plane (an expression which holds for rectangular or oblique axes). It is clear that the intercepts made by this plane on the axes x, y and z are $-D/A$, $-D/B$ and $-D/C$. But the first plane from the origin of the family

(hkl) makes intercepts on the axes equal to a/h, b/k and c/l, and it follows that

$$A = -Dh/a, \quad B = -Dk/b \quad \text{and} \quad C = -Dl/c$$

for this particular (hkl) plane. The direction ratios of the normal to the plane are thus

$$-Dh/a : -Dk/b : -Dl/c$$

and since the direction ratios of the line joining $(0, 0, 0)$ to (ua, vb, wc) are

$$ua : vb : wc$$

then these two lines are at right angles if

$$-\frac{Dh}{a}.ua - \frac{Dk}{b}.vb - \frac{Dl}{c}.wc = 0$$

that is, if $hu + kv + wc = 0$

This is also the condition for the line joining the origin to (ua, vb, wc) to be parallel to the plane (hkl), that is for $[uvw]$ to be the zone axis of $\{hkl\}$.

For two planes $(h_1k_1l_1)$ and $(h_2k_2l_2)$ of the same zone

$$h_1u + k_1v + l_1c = 0$$
$$h_2u + k_2v + l_2c = 0$$

since both planes must be parallel to the same zone axis. From these relations,

$$u = k_1l_2 - l_1k_2$$
$$v = l_1h_2 - h_1l_2$$
and $$w = h_1k_2 - k_1h_2$$

1.1.5   The stereographic projection

All the information about the external symmetry of a crystal can conveniently be displayed by stereographic projection. The projection is constructed by imagining the crystal to enclose the centre of a sphere and the crystal in its turn to be enclosed by the surface of the sphere. Normals are drawn from the centre to the crystal faces and are then produced to cut the sphere in what are termed the **spherical poles** of the corresponding faces (fig 1.6(a)). From each spherical pole in the upper, or northern, hemisphere a line is drawn to the south pole of the sphere to cut the equatorial plane in the corresponding stereographic pole (fig. 1.6(b)). A face of a crystal can thus be represented by its pole in the equatorial plane—the plane of stereographic projection. If we draw lines from the south pole to the spherical poles in the lower, or southern, hemisphere, then the resultant stereographic poles will be outside the circle, called the **primitive circle**, in which the sphere is cut by the equatorial plane. It is, however, usual to confine the complete projection within the primitive circle and to do this it is necessary to construct the projection of the poles on the southern hemisphere by drawing lines to these poles from the north pole. To distinguish between the two sets, stereographic poles projected from the northern hemisphere are shown as dots while those projected from the southern

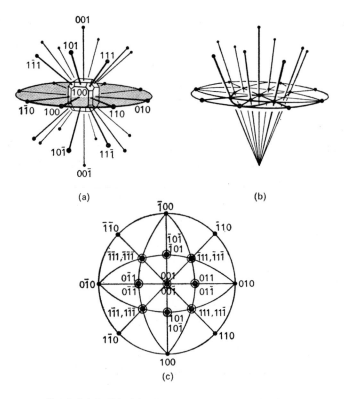

Fig. 1.6 (a), (b), (c)   The stereographic projection

hemisphere are shown as small open circles. The centre of the projection sphere is also the centre of the primitive circle on which the stereographic poles lie. If the crystal has a centre of symmetry (1.2.1) then every face will have a congruent face parallel to it and equidistant from, but on opposite sides of, this centre of symmetry, so that each stereographic pole represented by a dot will be paired with one represented by an open circle. The two poles will lie on a diameter of the primitive circle and will be equidistant from, but on opposite sides of, the centre.

A **great circle** on the surface of the reference sphere is produced by the intersection with the sphere of a plane which passes through its centre; if a plane does not pass through the centre but still cuts the sphere, it will do so in a **small circle**. The stereographic projection of a great circle, provided only one projection point is used, is also a circle, but if the two projection points are used—one for each hemisphere—then the great circle projects as two circular arcs of equal radii (fig. 1.7(a)). Similarly, a small circle will

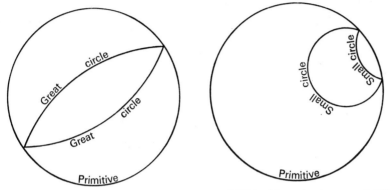

Fig. 1.7(a)   Two circular arcs of equal radii obtained when a great circle is projected from the north and the south poles on to the primitive

Fig. 1.7(b)   Two circular arcs of different radii obtained when a small circle lying in both hemispheres is projected from the north and the south poles on to the primitive

project as a circle if only one projection point is used, but if it lies in both hemispheres and two projection points are used then the stereogram will in general be two circular arcs of different radii (fig. 1.7(b)). In the special case where the circumference of the primitive circle lies on the diameter of the small circle, the two projected circular arcs will have the same radii and will lie one above the other.

One further property of the stereographic projection is that it preserves angular truth so that the angle between the projections of great circles is equal to the angle between the circles themselves. This is useful in representing zones, for a zone is a set of faces (or planes) which intersect in parallel edges (1.1.4) and hence the spherical poles of the faces of a zone all lie on the same great circle. Correspondingly, the stereographic poles will lie either on an arc of a circle or, if the great circle passes through the projection point, on the diameter of the primitive. The projection plane, and hence the primitive circle, is chosen to be parallel to an important face, say the (001) face of the cubic crystal shown in fig. 1.6(a). In this particular example both the spherical and the stereographic poles of the zone which has faces parallel to [001]—that is faces with $l = 0$—will lie on the primitive circle (fig. 1.6(c)); these poles can be plotted by measuring the angles be-tween the zone faces. In fig. 1.6(c) the directions of the axes are conven-tional with $+z$ upwards from the plane of the paper, $+y$ from left to right and $+x$ downwards.

For the spherical poles which do not lie on the primitive circle a vertical section is taken through the upper and lower projection points and the appropriate spherical pole. Such a section is drawn in the plane of the paper in fig. 1.8. The known angle θ between the normal to the face under con-sideration and the normal to the (001) face is marked on the great circle and

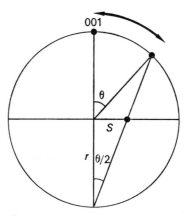

Fig. 1.8   Determination of the distance, S, of a stereographic pole from the centre of the primitive when the spherical pole does not lie on the primitive circle

the distance S is determined. This distance, equal to $r \tan \theta/2$ where r is the radius of the projection sphere, is that of the required pole from the centre of the primitive.

When the pattern is complete the symmetry and the orientation of the crystal can be established.

1.2 Symmetry of Crystals

1.2.1 Types of symmetry

We have so far discussed the external shapes of crystals in general terms only. If we look more closely into the arrangement of face normals of crystals it will be seen that, in all, only three types of symmetry are possible. These involve inversion through a point, reflection across a plane and rotation about an axis, and are examples of the symmetry elements dis-cussed in section 1.1.3. A point of inversion, or **centre of symmetry**, is present when for each face of the crystal there is a parallel, congruent face equidistant from, but on the opposite side of, this point. A **reflection** or **symmetry plane** divides the crystal into halves which, by reflection across the plane, are mirror images of each other. Finally, a **rotation** or **symmetry axis** brings each face of the crystal into coincidence with a similar, but not necessarily parallel face by a rotation of $2\pi/n$ about an axis; such an axis is designated n-fold. From the external or morphological aspect the crystal can have only one centre of symmetry, but it can have more than one symmetry plane and more than one rotation axis. However, the types of axis are limited by the geometry of the internal structure of the crystal to 1-, 2-, 3-, 4- and 6-fold. The last four are commonly called **diad**, **triad**, **tetrad** and **hexad** axes respectively.

1.2.2 Combinations of symmetry elements

The symmetry elements may, according to the nature of the crystal, appear in a limited number of combinations. The introduction of what are termed

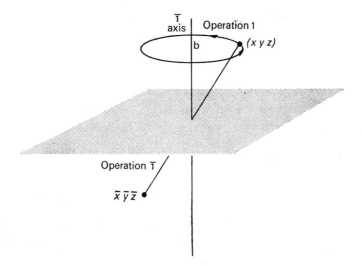

Fig. 1.9   Equivalence of $\bar{1}$ axis and centre of symmetry

inversion axes, or rotary inversion axes, has made it unnecessary to use the centre of symmetry as an independent symmetry element and the latter is no longer incorporated as such in the description of the symmetry of a crystal. Inversion axes are compound elements consisting of an axis of rotation combined with inversion across a point on the axis. The point itself is not necessarily a centre of symmetry. The axes are denoted by the symbols $\bar{1}, \bar{2}, \bar{3}, \bar{4}$ and $\bar{6}$; a centre of symmetry is equivalent to a $\bar{1}$ axis (fig. 1.9).

12   X-Ray Powder Diffraction

It is also possible to dispense with the symmetry plane (in fig. 1.10 it can be seen that it is equivalent to a $\bar{2}$ axis) in the description and classification of the external symmetries of all crystals and to use rotation and inverse rotation axes alone. In practice, however, groups of symmetry elements are

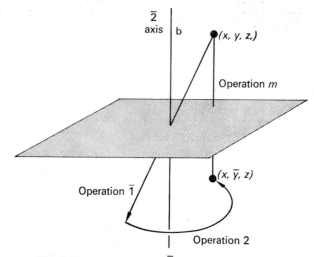

Fig. 1.10   Equivalence of $\bar{2}$ axis and mirror plane

not always unique and for convenience symmetry planes are, in fact, mentioned when they exist.

1.2.3   Crystal systems

By considering the limited number of different combinations of symmetry elements which occur in the description of the external symmetry of a perfect specimen, crystals are grouped into what are known as **crystal systems**. The seven possible systems are called **triclinic, monoclinic, orthorhombic, trigonal, hexagonal, tetragonal** and **cubic** and each is characterized by the number and type of symmetry axes present as shown in the following table.

System	Essential symmetry	Symbols
Triclinic	one-fold axis	1 or $\bar{1}$
Monoclinic	two-fold axis	2 or $\bar{2}$
Orthorhombic	three perpendicular two-fold axes	222, $\bar{2}\bar{2}\bar{2}$, $\bar{2}\bar{2}\bar{2}$
Trigonal	three-fold axis	3, $\bar{3}$
Tetragonal	four-fold axis	4, $\bar{4}$
Hexagonal	six-fold axis	6, $\bar{6}$
Cubic	four three-fold axes parallel to the diagonals of the cube	23, $\bar{2}3$

In all but the cubic system the essential symmetry axes may be the only ones present. These axes also represent the minimum external symmetry which must be displayed by a perfect crystal in order that the crystal should belong to the appropriate system. This does not apply to crystals of cubic symmetry because the presence of the characteristic four triad axes implies the existence of three mutually perpendicular diad axes; the combination of four triad and three diad axes therefore gives the minimum symmetry which can be exhibited by a cubic crystal. Just as there is a minimum symmetry required for each system, there is a maximum to the number and type of symmetry elements which can be present in each system. A crystal showing the full symmetry of the cubic system, for example, has three tetrad axes with reflection planes perpendicular to each, four triad axes, six diad axes with reflection planes perpendicular to each, and a centre of symmetry.

1.2.4   Crystal lattices

By (1.1.1) a crystal or space lattice has centres of symmetry; that is, it is **centrosymmetric**. Also, the environment of every point is exactly the same. When the vectors defining the lattice and its unit cell are the three shortest that can be drawn from the origin then the lattice is a primitive lattice and the unit cell is primitive, having lattice points at each of the eight corners only. In the extended crystal lattice each point at the corner of a given cell is shared by seven other cells so that associated with each primitive cell there is just one lattice point and, consequently, just one unit of pattern. As might be expected, because of the close relationship between the external symmetry of a crystal and the internal atomic structure, all possible space lattices defined by these primitive cells can be classified as triclinic, monoclinic, orthorhombic, trigonal, hexagonal, tetragonal or cubic, according to the type of symmetry they display. In order to qualify as a particular type, the lattice need have no more than the appropriate characteristic symmetry and, of course, any further symmetry elements that this may automatically generate. Thus a cubic lattice need have no more than the characteristic four triad axes and the associated three diad axes which are the minimum requirements for cubic symmetry.

Space lattices, however, are usually specified so that they display the full symmetry of the system to which they correspond; thus the unit cell of a cubic lattice is so chosen that the lattice, as defined by this cell, possesses maximum cubic symmetry (1.2.3). If the primitive unit cell produces by translations a lattice which does not display the maximum symmetry of the system in question, then an alternative unit cell which does so is selected. The new lattice may not be primitive and the new unit cell may then have more than one lattice point corresponding to more than one unit of pattern.

For example, it may be that a certain cubic lattice has been defined in terms of a primitive unit cell which has its three non-coplanar vectors, drawn from the origin, equal in length but inclined to each other at an angle of 60°. A larger unit cell can be chosen with equal non-coplanar vectors at right angles to each other as edges. Such a cell would have lattice points at the centres of each face (fig. 1.11) but it defines a lattice which will display the full symmetry of the cubic system whereas that defined by the primitive cell will not.

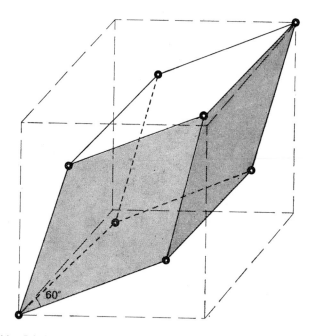

Fig. 1.11   Primitive cell (full lines) and alternative face-centred cell (broken lines) having full cubic symmetry

Bravais established that, in all, there are fourteen unique lattices which conform to the full symmetry requirements of the seven systems. These lattices are listed below, together with the angular and linear relations of the unit cells by which they are specified.

Because of the symmetry requirements of the appropriate system, the lengths shown equal are identically equal, and so are the angles. A tetragonal lattice may, for example, have $a = b = c$ accidentally, but it does not have the four triad axes which are characteristic of a cube and does not, therefore, have cubic symmetry.

The Fourteen Bravais Lattices

Symmetry	Linear relations	Angular relations	No. in system	Positions of lattice points	No. of points in unit cell	Lattice symbol
Cubic	$a = b = c$	$\alpha = \beta = \gamma = 90°$	3	Points at corners	1	P
				Points at corners and one at centre	2	I
				Points at corners and at middle of all faces	4	F
Tetragonal	$a = b \neq c$	$\alpha = \beta = \gamma = 90°$	2	Points at corners	1	P
				Points at corners and one at centre	2	I
Hexagonal	$a = b \neq c$	$\alpha = \beta = 90°$ $\gamma = 120°$	1	Points at corners	1	P
Trigonal	$a = b = c$	$\alpha = \beta = \gamma \neq 90°$	1	Points at corners	1	R
Orthorhombic	$a \neq b \neq c$	$\alpha = \beta = \gamma = 90°$	4	Points at corners	1	P
				Points at corners and one at centre	2	I
				Points at corners and points at middle of A (100) or B (010) or C (001) faces	2	A B C
				Points at corners and at middle of all faces	4	F
Monoclinic	$a \neq b \neq c$	$\alpha = \gamma = 90°$ $\beta \neq 90°$	2	Points at corners	1	P
				Points at corners and at middle of A (100) or B (010) or C (001) faces	2	A B C
Triclinic	$a \neq b \neq c$	$\alpha \neq \beta \neq \gamma$	1	Points at corners	1	P

Of the fourteen Bravais lattices seven are primitive (usually denoted by P), one to each crystal system, and seven are centred, two of which are cubic, one is tetragonal, three are orthorhombic and one is monoclinic. When the unit cell contains a lattice point at the intersection of the body diagonals it has two lattice points in all and the lattice is body-centered (I). A cell which has two of its opposite faces centred also has two lattice points and the corresponding lattice symbol is A, B or C according to whether the centred faces belong to the (100), (010) or (001) planes respectively. A lattice which is defined by a unit cell with all its faces centred is called a face-centred (F) lattice and there are four lattice points in the cell. The hexagonal lattice, which is primitive, was at one time labelled C; that exhibiting trigonal symmetry is based on a rhombohedral cell and, although primitive, is always denoted by R.

The vectors *a, b* and *c* conventionally define not only the unit cell and the space lattice, but also the origin and the direction of the crystallographic axes x, y and z. Examination of the unit cells of the Bravais lattices (fig. 1.12) shows that the environment of every lattice point is the same and that rotational symmetry of the lattice is limited to 1-, 2-, 3-, 4- and 6-fold. This restriction is imposed by the simultaneous rotational and translational symmetry requirements for an extended lattice. That is, the unit cells must produce by translation a continuous network and must therefore leave no gaps when translated in the directions of the crystallographic axes. In the hexagonal system it would appear that a suitable choice of unit cell would be a hexagonal prism (fig. 1.13(a)), but a smaller cell based on a 60°-angle rhombus is preferred (fig. 1.13(b)) although this cell, considered as a geometrical figure, does not display a six-fold axis. This latter cell is primitive (P). It is often convenient, so as not to obscure the symmetry, to refer the primitive lattice to three equal coplanar axes x, y and u (fig. 1.13(b)) spaced at angular intervals of 120° and at right angles to the fourth axis, z. Previously the lattice was called C because hexagonal structures so defined (by reference to the C face-centred cell shown in fig. 1.13c) have certain characteristics which appear similar to those of structures based on C face-centred lattices in other systems.

1.2.5   Crystal classes

Reference has been made (1.1.1) to the fact that the unit of pattern associated with each lattice point may itself possess symmetry and that this symmetry is closely related to that of the crystal as a whole. There are only thirty-two different combinations of the external symmetry elements of crystals; this total includes that of the asymmetric crystal which, except for a one-fold axis, possesses none of the elements of reflection across a plane, rotation about an axis, or inverse rotation about an axis which are

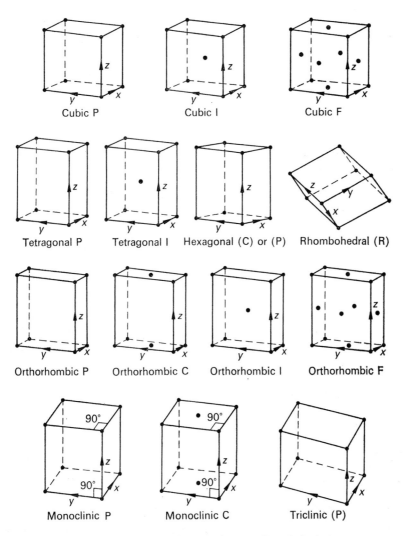

Cubic P   Cubic I   Cubic F

Tetragonal P   Tetragonal I   Hexagonal (C) or (P)   Rhombohedral (R)

Orthorhombic P   Orthorhombic C   Orthorhombic I   Orthorhombic F

Monoclinic P   Monoclinic C   Triclinic (P)

Fig. 1.12   The unit cells of the fourteen Bravais lattices

customarily used for the purposes of classification. These thirty-two sets of symmetry elements are called **point groups,** because all the elements of any one group pass through one point which is at the origin of crystallographic axes chosen in conformity with morphological conventions. Besides division into the seven systems, crystalline matter can be further divided into thirty-two **crystal classes** each of which is characterized by the symmetry of one of the thirty-two point groups.

18   X-Ray Powder Diffraction

In point-group nomenclature, X and $\overline{X}$ denote principal axes of rotation and inverse rotation respectively where X may be 2, 3, 4 or 6; when one or other of these axes is named, the appropriate symbol is placed first. The

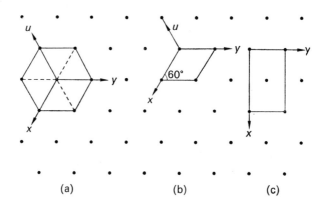

Fig. 1.13   Three possible unit cells of the hexagonal lattice. The z axis is upwards ⊥ to the plane of the paper

symbol m represents a symmetry plane and the only other symbols used are 2 for a diad perpendicular to a principal axis and 3 in a subsidiary position for the four triad axes of the cubic system. Some examples of point-group symbols, and their meanings are set out in the table.

Examples of Point-Group symbols (*after Henry, Lipson, and Wooster,* 1960)

Symbol	Meaning
X or $\overline{X}$	rotation or inverse axis by itself.
X2 or $\overline{X}$2	subsidiary rotation diad or diads perpendicular to X or $\overline{X}$.
X3 or $\overline{X}$3	the symbol 3 in the second position denotes the four triad axes of the cubic system.
Xm or $\overline{X}m$	symmetry plane or planes containing the axis X or $\overline{X}$.
X/m	symmetry plane perpendicular to axis X.
X/mm	symmetry plane or planes containing the axis X and one plane perpendicular to the axis.
mm	two mutually perpendicular symmetry planes; these intersect in a diad and hence the symbol = mm2 (2mm being sometimes used).
$\overline{X}$2m or $\overline{X}m$2	the symmetry planes contain the axis $\overline{X}$, while the diads are perpendicular to X.
X/m3m	the symbol 3 in a subsidiary position denotes the four triad axes of the cubic system; the first m denotes three symmetry planes, each of which is normal to one of the three crystallographic axes the symmetry of which is given by X; the second m denotes the six symmetry planes each containing one of the crystallographic axes and two triad axes.

All the information about the symmetry of a class may be obtained from the point-group symbol by starting with a general face—that is a face with no special relation to any symmetry element—and operating the symmetry elements represented by the symbol. This will produce a general form, defined as 'an assemblage of faces necessitated by the symmetry when one face in a general position is given' (Phillips, 1963). From the general form all the symmetry elements present in the crystal class, including those which it is not necessary to express in the point-group symbol, may be derived; the information can be presented in a convenient way by reference to a stereographic projection (1.1.5). The stereograms of the thirty-two point groups are shown in fig. 1.14 where the appropriate symmetry elements are shown in conventional orientation with the z axis upwards out of the plane of the paper and the y axis from left to right. A dot represents the pole of a face whose normal is directed upwards and a small open circle represents the pole of a face whose normal is directed downwards. The complete symmetry elements are shown but, where these are sufficient to determine the corresponding point-group symmetries unambiguously, the symbols are in shortened form. Thus the set of symmetry operations $\frac{6}{m}\,mm$ will generate two subsidiary rotation diads at right angles to the hexad axis so that the full symbol for the point group is $\frac{6}{m}\frac{2}{m}\frac{2}{m}$. If it is necessary to specify the orientation of the point group mm with respect to the lattice then the symbol mm is expanded to either $2mm$ or $mm2$ in order to show parallelism of the diad axis with the z axis; the symbols 42, 62 and 43 are often given in full as 422, 622 and 432 respectively to indicate that the diad is in the third position.

For the holosymmetric classes, those having the maximum symmetry of the system to which they belong, the point groups are $\bar{1}$, $\frac{2}{m}$, $mmm\left(\frac{2}{m}\frac{2}{m}\frac{2}{m}\right)$, $3m\left(\bar{3}\frac{2}{m}\right)$, $\frac{4}{m}mm\left(\frac{4}{m}\frac{2}{m}\frac{2}{m}\right)$, $\frac{6}{m}mm\left(\frac{6}{m}\frac{2}{m}\frac{2}{m}\right)$ and $m\,3m\left(\frac{4}{m}\bar{3}\frac{2}{m}\right)$. It is clear from the stereograms that eleven of the thirty-two classes are centrosymmetric and that the $\bar{1}$ and $\bar{3}$ operations add a centre of symmetry to the corresponding rotation axes; the $\bar{2}$, $\bar{4}$ and $\bar{6}$ axes do not add a centre of symmetry. The stereograms also show that every crystal class possesses at least the minimum symmetry of its system.

1.3 Symmetry of Crystal Structures

1.3.1 Glide planes and screw axes

Operation of any one of the symmetry elements of the point group will bring the crystal as a whole—that is the normals to the faces or the faces

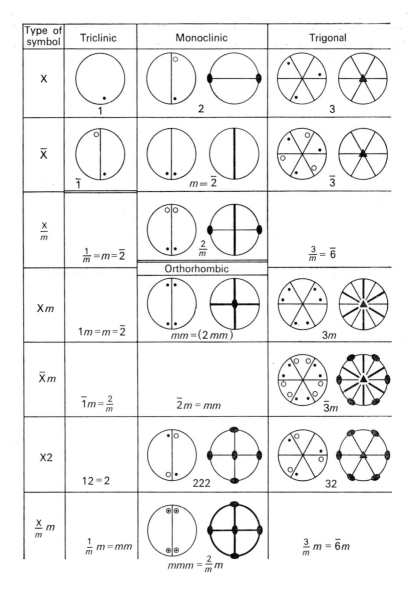

Fig. 1.14   Stereograms of the 32 point groups

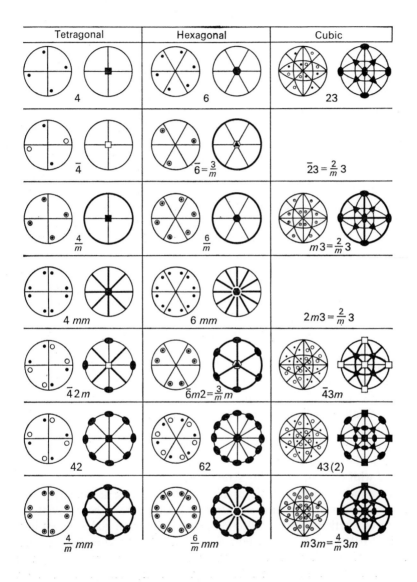

(Henry, Lipson and Wooster (1960))

themselves if they are fully developed—into a new orientation which is indistinguishable from the old; repeated operation will ultimately return the crystal to its initial orientation. This external symmetry displayed by the crystal is dependent upon the atomic arrangement within the crystal itself, which in its turn depends upon the symmetry displayed by the atoms and molecules which are associated with each lattice point and which comprise what we have called the unit of pattern. The unit of pattern may possess exactly the same combination of reflection planes, rotation axes and inverse rotation axes as does the crystal as a whole. When this occurs the point-group symbol serves to designate both the internal grouping of the atoms round the lattice point and the external symmetry of the crystal. However, the unit of pattern may possess symmetry elements which relate every point of the pattern with another by operations which involve translation; this type of symmetry will have no place in a collection of symmetry elements, such as a point group, which is not concerned with translations. There are two such additional symmetry elements, called **glide planes** and **screw axes**, in accordance with which the unit of pattern is unaltered by the operation of a reflection across a plane combined with a translation parallel to the plane (glide) or by a rotation combined with a translation in the direction of the rotation axis (screw). By *repeated* operation of the translational symmetry elements the unit of pattern is brought into coincidence, not with itself, but with the corresponding group in an adjoining unit cell, and so on into the next cell and the next through the crystal lattice.

If the unit of pattern associated with a lattice point does not possess a glide plane but has, however, a reflection plane, then the combination of this latter plane with a glide plane which is not a symmetry element of the unit of pattern will produce a centred lattice. This is one way, that is by combination of a reflection plane and a glide plane, in which a centred lattice can be generated, and it is illustrated in two dimensions in fig. 1.15. In a similar way a centred lattice is obtained when a rotation axis is combined with the corresponding screw axis.

Translations on the atomic, or microscopic, scale are not apparent on the macroscopic scale and there is no way of knowing from the morphological features whether external reflection planes arise from glides planes in the unit of pattern or from true reflection planes; for the same reason there is no way of knowing whether external rotation axes are produced by screw axes or by true rotation axes, internally. Thus any point group of external symmetry which includes reflection planes and, or, rotation axes may arise from any one of a combination of mirrors, rotations, glides and screws, and the net effect is to increase the total of different possible combinations of internal symmetry elements from 32 to 230.

There are three possible types of glide plane which correspond to translations parallel to the direction of a cell edge, a cell-face diagonal and, less commonly, to a cell-body diagonal. Reflection across a plane perpendicular to the cell edge, a, may be accompanied by a translation, parallel to the plane, of $\frac{1}{2}b$, $\frac{1}{2}c$, $\frac{1}{2}(b + c)$ or, occasionally, $\frac{1}{4}(b + c)$; these are called b, c, n and d glides respectively. Similarly, a, c, n and d glide planes may occur perpendicularly to the b axis and a, b, n and d perpendicularly to the c axis. In a few combinations, n glides and d glides may arise from reflection across a plane with simultaneous translations of $\frac{1}{2}(a + b + c)$ and $\frac{1}{4}(a + b + c)$ in the direction of the body diagonal of the unit cell.

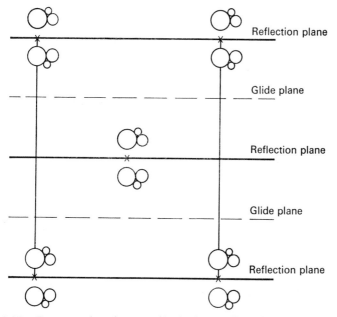

Reflection plane

Glide plane

Reflection plane

Glide plane

Reflection plane

Fig. 1.15   The generation of a centred lattice in two dimensions by a combination of a reflection plane and a glide plane

A screw diad axis combines a rotation of 180° with a translation of half the cell edge and is denoted by the symbol 2_1. With this and all other screw axes, when looking in the decreasing positive direction of the appropriate cell vector an anti-clockwise rotation is combined with a translation in the increasing positive direction. A screw triad, or 3_1, axis combines a rotation of 120° with a translation equal to one-third of a unit-cell edge; when the 120° rotation is combined with a translation of two-thirds the cell edge the screw triad is called a 3_2 axis. Similarly screw tetrad axes 4_1, 4_2, 4_3

combine 90° rotations with cell-edge translations of $\frac{1}{4}$, $\frac{2}{4}$ and $\frac{3}{4}$ respectively, and screw hexad axes 6_1, 6_2, 6_3, 6_4 and 6_5 combine rotations of 60° with corresponding translations of $\frac{1}{6}$, $\frac{2}{6}$, $\frac{3}{6}$, $\frac{4}{6}$ and $\frac{5}{6}$.

From fig. 1.16 it can be seen that a 3_2 axis is equivalent to a clockwise rotation of 120° combined with a one-third positive translation and hence that the two patterns, one showing a 3_1 axis and the other a 3_2 axis will be **enantiomorphously** related; that is they will be reflections of each other as are a left-handed and a right-handed glove. Similarly each of the pairs of patterns having 4_1 and 4_3, 6_1 and 6_5, 6_2 and 6_4 axes will be reflections of each other. However, if a unit of pattern possesses either a mirror or a

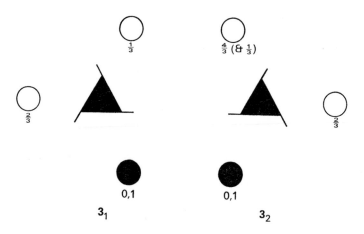

Fig. 1.16   The operations of axes 3_1 and 3_2

glide plane then it cannot be enantiomorphously related to any other unit of pattern. It follows that if a crystal belongs to a class whose point group does not have mirror planes then it is possible for that crystal to be enantiomorphously related morphologically to another crystal of the same class. A crystal class of which the point group includes either a reflection plane or a centre of symmetry is said to be non-enantiomorphous because the presence of these symmetry elements precludes the existence of a right- and left-hand relationship between crystals of the same class.

1.32   Space groups

The 230 different combinations of internal symmetry elements (1.3.1) constitute the 230 different kinds of extended three-dimensional framework into which the unit of pattern associated with each lattice point may be fitted in order to produce a periodic crystal lattice. They are called **space**

groups. Although the space group must account for the symmetery relations existing within the unit of pattern, it tells us nothing about the actual arrangement of the atoms; it could equally well apply to a collection of arrows or of flowers.

An example of a space group is shown in two dimensions in fig. 1.17. The design, which is one often used in wooden flooring, has a primitive, rectangular cell. Since the cell is primitive it has just one unit of pattern associated with it; the unit of pattern is shaded, the glide planes are indicated by broken lines, the diad axes by ellipses, and the unit cell is outlined by full lines. Inspection of the diagram will reveal that a reflection across a

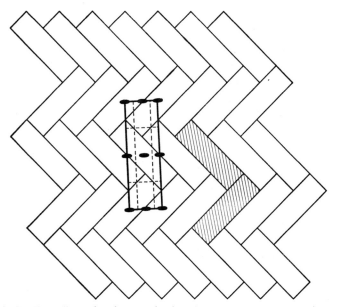

Fig. 1.17   Two-dimensional pattern having glide planes and diad axes. The unit of pattern is shaded

broken line followed by a translation of one half the unit-cell edge parallel to that line, and in the plane of the paper, will bring the whole pattern into self-coincidence. The same considerations apply to the diad axes.

To return to the study of crystals, each of the fourteen Bravais lattices has the full symmetry of the system to which the crystal as a whole belongs; in other words, it has the symmetry of the holosymmetric class of that system. If the atomic grouping associated with each lattice point does not possess this maximum symmetry (in the specification of the symmetry, glide planes are considered to be reflection planes and screw axes are counted as rotation axes) then the point-group symmetry of the crystal

2+

will be lower than that of the holosymmetric class. Each class of symmetry, apart from the asymmetric class, will include elements of symmetry other than a centre and because these elements may arise from elements of translational symmetry it follows that all such classes will include more than one space group. It is clearly advantageous that the procedure for the symbolic representation of space groups should follow closely that adopted for point-group symbols. In the first position is a capital letter which gives the lattice type and which will be chosen appropriately from the letters P, A, B, C, F, I and R (1.2.5). Next comes the symbol for the principal axis, if present, followed by the symbols for any secondary axes and planes. A reflection plane at right angles to an axis, X, whether the axis is principal or secondary, is written $\frac{X}{m}$ (or X/m), and a plane parallel to an axis is denoted by Xm. Thus P2/a is a space group based on a primitive lattice with an a-glide at right angles to the screw diad; the symbol Cm means that the lattice is C-face centred and that the space group includes a mirror plane.

Each space group is generally described in terms of the minimum number of symmetry elements essential to identify it but the minimum number chosen may not be the only one possible. As with point groups, the operation of these symmetry elements may automatically give rise to further elements which must necessarily also be present, but which need not be specified. For example, *Pmmm* must have two-fold axes (International Tables, Vol. I, 1952) at right angles to each of the planes and could be expressed in the unnecessarily elongated form P2/m 2/m 2/m. The space group of the two-dimensional pattern of fig. 1.17 can be specified as primitive with two glide planes at right angles to each other; the diad axes arise from the operation of the glide planes. (It is usual to refer to the space groups of two-dimensional patterns as **plane groups**. The axes, which may be oblique, rectangular, square or hexagonal, represent the four systems to which two-dimensional lattices can belong; of these, the rectangular lattice may be centred as well as primitive, but the other three lattices can be primitive only. The point groups and plane groups of two-dimensional systems are given in Vol. I of the International Tables, 1952.)

Additional symmetry elements may arise also from centering points when the lattice is other than primitive. The space group Cm, a portion of which is shown diagrammatically in fig. 1.18 in the conventional orientation with z axis upwards from the plane of the paper, y axis from left to right and x axis downwards, has the C face centred and will thus have repeated at this lattice point the unit of pattern which is associated with each of the corners of the unit cell. Because of this there is a glide plane parallel to the mirror plane, but it is not essential that the glide plane should be specified as well as the mirror plane.

Some crystals which morphologically have trigonal symmetry may be referred either to a primitive rhombohedral lattice or to a hexagonal lattice which is triply primitive; that is, the unit cell contains three lattice points. Other crystals of this system may, on the other hand, have an external structure which has a primitive hexagonal lattice or, a triply primitive rhombohedral lattice. For space-group representation the more convenient lattice is chosen and the corresponding symbol will be either R or P.

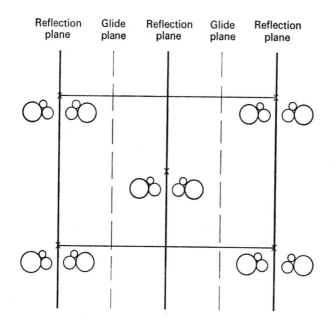

Reflection plane Glide plane Reflection plane Glide plane Reflection plane

Fig. 1.18   The space group *Cm*. It is not essential to specify the glide planes

1.3.3   Laue groups

The disposition of the atoms with respect to one set of crystollagraphic planes is not always the same as that with respect to another set of planes with the same spacing; this means that the intensities of the radiation diffracted by these two sets of planes will be different. Further, because the atoms are orientated differently with respect to the two sets of planes, the crystal faces which are parallel to these sets will grow, even under ideal conditions, at different rates; they will, therefore, belong to different forms and will not be related to each other by the operation of the symmetry elements of the point group. Just as faces belonging to different forms are parallel to internal planes which produce different diffracted intensities, so faces of the same form (that is, faces related by the point-group symmetry) are parallel to internal planes which, all other things being equal, produce the same diffracted intensities. There is thus a close relationship between the symmetry of the diffraction effects from the internal planes and the symmetry of the point group. The close relationship is sometimes, but not always, an exact one. The difference arises because, in general, according to Friedel's law (2.1.2), the diffracted intensity from the back of a plane of atoms is the same as that from the front so that hkl and $\bar{h}\bar{k}\bar{l}$ intensities are equal. This means that the pattern of intensities will show a centre of symmetry whether or not the point group is centrosymmetric. Diffraction symmetry is therefore the corresponding point-group symmetry to which, if noncentrosymmetric, a centre has been added. Eleven point groups are centrosymmetric (fig. 1.14) and when a centre has been added to each of the remaining twenty-one the resulting symmetry will be identical with that of one or other of the centrosymmetric groups; there are thus eleven distinguishable groups of diffraction symmetry. Since diffraction symmetry is best shown in Laue photographs (Henry, Lipson and Wooster, 1960) it is called Laue symmetry and the eleven distinguishable groups are called **Laue groups**.

The Laue-symmetry groups are shown in the table below which is taken from Vol. I of the International Tables (1952). Point groups belonging to the same Laue group are enclosed in a rectangle and the symbol of the Laue group is that of the centrosymmetrical point group which is given last within each rectangle.

The 11 Laue-symmetry Groups

Triclinic	Monoclinic	Tetragonal	Trigonal	Hexagonal	Cubic
1	2	4	3	6	23
	m	$\bar{4}$		$\bar{6}$	
$\bar{1}$	2/m	4/m	$\bar{3}$	6/m	m3

	Orthorhombic				
	222	422	32	622	432
		4mm	3m	6mm	
	mm2	$\bar{4}2m$		$\bar{6}m2$	$\bar{4}3m$
	mmm	4/mmm	$\bar{3}m$	6/mmm	m3m

CHAPTER 2

Geometry of X-ray Reflection

2.1 Diffraction by a Crystal

2.1.1 Diffraction from a three-dimensional grating

A crystal is built up of a more or less complex unit-of-pattern of atoms which is repeated in three dimensions by regular linear translations, and when each unit of pattern is represented by a point the resulting regular array is called a lattice (1.1.1). If we now imagine that each point of the lattice is capable of scattering incident X-radiation, then it is possible to determine the conditions under which the scattered waves from every point can combine to form a strongly diffracted beam, just as the condition can be determined that the light waves scattered by the separate lines on a ruled grating should reinforce each other to produce spectra. Radiation imagined scattered by the lattice points will be reinforced in the same directions as radiation scattered by corresponding atoms in each unit of pattern so that each array of corresponding atoms will produce strong diffraction effects in identical directions. Although the *directions* of the diffraction spectra produced by the crystal thus depend only on the geometry of the lattice, the actual diffracted intensities depend on the way in which the effects from the individual atoms combine and therefore upon the nature of the unit of pattern itself.

The condition for additive interference of X-rays supposed scattered from a row of points in the lattice may be determined by reference to fig. 2.1, in which A and B are successive points on the lattice line. Incident radiation travelling in the direction PA makes an angle ϕ with the line which joins AB and passes through the row of points. Diffracted radiation travelling in the direction AQ makes an angle ψ with the same line. The path difference between the rays PAQ and RBS is $BC-AD$ where the angles ACR and BDQ are both $90°$ and this difference is clearly given by

$$AB(\cos \phi - \cos \psi)$$

For additive interference between the diffracted rays AQ and BS this path difference must be equal to a whole multiple of the wavelength, λ, of the radiation; the condition is, therefore,

$$n\lambda = AB(\cos \phi - \cos \psi) \qquad \ldots 2.1$$

where n is integral.

In order to specify the conditions for additive interference of the diffracted radiation in three dimensions it is sufficient to choose three non-coplanar lattice rows, not necessarily mutually perpendicular, and to write down the equations corresponding to (2.1) for each of them. If the repeat

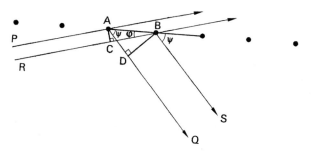

Fig. 2.1   Diffraction by a row of scattering points

distances of the points along the three non-coplanar rows are a, b, c, and if the angles which the directions of the incidence and the diffracted radiation make with these rows are ϕ_a, ϕ_b, ϕ_c and ψ_a, ψ_b, ψ_c respectively, then the conditions for additive interference are

$$h\lambda = a(\cos \phi_a - \cos \psi_a)$$
$$k\lambda = b(\cos \phi_b - \cos \psi_b)$$
$$l\lambda = c(\cos \phi_c - \cos \psi_c)$$

These are known as the three **Laue equations** and they must be satisfied simultaneously. The integers h, k, l are the numbers of wavelengths in the path differences between the rays diffracted from consecutive points along the three lattice rows defined by a, b and c; that is, they give the order of the spectrum.

Suppose now that one of the lattice points is chosen as the origin, O, of axes x, y, z, each of which lies in the direction of one of the three non-coplanar rows of points so that the first points A, B. C along these axes are distant a, b and c respectively from the origin (fig. 2.2). From the first of the Laue equations the path difference between the radiation diffracted from O and that diffracted from A is h wavelengths. The distance from O along the x axis of the point from which the diffracted radiation has a path difference of one wavelength from that diffracted from O would therefore be a/h. Similarly points at the origin and at a distance b/k along the y axis would produce diffracted waves having one wavelength path difference, and the corresponding distance along the z axis would be c/l. These three points would therefore scatter waves which would have zero path difference and they define one of the family of parallel equidistant planes having indices (hkl).

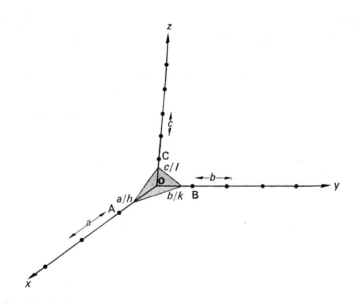

Fig. 2.2   Points distant a/h, b/k, c/l from the origin of a three-dimensional lattice. Radiation scattered from these points is in phase

Although the radiation scattered by each of these three points might have zero path difference, there will only be reinforcement of this scattered radiation, and hence a diffracted ray, for a direction in space for which the three Laue equations are satisfied simultaneously. If this condition holds then the atoms at all the lattice points on the plane defined by the three points will produce a diffracted beam in that particular direction. Moreover, atoms at the lattice points on the parallel plane through the origin will also scatter radiation with zero path difference in this direction. By definition, there is one wavelength path difference between the radiation scattered from the origin and from each of the three points distant a/h, b/k, c/l along the corresponding axes, so that the beams diffracted by the two planes will reinforce one another. Similarly, the path difference between rays scattered from adjacent planes of the family (*hkl*) is one wavelength and hence all the planes of this family will scatter X-rays additively in the direction defined by the three Laue equations. The **spacing** d_{hkl} between the planes is equal to the perpendicular distance from the origin to the plane which makes intercepts a/h, b/k, c/l on the x, y and z axes.

2.1.2 Bragg's law

Bragg's law is an alternative way of expressing Laue's equations and in many ways makes it easier to understand diffraction by a crystal. Again we consider a regular three-dimensional array either of atoms or of points capable of diffracting X-radiation. In this array (fig. 2.3) we choose a family

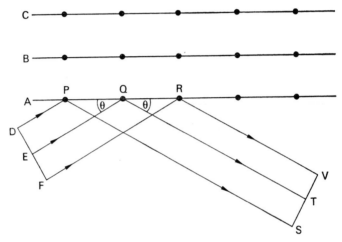

Fig. 2.3   Radiation scattered by a plane of atoms

of parallel equidistant planes of atoms A, B, C for which the indices are (*hkl*) and suppose that the orientation of these planes relative to the direction of an incident X-ray beam is such that they scatter the X-rays additively. The three appropriate Laue equations will then be satisfied simultaneously and this means that there is zero path difference between the rays scattered by the atoms in any one of these planes. Thus if DEF is a part of a plane wavefront incident on, say, the plane A of atoms, and if STV is part of the corresponding wavefront diffracted by the atoms P, Q, R, then each of the paths DPS, EQT and FRV must be equal in length. This is precisely the geometrical condition for the reflection of light by a plane mirror so that the diffraction of an X-ray beam from the plane A of atoms is geometrically equivalent to the reflection of the beam. Thus the angle θ which the incident beam makes with the plane of the atoms must be equal to the angle which the diffracted beam makes with that plane. It may be noted *when* the reflection condition is satisfied then all the paths between the incident and the diffracted wavefronts via the plane are equal in length. This means that for 'reflection' of X-rays to take place from the plane A it is not necessary that all the atoms in that plane should be in a regular two-dimensional array (fig. 2.4). If, however, the atoms were irregularly distributed over the plane then that plane could not be one of a family of

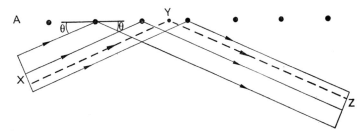

Fig. 2.4   Radiation 'reflected' from Y, which is not part of the regular array, has the same path length as that 'reflected' from the atoms in the array

equivalent (*hkl*) planes each member of which reflects X-rays at the glancing angle *θ*.

The same argument applies to the other planes B, C, of the family (*hkl*) so that when the three Laue conditions are satisfied then the parallel planes B, C, will reflect X-rays when the incident beam strikes each of the planes at the glancing angle *θ*. If now the three-dimensional array of atoms is to produce a diffracted beam from the family of planes (*hkl*) then the separate rays diffracted by each of the planes must all be in phase. But from fig. 2.5 the path difference between rays reflected from adjacent planes A and B is

$$DF + FG = d \sin \theta + d \sin \theta$$
$$= 2d \sin \theta$$

where *d* is the perpendicular distance between the planes. If this path difference is an integral number of wavelengths then the waves from each plane reinforce one other. The planes (*hkl*) were, however, so defined that when the three Laue equations were simultaneously satisfied the path difference between rays diffracted from successive planes of atoms was one wavelength, λ. Thus the condition that the planes should diffract X-rays at the angle *θ* to the planes is λ = 2*d* sin *θ*. This condition, called **Bragg's law** is equivalent to that laid down by the three Laue equations.

It is clear that for radiation of a particular wavelength, the planes with an interplanar spacing *d* can reflect X-rays for only one value of *θ*. This is because the constructive-interference conditions must be satisfied in three

2*

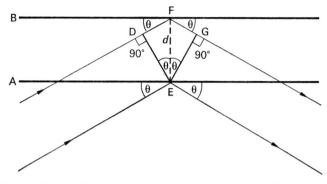

Fig. 2.5   Reflection of rays from two successive planes, A and B, of a family of planes

dimensions; in the two-dimensional examples of reflection of X-rays from a single plane of atoms and of reflection of light from a plane mirror, the value of θ is continuously variable.

The above is an elementary derivation of Bragg's law and for a more rigorous and elegant treatment vectorial methods are best used. In order to express the three Laue equations (2.11) in vector notation, consider in turn the three lattice points at vector distances a, b and c from the origin. Let i and f be unit vectors along the positive directions of the incident and diffracted rays and suppose these vectors make angles ϕ_a and ψ_a respectively with the vector a (fig. 2.6).

The scalar (dot) product $a \cdot i = ai \cos \phi_a$ by definition.
But i is of unit length so that

$$a \cdot i = a \cos \phi_a$$

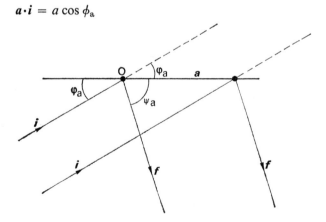

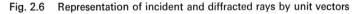

Fig. 2.6   Representation of incident and diffracted rays by unit vectors

Similarly $a \cdot f = af \cos \psi_a = a \cos \psi_a$
since f is of unit length.

Also $\qquad\qquad\qquad b \cdot i = b \cos \phi_b; \qquad b \cdot f = b \cos \psi_b$

and $\qquad\qquad\qquad c \cdot i = c \cos \phi_c; \qquad c \cdot f = c \cos \psi_c$

The three Laue equations can now be written

$$h\lambda = a \cos \phi_a - a \cos \psi_a = a \cdot i - a \cdot f = a \cdot (i - f)$$
$$k\lambda = b \cos \phi_b - b \cos \psi_b = b \cdot i - b \cdot f = b \cdot (i - f)$$
$$l\lambda = c \cos \phi_c - c \cos \psi_c = c \cdot i - c \cdot f = c \cdot (i - f)$$

Relations of the type $a \cdot i - a \cdot f = a \cdot (i - f)$ are a consequence of the definitions of addition and subtraction of scalar products of vectors.

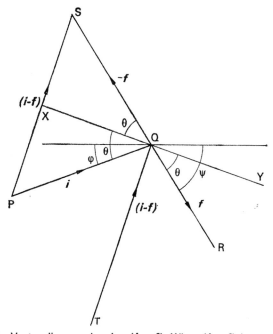

Fig. 2.7   Vector diagram showing $(i - f)$. When $(i - f)$ is perpendicular to XY then the unit vectors PQ and QR, in the directions of the incident and diffracted rays, make equal angles, θ, with XY

By reference to fig. 2.7 consider the vector difference $(i - f)$ between the unit vectors i and f. If the direction of f is reversed then $(i - f)$ is the vector sum of the unit vectors PQ and QS. That is, $(i - f) = $ PS and if we draw TQ equal and parallel to PS then TQ also represents $(i - f)$. Because $|i| = |f|$ the vector $(i - f)$, or TQ, bisects the angle between i and f, or PQ and QR, and therefore i and f make equal angles, θ, with the plane, XY, perpendicular to **TQ**. The magnitude of $(i - f)$ is given by the magnitude of **PS** and is equal to $2 \sin \theta$.

Let us now rewrite the Laue equations in the forms

$$1 = \frac{a}{h\lambda} \cdot (i - f) \qquad \qquad \dots 2.2$$

$$1 = \frac{b}{k\lambda} \cdot (i - f) \qquad \qquad \dots 2.3$$

$$1 = \frac{c}{l\lambda} \cdot (i - f) \qquad \qquad \dots 2.4$$

If a three-dimensional array of scattering points produces a diffracted beam, the equations must be satisfied simultaneously and in this case, we can write, by subtracting 2.3 from 2.2 and 2.4 from 2.2,

$$0 = \left(\frac{a}{h\lambda} - \frac{b}{k\lambda}\right) \cdot (i - f)$$

and
$$0 = \left(\frac{a}{h\lambda} - \frac{c}{l\lambda}\right) \cdot (i - f)$$

These expressions are of the form

$$\mathbf{A} \cdot \mathbf{B} = 0$$

i.e. $AB \cos \epsilon = 0$ where ϵ is the angle between the positive directions of **A** and **B**, i.e. $\cos \epsilon = 0$ and $\epsilon = 90°$ provided that neither **A** nor **B** is zero. Thus the vectors

$$\left(\frac{a}{h\lambda} - \frac{b}{k\lambda}\right) \quad \text{and} \quad \left(\frac{a}{h\lambda} - \frac{c}{l\lambda}\right)$$

are both perpendicular to the vector $(i - f)$. A diffracted beam is therefore obtained when the vector $(i - f)$ is perpendicular to the plane defined by the vectors

$$\left(\frac{a}{h\lambda} - \frac{b}{k\lambda}\right) \quad \text{and} \quad \left(\frac{a}{h\lambda} - \frac{c}{l\lambda}\right)$$

i.e. by

$$\frac{1}{\lambda}\left(\frac{a}{h} - \frac{b}{k}\right) \quad \text{and} \quad \frac{1}{\lambda}\left(\frac{a}{h} - \frac{c}{l}\right)$$

Since $1/\lambda$ is a factor common to both vectors and affects only their magnitudes and not their directions it follows that $(a/h - b/k)$ and $(a/h - c/l)$ define the plane to which $(i - f)$ must be perpendicular if a beam is to be strongly diffracted from that plane.

From fig. 2.8 it is evident that the vectors $(a/h - b/k)$ and $(a/h - c/l)$ define the plane PQR which passes through the points distant a/h, b/k, c/l from the origin along the x, y and z axes, so that $(i - f)$ must be perpendicular to PQR for diffraction to occur. But we have already seen that when $(i - f)$ is perpendicular to a given plane then the vectors i and f must make equal angles, which we have called θ, with that plane (fig. 2.7), and therefore when diffraction occurs from PQR the incident and diffracted beams are in the directions of i and f respectively and make equal angles with PQR. The plane PQR is one of the family (hkl) and hence the incident and diffracted beams make equal angles with all these planes.

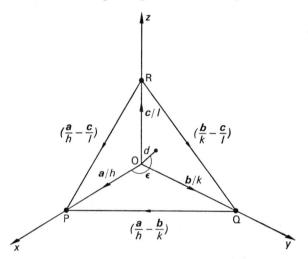

Fig. 2.8   The plane PQR defined by $(a/h - b/k)$ and $(a/h - c/l)$

Now let d be the length of the normal from the origin O to the (hkl) plane which passes through a/h, b/k, c/l (fig. 2.8). O lies on the adjacent (hkl) plane, so that d is also the perpendicular distance between successive planes of the family. The vector $(i - f)$ is perpendicular to the family of planes (hkl) and is therefore parallel to the direction of the normal. Let the positive direction of $(i - f)$, and therefore also of the normal, make an angle ϵ with the positive direction of a/h. Then

$$\frac{a}{h} \cdot (i - f) = \frac{a}{h} \times \text{length of } (i - f) \times \cos \epsilon$$

by the definition of a scalar product.

But $(a/h) \cos \epsilon = d$ and we have already shown that the magnitude of $(i - f)$ is $2 \sin \theta$, therefore

$$\frac{a}{h} \cdot (i - f) = 2d \sin \theta$$

From the first of Laue's equations

$$h\lambda = a\cdot(i - f)$$

so that

$$\lambda = \frac{a}{h}\cdot(i - f)$$

$$= 2d \sin \theta$$

which is Bragg's law.

The value of d depends upon the value of $\cos \epsilon$ which in turn depends upon the lengths a, b, c of the unit-cell edges and the angles α, β, γ between b and c, c and a, a and b respectively. The value of d, and consequently that of $\sin \theta$, will thus be different for different systems and the expressions for these two quantities will become increasingly complicated for lattices of lower and lower symmetry. Calculations are easier if values of either $1/d^2$ or $\sin^2 \theta$ rather than values of d or of $\sin \theta$ are available. The choice between $1/d^2$ and $\sin^2 \theta$ may depend on whether an expression is required which is independent of the wavelength λ of the X-radiation; $1/d^2$ is independent of λ whereas $\sin^2 \theta$ is not. In the following table (p. 39) expressions for $\sin^2 \theta$ are given but if desired they can be converted to $1/d^2$ using Bragg's law.

The indices of planes referred to a rhombohedral lattice are usually denoted by p, q and r; if the indices are transformed to those corresponding to an appropriate hexagonal cell (5.2.2) they are denoted by h, k and l. The constants a^*, b^*, c^*, α^*, β^* and γ^* are the constants of a unit cell in reciprocal space and they are defined in section 2.2.1.

In the discussion leading to Bragg's law it was assumed implicitly that reflection of X-rays from opposite sides of a parallel plate of crystal must occur at the same angles. It is usually also true that the intensities reflected from two sides of such a plate under comparable conditions are equal; this is known as **Friedel's law**. It may happen that when the wave-length of the incident radiation is close to a resonance level for one kind of atom in the crystal, there is a differential phase change on scattering which enables hkl to be distinguished from $\bar{h}\bar{k}\bar{l}$ in a non-centrosymmetric crystal. We shall assume that Friedel's law holds over the field covered, and therefore that the hkl and $\bar{h}\bar{k}\bar{l}$ reflections from a given set of parallel planes have equal intensities.

2.1.3   Orders of reflection

Bragg's law is generally stated in the form

$$n\lambda = 2d \sin \theta$$

where n is the order of the reflection. That is, for path differences of $n\lambda$ between rays scattered from adjacent planes, the reflection nh, nk, nl is

Expressions for $\sin^2 \theta$

Lattice	$\sin^2 \theta$
Cubic	$\dfrac{\lambda^2}{4a^2}(h^2 + k^2 + l^2)$
Tetragonal	$\dfrac{\lambda^2}{4a^2}(h^2 + k^2) + \dfrac{\lambda^2}{4c^2}l^2$
Hexagonal	$\dfrac{\lambda^2}{3a^2}(h^2 + hk + k^2) + \dfrac{\lambda^2}{4c^2}l^2$
Orthorhombic	$\dfrac{\lambda^2}{4a^2}h^2 + \dfrac{\lambda^2}{4b^2}k^2 + \dfrac{\lambda^2}{4c^2}l^2$
Rhombohedral	$\dfrac{\lambda^2}{4a^2}\left[\dfrac{(p^2 + q^2 + r^2)\sin^2\alpha + 2(pq + qr + rp)(\cos^2\alpha - \cos\alpha)}{1 + 2\cos^3\alpha - 3\cos^2\alpha}\right]$
Monoclinic	$\dfrac{\lambda^2}{4a^2\sin^2\beta}h^2 - \dfrac{\lambda^2\cos\beta}{2ac\sin^2\beta}hl + \dfrac{\lambda^2}{4c^2\sin^2\beta}l^2 + \dfrac{\lambda^2}{4b^2}k^2$
Triclinic	$\dfrac{\lambda^2}{4}[h^2a^{*2} + k^2b^{*2} + l^2c^{*2} + 2klb^*c^*\cos\alpha^*$ $+ 2lhc^*a^*\cos\beta^* + 2hka^*b^*\cos\gamma^*]$

where

$$a^* = \frac{bc}{V}\sin\alpha; \quad \cos\alpha^* = \frac{\cos\beta\cos\gamma - \cos\alpha}{\sin\beta\sin\gamma}$$

$$b^* = \frac{ca}{V}\sin\beta; \quad \cos\beta^* = \frac{\cos\gamma\cos\alpha - \cos\beta}{\sin\gamma\sin\alpha}$$

$$c^* = \frac{ab}{V}\sin\gamma; \quad \cos\gamma^* = \frac{\cos\alpha\cos\beta - \cos\gamma}{\sin\alpha\sin\beta}$$

The volume, V, of the unit cell is equal to
$abc(1 + 2\cos\alpha\cos\beta\cos\gamma - \cos^2\alpha - \cos^2\beta - \cos^2\gamma)^{1/2}$.

regarded as the nth order of the hkl reflection. This is not really necessary since the nh, nk, nl reflection is that of the first order from the planes ($nh\ nk\ nl$). In this connection it is conventional to distinguish the hkl reflection from the (hkl) planes by omission of the brackets.

2.2 The Reciprocal Lattice

2.2.1 The nature of the reciprocal lattice

Some of the characteristics of a crystal such as its morphological features and the intensity of the reflections, are most simply considered in terms of the crystal, or direct, lattice. Other features, including many diffraction effects, are however, more easily understood by the use of a concept known as the **reciprocal lattice**. In this lattice, each point, hkl, represents a family of planes (hkl) in the direct lattice and is situated on the line drawn through the origin at right angles to the planes. The distance between the point and the origin along this line is inversely proportional to the spacing of the planes concerned. Distances in reciprocal space may be denoted by an asterisk, so that the relation between the distance, d^*_{hkl}, of a reciprocal point from the origin, and the spacing d_{hkl} of the corresponding family of planes, is of the form

$$d^*_{hkl} = K/d_{hkl}$$

where K is a constant which is sometimes made equal to unity and at other times to the wavelength λ of the X-radiation. Similarly, the axes of the reciprocal lattice are labelled x^*, y^*, z^* and are perpendicular to the yz, zx and xy planes of the direct lattice respectively. Hence for the three crystal systems—cubic, tetragonal and orthorhombic—which are referred to orthogonal axes, the two sets of coordinate axes coincide; but when the direct lattice has an axis which is not at right angles to the plane of the other two then this axis cannot be collinear with the corresponding axis in reciprocal space. Thus in the monoclinic system, the axes of which are shown in fig. 2.9 with the x and z axes in the plane of the paper and the y axis at right angles to this plane, neither the x and x^* nor the z and z^* axes coincide; the angle β^* between the x^* and z^* axes is the supplement of the monoclinic angle β between the x and z axes. Similarly, as shown with the hexagonal system in the same figure, the angle x^*Oy^* is 60° and is the supplement of the angle xOy in the direct lattice; the unique z axis and its reciprocal z^* axis are both directed upwards at right angles to the plane of the paper.

To construct the reciprocal lattice of, say, a monoclinic crystal, the points are plotted as illustrated in fig. 2.10. The repeat distance a^* along the x^* axis is inversely proportional to the spacing $a \sin \beta$ of the d_{100} planes in the direct lattice, and 100, the first point out from the origin along Ox^*, represents the family of (100) planes parallel to the plane yz. At a distance $2a^*$ from the origin is the second point 200 corresponding to the planes (200) with a spacing $\frac{1}{2} a \sin \beta$; next, 300 represents the (300) planes with spacing $\frac{1}{3} a \sin \beta$ and so on. Points 001, 002, 003, . . . along Oz^* with a repeat distance $c^* = k/c \sin \beta$ are derived from the spacings, $\frac{1}{2} c \sin \beta$ . . . of the (001), (002), . . . planes. Similarly points 010, 020, 030—spaced at intervals $b^* = K/b$—are obtained along the y^* axis at right angles to the plane of the paper. The reciprocal-lattice point 101 is on the line through the origin at right angles to the (101) planes of the direct lattice at a dis-

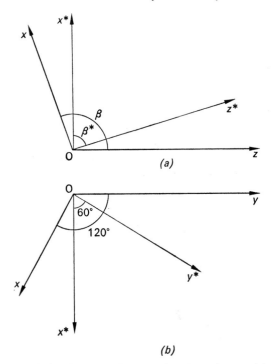

(a)

(b)

Fig. 2.9   The relation between direct and reciprocal axes (*a*) monoclinic
system (*b*) hexagonal system

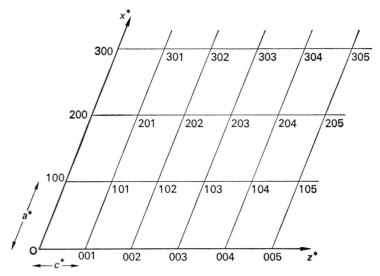

Fig. 2.10   The *h0l* plane of the reciprocal lattice of a monoclinic crystal

tance from the origin equal to K/d_{101}. If the plane x^*Oz^* is considered to be horizontal the 111 point will be at a vertical height K/b above the point 101. Thus with origin O, a unit cell with edges of lengths a^*, b^*, c^* can be drawn, and the reciprocal lattice can be built up by repeated three-dimensional translations of the cell, just as the direct lattice is constructed by the repeated translations of the direct-lattice cell.

It can be proved as follows for the general case that the points plotted as indicated above do, in fact, form a lattice. This means that the point hkl plotted along the perpendicular through the origin to the (hkl) planes at a distance $d^*_{hkl} = K/d_{hkl}$ from the origin is the same as that which would be obtained by successive translations ha^*, kb^* and lc^* along the x^* axis and then along lines parallel to y^* and z^* axes. Now of the family of planes with indices (hkl), the plane nearest the origin cuts the direct-lattice x, y and z axes at A, B and C respectively (fig. 2.11a) and the vectors from the

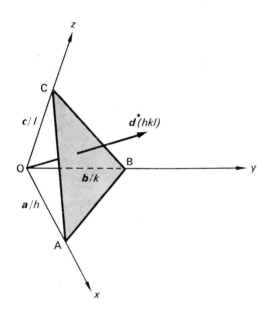

Fig. 2.11(a)   The reciprocal vector d_{hkl} corresponding to the family of planes (hkl)

origin to these three points are consequently a/h, b/k and c/l. Suppose that the vector d^*_{hkl} from the origin to the corresponding reciprocal-lattice point hkl is given by

$$d^*_{hkl} = ha^* + kb^* + lc^*.$$

That is we are assuming that the point hkl in reciprocal space, corresponding to the family of planes (hkl) in a real or direct space, is reached by successive translations ha^*, kb^* and lc^* from the origin in the directions Ox^*, Oy^* and Oz^*. We must therefore establish both that the vector d^*_{hkl} is perpendicular to the plane of A, B, C and that its length is given by $d^*_{hkl} = K/d_{hkl}$.

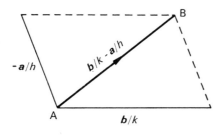

Fig. 2.11(b)   Construction of the vector **AB**

In the direct lattice the length and direction of the vector **AB** is given by the reversing the direction of a/h and adding this vector to b/k (fig. 2.11b); hence $\mathbf{AB} = (b/k - a/h)$.

The scalar product of d^*_{hkl} and **AB** is given by

$$d^*_{hkl} \cdot \mathbf{AB} = (ha^* + kb^* + lc^*) \cdot (b/k - a/h)$$

and since, by definition, the right-hand side is evaluated by term-by-term multiplication, the relation becomes

$$d^*_{hkl} \cdot \mathbf{AB} = \frac{h}{k}(a^* \cdot b) + \frac{k}{k}(b^* \cdot b) + \frac{l}{k}(c^* \cdot b)$$

$$- \frac{h}{h}(a^* \cdot a) - \frac{k}{h}(b^* \cdot a) - \frac{l}{h}(c^* \cdot a)$$

Because of the relationship between the reciprocal and the direct lattices a^* is perpendicular to b, c^* to b, b^* to a and c^* to a. Since the angle between each of these pairs of vectors is 90° the scalar products $a^* \cdot b$, $c^* \cdot b$, $b^* \cdot a$ and $c^* \cdot a$ are all zero.

Also, by the relationship between the lattices, the scalar products $a^* \cdot a$, $b^* \cdot b$, $c^* \cdot c$ are all equal to K, the constant of proportionality between the two lattices. It follows that $d^*_{hkl} \cdot \mathbf{AB} = K - K = 0$ and therefore d^*_{hkl} is normal to **AB**. Similarly d^*_{hkl} is normal to **BC** (and to **CA**) and so the vector d^*_{hkl} must be at right angles to the plane ABC.

Further, the projection of the length of a/h (or of b/k or of c/l) onto the normal from the origin to the nearest of the (hkl) planes is equal to d_{hkl}, the the perpendicular distance from the origin to this plane. We can therefore write

$$d_{hkl} = a \cos \epsilon / h = a \cdot n / h$$

where ϵ is the angle between a/h and the normal to the plane; n is a unit vector in the direction of the normal, that is, in the direction also of d^*_{hkl}.

But $d^*_{hkl} n = d^*_{hkl}$, i.e.

$$n = \frac{d^*_{hkl}}{d^*_{hkl}}$$

and therefore from $d_{hkl} = (a \cdot n / h)$ we have

$$d_{hkl} = \frac{a}{h} \cdot \frac{d^*_{hkl}}{d^*_{hkl}}$$

i.e.

$$d_{hkl} = \frac{a}{h} \cdot \frac{(ha^* + kb^* + lc^*)}{d^*_{hkl}}$$

$$= \frac{h(a \cdot a^*) + k(a \cdot b^*) + l(a \cdot c^*)}{h d^*_{hkl}}$$

$$= \frac{K}{d^*_{hkl}}$$

since $(a \cdot a^*) = K$ and $(a \cdot b^*)$ and $(a \cdot c^*)$ are both zero.

It follows therefore that the reciprocal points form a lattice with repeat distances a^* $(= K/d_{100})$, b^* $(= K/d_{010})$ and c^* $(= K/d_{001})$ along Ox^*, Oy^* and Oz^* which are perpendicular to the planes yz, zx and xy respectively, and that any point in this lattice with coordinates ha^*, kb^*, lc^* is the reciprocal point of the family of planes (hkl) drawn in the direct lattice.

Although for convenience both the reciprocal and the direct lattices have been considered to share a common origin this is purely conventional, since they exist in different kinds of space. Also, the scales to which the two lattices are drawn are independent and they can have any value. In fact, when the reciprocal lattice is used as a device for the solution of problems in diffraction, it is quite unnecessary to construct the direct lattice.

As an example of the simplification which is introduced by applying the reciprocal-lattice concept it is instructive to derive the relation between the zone of planes $\{hkl\}$ and the associated zone axis $[uvw]$, and to compare the derivation with that already given in 1.1.4. This axis is parallel to all the planes in the zone and therefore it will be at right angles to the normals of all these planes; in particular, if we consider the planes (hkl) then $[uvw]$ will be at right angles to d^*_{hkl} which is normal to (hkl).

In vector notation, $d^*_{hkl} = ha^* + kb^* + lc^*$ and the vector representing the zone axis is $ua + vb + wc$. Since these vectors must be mutually perpendicular, their scalar product is zero, and hence

$$(ha^* + kb^* + lc^*) \cdot (ua + vb + wc) = 0$$

i.e.

$$hu(a^* \cdot a) + ku(b^* \cdot a) + lu(c^* \cdot a) + hv(a^* \cdot b) + kv(b^* \cdot b) + lv(c^* \cdot b)$$
$$+ hw(a^* \cdot c) + kw(b^* \cdot c) + lw(c^* \cdot c) = 0$$

All the products are zero with the exception of those involving $(a^* \cdot a)$, $(b^* \cdot b)$ and $(c^* \cdot c)$ which are each equal to K and therefore

$$K(hu + kv + lw) = 0$$

or

$$hu + kv + lw = 0$$

2.2.2 Types of reciprocal lattice

The symmetry of any reciprocal lattice is identical with that of the direct lattice from which it is derived. There are thus seven different types of reciprocal lattice each of which conforms to the full symmetry requirements of the system to which it belongs—cubic, tetragonal, hexagonal, trigonal, orthorhombic, monoclinic or triclinic.

Considered as a geometrical construction, the reciprocal lattice is determined solely by the size and shape of the unit cell in real space; that is, it is formed by the repeated translations of lengths a^*, b^*, c^* in three dimensions of a cell which has a volume V^*, and which is defined by the vectors a^*, b^* and c^*. The lattice is therefore primitive whether or not the direct lattice is centred and it is, of course, independent of the atomic arrangement within the crystal itself. If it is convenient to consider the reciprocal lattice as other than a geometrical tool and to invest it with the diffracting properties of the crystal, then each point can be 'weighted' by an amount which is proportional to the intensity of the reflection from the corresponding family of planes. For example, when a lattice is non-primitive, certain general reflections hkl are absent and the intensity of these reflections is identically zero; the corresponding reciprocal-lattice point can be assigned zero weight. The reason for such absences can be illustrated by reference to a body-centred lattice considered to have identical diffracting material at all the lattice points. Now a family of (hkl) planes is defined with respect to the unit-cell dimensions in such a way that it does not matter whether or not the cell is centred, and it may well be that planes passing through corners of the unit cell alternate with planes which pass through the centre of the cell. In a given direction, diffraction effects from the planes through corners will be exactly out of phase with those from the planes through centres, and since there are as many centres as there are corners, the two diffraction effects will cancel, and there will be zero diffracted intensity in the given direction. The condition that this should happen with a body-centred lattice is that $h + k + l$ should be odd. When a lattice is C-face centred, hkl reflections are absent when $h + k$ is odd, and of a lattice which is based on a unit cell in which all the faces are centred, the hkl points in reciprocal space will have zero weight when $h + k$ is odd, $k + l$ is odd and $l + h$ is odd.

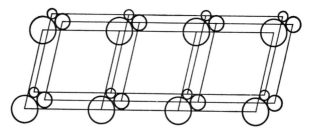

Fig. 2.12   Lattices formed by joining different corresponding atoms in the unit of pattern

As well as the systematically absent reflections which arise from the lattice type and to whose reciprocal points are assigned zero weights, there are also variations in the intensities of reflections which result from the atomic arrangement within the unit of pattern associated with each point of the direct lattice. The weights given to the corresponding reciprocal-lattice points are made proportional to the intensities of the reflections. If the unit of pattern possesses neither glide planes nor screw axes then the variation of intensity arising from the atomic arrangement is one which affects all hkl reflections in general. If however, the unit of pattern does possess elements of translational symmetry, then superimposed on the general variation there will be interference effects resulting from the operation of these symmetry elements which will cause certain types of reflection to be absent.

The effect of the atomic arrangement on the intensity of the general hkl reflections can be understood by reference to the direct lattice. The crystal structure is built up by repeated bodily translations of the unit of pattern (1.1.1) so that the direct lattice can be formed by joining all corresponding atoms in successive units of pattern (2.1.1). In this way a motif of m atoms will give rise to m direct lattices, each of which will be in the same angular orientation in space but displaced bodily relative to one another (fig. 2.12).

From the condition

$$\lambda = 2d_{hkl} \sin \theta$$

for a reflection *hkl* to occur at the Bragg angle θ, it is clear that for the same value of λ, θ depends only on the lattice constants, and hence from each of the *m* lattices there will be a diffracted beam in the same direction. However, the diffracted radiation from these separate lattices of atoms will not necessarily bear the same phase relationships to one another for different values of the common direction θ, and thus the resultant intensity will depend on the direction of the diffracted beam. That is, the intensity will depend upon the indices *hkl* of the reflection. Also, if an atom in the unit of pattern is placed in another position then this will modify the resultant intensity in any particular direction, because the phase relationships between the contribution from this atom and those from the other atoms in the pattern will change.

The effect of the presence of translational symmetry elements in the unit of pattern on the intensities of special types of reflection is somewhat similar to that produced by centering of the lattice, and the general principle can be illustrated by considering the operation of a glide plane. Fig. 2.13a shows the projection on a plane perpendicular to the *c* axis, of a primitive unit cell, and fig. 2.13b shows what happens when a glide plane is introduced. This particular plane produces a *b*-glide parallel to the plane of the paper so that now, in projection, molecules appear at E and F as well as those already present at A, B, C and D. These six molecules will be indistinguishable in projection and since E and F are exactly halfway along the *b* edges of the cell they effectively halve the *b* dimension in projection. This halving causes all *hk*0 reflections with *k* odd to be absent because in this zone, when *k* is odd, the waves diffracted from E and F will be exactly out of phase with those from A, B, C and D, and since molecules of the type E and F are equal in number to those of type A, B, C and D, the total diffracted intensity in the given direction will be zero.

Another factor which affects the intensities of X-ray reflections, and hence the weighting of the reciprocal lattice-points is the different diffracting powers of atoms in different directions. The intensity of the radiation diffracted by an atom depends upon the number of electrons in the atom. If the atom were a mathematical point the intensity would not vary with direction, but since it is finite the scattering power depends upon the angle θ and decreases as θ increases. This will reduce the intensities of reflections with high *h, k, l* values relative to those of reflections with low values of *h, k* and *l*. A fuller discussion of these and other factors which affect the intensities of the reflections from a crystal appears in chapter 7.

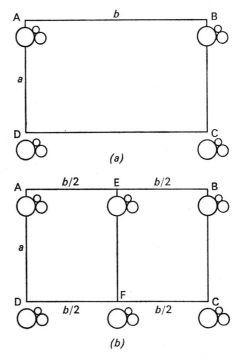

Fig. 2.13   Projection of unit-cell contents along the *c* axis, perpendicular to the plane of the paper: (*a*) with no glide planes present; (*b*) with a *b*-glide plane parallel to the plane of the paper

2.2.3   Use of the reciprocal lattice in diffraction studies

An interpretation of Bragg's Law was proposed by Ewald (1921) whereby the condition $\lambda = 2d_{hkl} \sin \theta$ for an X-ray reflection from a set of planes (*hkl*) could be expressed by a geometrical construction. Suppose that the orientation of a set of planes, spacing d_{hkl} (fig. 2.14), with respect to the incident X-ray beam AO, is such that the reflection *hkl* is obtained. The direction of reflection will be OB where the directions OA and OB each make an angle θ with the planes. The reciprocal-lattice point, P, corresponding to the planes (*hkl*) will lie on the normal to the planes at a distance K/d_{hkl} from the origin O. Now construct a sphere, with centre I on the direction AO of the incident X-ray beam, such that the radius IO is equal to K/λ where λ is the wavelength of the radiation concerned. The particular great circle of this sphere which is shown in the diagram is one on which the point P is assumed to lie; it must, by definition, pass through O and have radius IO and diameter OX. If P does lie on this circle then the angle OPA is a right angle and therefore, since OP is normal to the reflecting planes, angle OXP $= \theta$. It follows from this that angle OIP $= 2\theta$.

Thus $\qquad\qquad$ OP/OX $= \sin \theta$

But OP $= K/d_{hkl}$, and OX $= 2K/\lambda$ by construction, and therefore

$$OP/OX = \lambda/2d_{hkl}$$

i.e. $\qquad\qquad \lambda/2d_{hkl} = \sin \theta \quad \text{or} \quad \lambda = 2d_{hkl} \sin \theta$

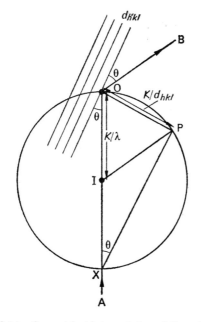

Fig. 2.14   Geometrical interpretation of Bragg's law

Thus when the reciprocal point P lies on the surface of the sphere of diameter OX, the Bragg condition for reflection from the corresponding (hkl) planes is obeyed. Further, IP is parallel to OB and its direction is that of the reflected beam. Since the condition that a family of planes shall be in the correct orientation to reflect X-rays is that the associated reciprocal point should lie on the surface of the sphere as defined above, the sphere is called the **reflecting sphere**.

The two methods of X-ray diffraction are that in which the orientation of a single crystal relative to the direction of the beam is kept fixed and white radiation is used, and that in which a specimen is either rotated or oscillated and the radiation is monochromatic. In the former (the Laue) method the surface of the reflecting sphere is brought into contact with a reciprocal lattice point by the variation of λ and hence by variation of the radius of the sphere; to satisfy this requirement K is chosen equal to unity making the radius of the sphere $1/\lambda$ and therefore variable. When the specimen is either rotated or oscillated in monochromatic radiation, as in both the single-crystal oscillation and the powder methods, the reciprocal-lattice points are brought into contact with the surface of a sphere of fixed radius either by oscillations or rotations of the reciprocal lattice itself; under these circumstances K is chosen equal to λ and the radius of the sphere is therefore constant and equal to unity. In practice the reciprocal-lattice oscillations or rotations are about an axis *through O*.

CHAPTER 3

X-rays

3.1 Nature of X-radiation

3.1.1 Introduction

The complete spectrum of electromagnetic radiation covers the range of wavelengths from infinity down to less than 10^{-13} m. At the longer-wavelength end of the spectrum are the radio waves and at the short-wavelength end are the gamma and cosmic radiations. X-radiation immediately precedes gamma radiation and follows the ultra-violet in the table of decreasing wavelengths, but there are no sharp demarcation lines between the three as far as their wavelengths are concerned. The distinction lies in the manner of their production, and the distinguishing feature about X-rays is that they are produced by the bombardment of matter by high-speed electrons. Wavelengths used in X-ray diffraction are of the order of 10^{-10} m (or 1 Å) because this is the order of magnitude of the distance between atoms in crystalline matter; fortunately there are readily available sources of radiation of about these wavelengths.

3.1.2 White radiation

X-rays can be generated by bombarding a relatively massive target with fast-moving electrons. Electrons can be accelerated between a pair of electrodes kept at a potential difference of upwards of a few thousand volts. For X-ray diffraction studies a typical value of this potential difference is 50 000 volts. Provided that the negative electron does not suffer any loss of energy in its passage from the negative cathode to the positive anode, the energy of the electron on reaching the anode will be eV, where e is the electronic charge and V is the potential difference applied to the electrodes. On striking the target anode the electron will lose energy $E_1 - E_2$, where E_1 and E_2 are the energies of the electron before and after collision respectively. If the resulting deceleration is sufficiently rapid then this loss of energy will be converted into radiation according to the quantum law

$$hv = E_1 - E_2$$

where h is Planck's constant and v is the frequency of the emitted X-radiation.

If an electron loses the whole of its energy in a single collision then the maximum frequency emitted is given by $hv_{max} = eV$. Since $v = c/\lambda$, where c is the velocity of propagation and λ is the wavelength of the radiation, it follows that the minimum value of the wavelength is

$$\lambda_{min} = hc/eV$$

If V equals 50,000 volts, the corresponding value for λ_{min} is approximately 25 pm or 0·25 Å.

It is only rarely that an electron loses the whole of its energy in one impact; more usually it collides successively with several atoms, losing a fraction of its energy on each collision and thus producing several photons

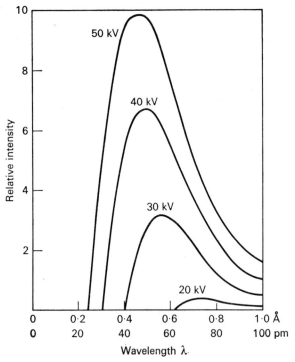

Fig. 3.1   Continuous spectrum from a tungsten target (after Ulrey)

each of which corresponds to a longer wavelength than the minimum, λ_{min}. Hence a continuous spectrum, called **white radiation,** is emitted which is sharply cut-off at the shorter-wavelength end but which tails off gradually at the longer-wavelength end.

As the applied voltage increases, not only will the energy eV of each electron increase but there will also be more electrons striking the target in unit time. It follows that, as well as a decrease in the value of the minimum wavelength emitted, there will be a general all-round increase in the rate of production of photons, with a consequent increase in the intensity of the radiation at all values of the wavelength. The family of curves showing the variation of intensity with wavelength for different applied voltages is shown in fig. 3.1; they are not unlike the curves showing the energy distribution for different temperatures in the spectrum of black-body radiation.

There is a high probability that the electrons will not decelerate sufficiently rapidly to cause the emission of X-radiation, in which case the energy of the electrons will contribute to the internal energy of the target and will be dissipated in the form of heat. In fact less than 1 per cent of the electronic kinetic energy is actually converted into X-radiation. The efficiency of the conversion depends upon the material of the target and increases as Z, the atomic number of the target atoms, increases. When this effect is combined with the effect of increasing applied voltage, V, the total intensity of the X-radiation is found to be roughly proportional to ZV^2.

3.1.3 Characteristic radiation

As well as the continuous spectrum, which depends on the direct loss of energy of the incident electrons, sharp line spectra may also be produced which depend on the material of the anode; these line spectra constitute what is called **characteristic radiation** (fig. 3.2). The intensity of these lines can be a hundred times that in a similar wavelength interval in the continuous spectrum. Characteristic radiation is produced when an incident electron has sufficient energy to dislodge an electron from one of the inner electron shells of a target atom; an electron from a shell of higher energy then drops into the vacant place and the energy is released as radiation. The difference in energy between two given shells will determine the emitted wavelength, and so while an increase in voltage might increase the intensity it does not alter the wavelength of the characteristic lines.

The characteristic-radiation spectra are very simple and are classified in order of increasing wavelength as K, L, M, . . . lines according to the level from which an electron is dislodged. Thus the K lines are produced when an electron is dislodged from the K (the innermost) shell and the gap so created is filled by an electron from a shell of higher level, such as the L or the M shell. If an electron is removed from the next innermost, the L, shell, and is replaced by one from a shell such as the M or the N, then the L lines result.

In accordance with the Pauli exclusion principle, four quantum numbers are required to specify the state of each of the Z electrons in an atom and no two sets of four can be alike. The principal quantum numbers $n = 1, 2, 3 \ldots$ refer to the energy levels K, L, M, $\ldots$ shells respectively. The second

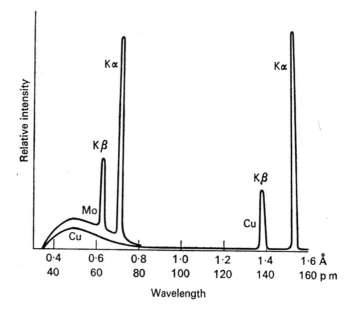

Fig. 3.2   The K spectra of Mo and Cu at 35 kV. The α-line is a doublet

quantum number $l = n - 1, n - 2, n - 3, \ldots , 0$ is related to the orbital angular momentum of each electron in a given level, and the third quantum number j is given by the orbital and spin angular momenta of the electron concerned. To determine the possible values which j can take, reference must be made to the coupling between the orbital angular momentum vector $\boldsymbol{l}$ and the spin angular momentum vector $\boldsymbol{s}$ of the electron. As a consequence of this coupling the total angular momentum vector $\boldsymbol{j}$ of the electron is given by

$$j = l + s$$

and hence from the vector inequality

$$|l + s| \geq \left||l| - |s|\right|$$

it follows that

$$|j| \geq \big||l| - |s|\big|$$

In terms of the orbital angular momentum number l, and spin quantum numbers $\pm\frac{1}{2}$ it can be shown (Eisberg, 1961) that this leads to

$$\sqrt{[j(j+1)]} \geq |\sqrt{[l(l+1)]} - \sqrt{(\frac{1}{2} + 1)}|$$

from which it can be deduced that the only possible values of j which satisfy this inequality are

$$j = l \pm \tfrac{1}{2} \quad \text{when} \quad l \neq 0$$

and

$$j = \tfrac{1}{2} \quad \text{when} \quad l = 0$$

Finally there is the magnetic quantum number $m = \pm j, \pm(j-1), \ldots$, which is the only one that can take negative values. There are thus only two possible sets of quantum numbers for electrons in the K shell, namely $n = 1$, $l = 0$, $j = \frac{1}{2}$, $m = \pm\frac{1}{2}$, and there cannot, therefore, be more than two K electrons. Both electrons will have the same numerical value for their energy so that the K level will be two-fold degenerate (fig. 3.3).

For the L shell, the possible sets of quantum numbers are $n = 2; l = 1, 0;$

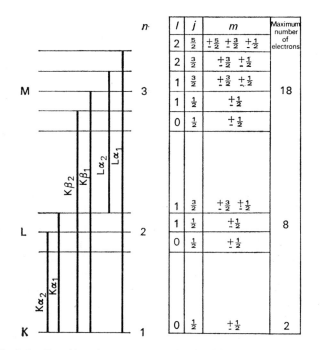

n	l	j	m	Maximum number of electrons
	2	$\frac{5}{2}$	$\pm\frac{5}{2}$ $\pm\frac{3}{2}$ $\pm\frac{1}{2}$	
	2	$\frac{3}{2}$	$\pm\frac{3}{2}$ $\pm\frac{1}{2}$	
3	1	$\frac{3}{2}$	$\pm\frac{3}{2}$ $\pm\frac{1}{2}$	18
	1	$\frac{1}{2}$	$\pm\frac{1}{2}$	
	0	$\frac{1}{2}$	$\pm\frac{1}{2}$	
	1	$\frac{3}{2}$	$\pm\frac{3}{2}$ $\pm\frac{1}{2}$	
2	1	$\frac{1}{2}$	$\pm\frac{1}{2}$	8
	0	$\frac{1}{2}$	$\pm\frac{1}{2}$	
1	0	$\frac{1}{2}$	$\pm\frac{1}{2}$	2

Fig. 3.3   Transitions between energy levels giving rise to X-ray spectra

$j = \frac{3}{2}; m = \pm\frac{3}{2}, \pm\frac{1}{2}$ as shown in fig. 3.3. There are three different numerical values of the energy associated with the L shell and no more than eight electrons can be distributed among these energies. Similarly there are five different numerical values of the M-shell energies and the shell has a possible population of 18 electrons.

Just as the population of the various energy levels is limited by the exclusion principle, so the number of lines in the X-ray characteristic spectrum is limited by the selection rules of quantum mechanics. Electric-dipole radiation is responsible for the most intense lines, and for this type of radiation the selection rules for possible transitions are:

The change in the value of n can take all values
The change in the value of L can be ± 1 only
The change in the value of J can be ± 1 only

where $L = 0, 1, 2, \ldots$ and is the quantum number for the resultant of all the electron orbital motions in the atom. $J = L + S, \ldots L - S$, is the total quantum number involving both orbital and spin motions of the electrons in the atom. The quantum number S is that for all the electronic spins and takes the values $0, \frac{1}{2}, 1, \frac{3}{2} \ldots$. When applied to X-ray spectra these conditions are expressed in the form that l can change by ± 1 only and j can change by $\pm 1, 0$ only. Thus of the three energy differences between the L and the K levels (fig. 3.3) there can be transitions from the L to the K level which involve only two of these—namely those from the L-level energies for which $l = 1$ and which produce the $K\alpha_1$ and the $K\alpha_2$ lines.

Similarly transitions from the M to the K level are possible for only those two of the five M-levels for which $l = 1$; these give the $K\beta_1$ and the $K\beta_2$ lines. Transitions from M to L produce the Lα lines, the two strongest of which, $L\alpha_1$ and $L\alpha_2$, are shown in fig. 3.3. Clearly these are not the only lines allowed by the selection rules.

If an incident electron is to dislodge an electron from, say, the K shell of an atom, then it must have an energy greater than a certain minimum value. The minimum applied potential difference necessary to generate this energy is called the **excitation voltage** (table 1); it varies from atom to atom and depends on how tightly the K electrons are bound. The K excitation potential for copper for example is 9 kV whereas that for molybdenum is 20 kV. The higher excitation voltage required for molybdenum and the fact that molybdenum has the higher atomic number and therefore converts incident energy more efficiently into white radiation (3.12) mean that the white radiation from a target of this material is more intense than that from copper. This results in a higher background from molybdenum. There is an optimum value of the applied potential difference which will give a maximum value of the ratio of the characteristic intensity to the white-radiation intensity. Suitable values of this potential for copper and molybdenum are 50 kV and 80 kV respectively (table 1), but these values are not critical.

Since the K electrons are the most tightly bound in the atom, it follows that if sufficient energy is available to dislodge a K electron then as well as the Kα and Kβ lines the L, M, . . . lines will be present. On the other hand, the latter lines may be present without the K lines.

3.2 Refraction of X-rays

The wavelengths of X-rays are altered when they pass from one medium to another and published wavelength tables refer to the values in vacuo. The difference between the wavelengths in a crystal and in vacuo will be slight because the refractive indices of ordinary crystalline substances for X-rays differ very little from unity; they are, in fact, slightly less than unity and can be calculated from the formula

$$1 - n = \frac{Ne^2\lambda^2\rho \sum Z}{10^7\, 2\pi m \sum W} = \delta \text{ (say)} \qquad \ldots 3.1$$

which has been derived on the assumption that the natural frequency of the scattering electrons is negligible compared with the frequency of the incident X-radiation. In the expression for δ,

n is the refractive index of the crystal
N is Avogadro's number
e is the electronic charge
m is the mass of the electron
λ is the wavelength of X-rays in vacuo
ρ is the density of the diffracting material
$\sum Z$ is the sum of the atomic numbers of the constituent elements per unit cell
$\sum W$ is the sum of the atomic weights of the constituent elements per unit cell
Values of Z and W for the elements are given in table 2.

It is shown in 3.3.1 that

$$N\rho = \frac{\sum W}{\text{volume of unit cell}}$$

i.e.
$$N\rho \frac{\sum Z}{\sum W} = \frac{\sum W \times \sum Z}{\sum W \times \text{volume of unit cell}}$$

$$= \frac{\sum Z}{\text{volume of unit cell}}$$

3+

But $\sum Z$ is the number of electrons in the unit cell, therefore $N\rho \sum Z/\sum W$ is the number of electrons per unit volume of the unit cell, and hence of the diffracting material.

Thus the expression for $1 - n$ can be rewritten as

$$1 - n = \frac{e^2\lambda^2}{2\pi m \times 10^7} \times \text{ number of electrons per unit volume of material.}$$

Returning to the original expression (3.1) for $1 - n$, substitution of numerical S.I. values gives

$$1 - n = 2\cdot71 \times 10^{11} \lambda^2\rho \frac{\sum Z}{\sum W}$$

or, since $\sum Z/\sum W \simeq \frac{1}{2}$,

$$1 - n \simeq 1\cdot35 \times 10^{11} \lambda^2\rho$$
$$\simeq 7 \times 10^{-6}$$

when $\lambda = 7\cdot1 \times 10^{-11}$ m (for molybdenum Kα radiation) and $\rho = 10^4$ kg m^{-3}.

Refraction effects, if taken into account, will modify Bragg's equation slightly. Inside the crystal, Bragg's law must be modified in the form

$$\lambda' = 2d \sin \theta'$$

where λ' and θ' have appropriate values for the interior of the crystal. When they penetrate the crystal, X-rays are refracted away from the normal, since n is less than unity, so that by Snell's law of refraction (fig. 3.4)

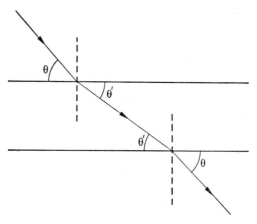

Fig. 3.4   Refraction of X-rays by a crystal

$$n(= 1 - \delta) = \frac{\cos \theta}{\cos \theta'}$$

$$= \frac{\text{velocity of X-rays in air}}{\text{velocity of X-rays in medium}}$$

$$= \frac{\nu\lambda}{\nu\lambda'} = \frac{\lambda}{\lambda'}$$

By writing

$$\frac{\cos^2 \theta}{\cos^2 \theta'} = \frac{1 - \sin^2 \theta}{1 - \sin^2 \theta'}$$

it can be shown that, approximately,

$$\frac{\sin \theta'}{\sin \theta} = \frac{1}{n}\left\{1 - \frac{1 - n}{\sin^2 \theta}\right\}$$

$$= \frac{\lambda'}{\lambda}\left\{1 - \frac{\delta}{\sin^2 \theta}\right\}$$

so that

$$\lambda = \frac{\lambda'}{\sin \theta'}\left\{1 - \frac{\delta}{\sin^2 \theta}\right\} \sin \theta$$

$$= 2d\left(1 - \frac{\delta}{\sin^2 \theta}\right) \sin \theta$$

from the equation $\lambda' = 2d \sin \theta'$

As we have already seen, δ is of the order of a few parts per million so that the effect of refraction is not serious.

3.3 X-ray Wavelengths

3.3.1 Units

The wavelength of X-rays diffracted by crystalline matter can be determined from Bragg's law, $\lambda = 2d \sin \theta$, provided that both sin θ and d, the spacing of the planes which produce the given reflection, can be determined. Sin θ can be measured experimentally but d must be calculated from a know-ledge of certain constants by the following well-established method.

The average density of a crystalline substance, that is the density deter-mined by experiment, is given by

$$\text{density} = \frac{\text{total mass of the atoms in the unit cell}}{\text{volume of the unit cell}}.$$

Now the atomic weight of an element is the mass of an atom of that element multiplied by Avogadro's number N; therefore

$$\text{mass of one atom} = \frac{\text{atomic weight}}{N}$$

$$\text{mass of } n \text{ atoms} = \frac{n}{N} \times \text{atomic weight}$$

Thus if in a unit cell there are n_1 atoms each of atomic weight M_1, n_2 atoms of atomic weight M_2, ... then

$$\text{total mass of the atoms in the unit cell} = n_1 M_1/N + n_2 M_2/N + \ldots$$

$$= (1/N) \sum n_i M_i$$

where n_i and M_i refer to the ith kind of atom.

Thus,

$$\text{density of the substance} = \frac{\sum n_i M_i}{N \times \text{Volume of cell}}$$

Hence the volume of the unit cell can be calculated. If the cell is cubic, then the edge a is known and so the value of any required interplanar spacing, d, is also known (5.2.1). The accuracy to which d can be determined depends upon the accuracy to which the density, the relevant atomic weights and N are known, and the greatest uncertainty in these quantities lies in the value of N. An accurate knowledge of N depends in turn upon the precision with which the electronic charge, e, can be measured and thus, finally, the accuracy of the value of d, and therefore of λ, depends largely upon the precision to which e is known.

To eliminate these uncertainties Siegbahn arbitrarily defined a unit of length which was based upon the 200 spacing of rock salt. From the data available at the time the value of this spacing was taken to be 2·814 Å and the 200 spacing of calcite was determined using this as a standard. Siegbahn estimated that spectrometer measurements relative to a fixed standard could be made to six significant figures and he therefore defined the d_{200} rock-salt spacing as 2814·00 units. Using the $K\alpha_1$ radiations of copper and iron, and the $L\alpha_1$ radiation of tin, he obtained three separate values for the 200 spacing of calcite from a comparison of the $\sin \theta$ values for the reflections from the (100) planes of calcite and of rock salt. The average of these three values was 3029·04 at 18°C in units which he called X units. Within the limits of accuracy of the figure 2·814 Å for the 200 spacing of rock salt, the X unit was equal to 10^{-3} Å.

The value of 3029·04 X units for the 200 spacing of calcite at 18°C has been universally adopted as a standard in terms of which X-ray wavelengths may be determined from measurements of the second-order reflections from the (100) planes of calcite. This value takes into account the effect of the refraction of X-rays by calcite so that the X-ray wavelengths derived from the use of this standard are 'true' wavelengths, that is, wavelengths in vacuo. The actual spacing of the (200) planes in calcite, obtained by multiplying 3029·04 by the appropriate correcting factor for refraction, is 3029·45 X at 18°C; this is also a definition of the X unit.

At the present time the wavelength of copper $K\alpha_1$ radiation is taken to be 1537·400 X units and that for molybdenum $K\alpha_1$, 707·831 X units. With the increased accuracy in the absolute determination of spacings now possible, the value of the multiplying factor for conversion from kX units (X units $\times$ 10^3) to Å is no longer unity, and the modified values which have variously been quoted range from 1·00202 (Bragg, 1947) to 1·00206 (DuMond and Cohen, 1953).

The conversion factor can be determined in one of three different ways. The first way is to measure an X-ray wavelength with a ruled grating, and divide this value by the known value of the wavelength in kX units. According to Bearden (1967) only two such ruled-grating determinations have been made with sufficient accuracy; one of these was by Bearden (1931) himself and the other was by Edlen and Svensson (1965). Bearden measured the conversion factor using a plane grating and the α and β lines of both copper and chromium radiations. Edlen and Svensson used a concave grating with AlKα radiation.

The second method by which the conversion factor can be determined has already been outlined, and involves calculating for example, the (100) spacing of a crystal such as calcite from a knowledge of the density and molecular weight of the material and of Avogadro's number; this is then compared with the known value of the spacing in kX units. Alternatively, once the spacings of a suitable material have been determined in this way, the wavelength of CuKα_1 can be found by powder methods and then compared with the known value in kX units. Unless there is a major error in Avogadro's number, the value of the conversion factor calculated in this way is the most accurate available.

The third method is described by Bearden (1967) and involves the high-frequency limit of the continuous X-ray spectrum. If the applied potential difference between the anode and cathode of an X-ray tube is V, then the minimum wavelength emitted is given (in S.I. units) by

$$\lambda_{min} = hc/Ve$$

Therefore $\lambda_{min}(\text{Å}) = (hc \times 10^{10})/Ve$ and the conversion factor $X(\text{Å})\lambda$ (kX units) $= (hc \times 10^{10})/eV\lambda$, where λ is the value of the minimum wavelength in kX units and can be measured (Spijkerman and Bearden, 1964). Since V is known, the conversion factor can be calculated.

From an examination of the published values obtained by the three methods Bearden (1967) has concluded that the best value for the conversion factor is 1·002056 ± 5 p.p.m. Thus the wavelength of CuKα_1 radiation is 1537·40 X units or 1·540562 Å ± 5 p.p.m.

The one flaw in the definition of the X unit is that every good calcite crystal is assumed to have the same value for corresponding interplanar spacings. This is, in fact, not so and although the variations are small and often less than the errors of measurement, the improved techniques now available mean that these variations remain a problem. It has been proposed that one of the characteristic radiations would be a more suitable basis for the definition of a length than would a crystalline material, samples of which are not all alike on the atomic scale. After an extensive research programme on the evaluation of X-ray wavelengths Bearden, Henins, Marzolf, Sauder and Thomsen (1964) recommended that such a standard should be the wavelength of the peak of the spectral line of tungsten Kα_1. The wavelength of CuKα_1 was assumed to be 1·540562 Å ± 5 p.p.m. and by comparison with this wavelength using five selected crystals of which one was calcite, two were quartz and two were silicon, the value of the wavelength of tungsten Kα_1 radiation was found to be

$$0·2090100 \text{ Å} \pm 5 \text{ p.p.m.}$$

The standard recommended by Bearden and his co-workers is therefore 0·2090100 Å for the wavelength of tungsten Kα_1 radiation.

3.3.2 Accuracy of measurement

When measurements of wavelengths are required to high accuracy certain properties of the spectra and of the diffracting matter have to be taken into account. The spectral lines themselves are not strictly monochromatic and often they are asymmetrical so that wavelengths corresponding to the peaks do not coincide with the wavelengths corresponding to the centres of gravity; lack of symmetry in the line thus leads to differences in the wavelength as determined by various instrumental methods. Refraction and other effects arising from crystal imperfections must be considered.

Bearden *et al.* include in their reasons for their choice of tungsten as a wavelength standard the facts that the line is narrow and highly symmetrical and that it requires a minimum of crystal corrections.

A list of the more useful wavelengths is given in table I.

3.4 Absorption of X-rays

3.4.1 Absorption coefficients

X-rays are partly absorbed and partly transmitted by crystals. The absorption of X-rays reduces the intensity of the primary beam from its initial value of I_0 to a value I according to the law

$$I = I_0 e^{-\mu t}$$

where I is the intensity after the beam has traversed a thickness t of matter. The coefficient μ is called the **linear absorption coefficient** for a homogeneous substance.

Factors which affect the value of μ are the wavelength of the incident X-radiation and the density of the scattering material. The density depends on whether the absorbing substance is in the solid, liquid or gaseous phase, and whether it is composed of heavy or light atoms. The absorbed energy may be transformed in several processes:

(1) Production of heat.
(2) Emission of X-rays of longer wavelength called **fluorescent radiation** which are characteristic of the scattering material.
(3) Emission of photo electrons; this accompanies the emission of fluorescent radiation.
(4) Production of X-rays of lower energy and longer wavelength (Compton effect).
(5) Scattering of X-radiation which has not suffered any change in wavelength; this is coherent radiation and is responsible for the phenomena with which this book is concerned.

Although μ has been measured for most elements for the wavelengths which are normally used in X-ray diffraction, these values are not particularly useful for calculating μ for other substances. More useful parameters are those which do not depend on the physical state of the element concerned—as does μ—but which for a given wavelength are constant for a given element. Generally two quantities are tabulated, namely, the mass absorption coefficient, μ_m, and the gram-atomic absorption coefficients, μ_g.

The mass absorption coefficient (table 3) is defined as μ/ρ and its units are $m^{-1}/kg\ m^{-3}$ or $m^2\ kg^{-1}$. To work out this coefficient for a substance composed of several elements the weighted mean of the mass absorption coefficients of the constituent elements is obtained for the wavelength involved. Thus μ/ρ for a substance comprising masses $m_1, m_2, m_3, \ldots$ of elements having mass absorption coefficients $(\mu/\rho)_1, (\mu/\rho)_2, (\mu/\rho)_3, \ldots$ is given by

$$\frac{\mu}{\rho} = \frac{m_1}{M}\left(\frac{\mu}{\rho}\right)_1 + \frac{m_2}{M}\left(\frac{\mu}{\rho}\right)_2 + \frac{m_3}{M}\left(\frac{\mu}{\rho}\right)_3 + \ldots$$

where M is the total mass of the substance, i.e.

$$\mu = \rho \sum p_i(\mu_m)_i$$

where p_i is the fraction by weight of the ith element and $(\mu_m)_i$ is the mass absorption coefficient of that element.

In X-ray diffraction work it is more usual for the relative numbers of the different atoms present to be known rather than the fractions by weight of the elements; under these circumstances it is more convenient to calculate μ in terms of the gram-atomic absorption coefficients, μ_g, of each element involved. The method is described by Henry, Lipson and Wooster (1960).

A fourth absorption coefficient may be used if both the volume and the content of the unit cell are known. This coefficient is designated μ_a and is defined by

$$\mu_a = \frac{\mu_m W}{N}.$$

In terms of this coefficient

$$\mu = \frac{\sum \mu_a}{V}$$

where the summation is over all the atoms in the unit cell and V is the volume of the cell.

Except at what are called absorption edges (3.4.2), the mass absorption coefficient is roughly proportional to $(\lambda Z)^3$ where Z is the atomic number of the absorbing atoms and λ is the wavelength of the incident radiation. Thus for a given element, μ/ρ is roughly proportional to λ^3, and if a mass absorption coefficient is required for a wavelength which is not listed in the tables the μ_m is first obtained for the nearest wavelength for which mass absorption coefficients are known. The unknown coefficient $(\mu_m)_x$ can then be obtained from

$$\frac{(\mu_m)_x}{\mu_m} = \left(\frac{\lambda_x}{\lambda}\right)^3$$

This relationship is not precise and is valid only for wavelengths which are nearly equal.

3.4.2   Absorption edges and fluorescent radiation

In the visible region of the electromagnetic spectrum there are marked similarities between emission and absorption spectra. With X-rays, however, the emission and absorption spectra are entirely different. Characteristic X-ray line emission spectra are produced by the ejection of electrons from the inner levels of the atom, which are then refilled by an electron from a level of higher energy (3.1.3); if X-rays constitute the exciting radiation then the incident photons must have sufficient energy to dislodge these inner electrons. The X-ray absorption spectrum however, consists not of a series of sharp lines as does the emission spectrum, but of one or more

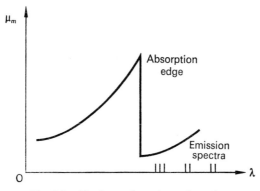

Fig. 3.5   The form of an absorption edge

absorption edges like that shown in fig. 3.5. In the figure the relation between the mass absorption coefficient of the material and the wavelength of the incident radiation is shown. As indicated in section 3.4.1 the value of the mass absorption coefficient is roughly proportional to λ^3 so that, in general, as λ increases so does the absorption of the radiation. However, as the wavelength increases, sudden falls in absorption occur at sharply defined values of the wavelength. The points on the wavelength scale at which these sudden changes take place are called absorption edges.

At wavelengths greater than that of the absorption edge (table 2) the X-ray quantum, hc/λ, is too small to dislodge an inner electron from a target atom. At sufficiently short wavelengths, the incident quanta are able to dislodge electrons from the inner shells of the target atoms and, in doing so, are absorbed; this results in a sudden increase in the value of μ_m and in the appearance of an absorption edge. Once an electron has been removed from an inner shell, the gap left is filled by an electron from a shell of higher energy and one of the characteristic radiations of the absorbing element is

3*

emitted; the emission is accompanied by the production of a photoelectron. The wavelength of the fluorescent radiation (3.4.1) is longer than that of the incident radiation because the exciting photon must have sufficient energy to remove an electron from an inner shell but the energy of a photon of the fluorescent radiation is the difference between the energies of two shells and is therefore smaller. Thus, to produce fluorescent radiation the wavelength of the incident radiation must be less than or equal to the wavelength of the absorption edge which, in turn, will be less than the wavelength of the emission lines of the absorbing substance.

Fluorescent radiation is emitted in all directions and the scattering is incoherent.

3.4.3 Modified and unmodified scattered radiation

As well as the reduction in intensity of the primary beam due to the absorption edges, there is also a reduction in intensity caused by modified and unmodified scattering of the incident radiation.

When the primary beam encounters either very loosely bound or free electrons, these electrons recoil and take up energy from the X-ray beam so that lower-energy, longer-wavelength X-rays emerge. These secondary X-rays are the primary X-rays which have been modified by the Compton effect (3.4.1); they are scattered in all directions and because the phase relationships of the scattered X-rays are random, the radiation is incoherent. The increase in wavelength of the secondary X-rays depends on the scattering angle—the larger the value of θ, up to 90°, the longer the wavelength of the scattered radiation. Also the lighter the scattering atoms, the more loosely bound are their electrons and the greater is the intensity of the scattered radiation.

When the primary beam encounters the electrons of the target, yet another type of interaction is possible. Instead of being ejected or recoiling, some electrons may simply cause the photons to bounce off without any transfer of energy; the scattered radiation is therefore unchanged in wavelength. This is coherent radiation and is responsible for the diffraction spectra normally observed in X-ray diffraction experiments.

3.4.4 Choice of radiation and reduction of background

On any powder photograph, the line spectra produced by diffraction of the characteristic radiation will be superimposed on a continuous background and under many circumstances, for example with weak lines, it is necessary to reduce the background intensity as much as is possible. Types of radiation contributing to the background are the coherently scattered white radiation and the incoherent fluorescent and modified radiations (3.4.1).

The intensity of white radiation relative to that of characteristic radiation is greater the heavier the target element (3.1.3). Hence one way of reducing the white-radiation background, whether the scattering is coherent or incoherent, is to reduce the general level of white radiation by using copper rather than, say, molybdenum radiation. Fluorescent radiation will be produced by absorption of both white and characteristic radiation. The absorption-edge effect on white radiation may be reduced by minimizing the general level of this radiation by suitable choice of target as just described, but reduction of the potentially more serious effect on characteristic radiation may require greater discrimination in the choice of radiation. If the characteristic wavelength is just greater than that of the absorption edge then it presents no problem, but if it is just less then there will be intense fluorescent radiation producing a very heavy background. The further the wavelength is reduced beyond the absorption edge the less will be the fluorescent effect because, away from the absorption edge, the higher the X-ray energy the less it is absorbed (3.4.1). CuKα radiation of wavelength 1·54 Å is totally unsuitable for taking photographs of specimens which contain a significant quantity of iron, the K absorption edge of which has wavelength 1·74 Å.

If fluorescent radiation is present, the relative background intensity produced by the absorption of the incident characteristic radiation may be minimized by interposing a metallic-foil screen between the specimen and the intensity-recorder. Both the incoherent fluorescent and the coherent diffracted radiation will be absorbed by the screen, but because the wavelength of the fluorescent radiation is necessarily longer than that of the incident radiation it will be preferentially absorbed (3.4.1). The longer the fluorescent-radiation wavelength compared with that of the exciting wavelength the greater will be its relative absorption; if the fluorescent wavelength is sufficiently long, such as AlKα = 8·34 Å, the air in the camera will have the desired effect.

Screens will clearly not be effective when the wavelength of the incoherent radiation is only slightly longer than that of the incident radiation; thus they cannot be used to reduce the effect of Compton-modified characteristic radiation, nor are they effective against either Compton-modified white radiation or coherent white radiation.

Methods of reducing background intensity relative to the line intensity by the use of monochromators, larger-diameter cameras and certain types of counter are discussed in sections 3.5.1, 4.4.1 and 4.5.1 respectively.

In order to simplify X-ray diffraction photographs, the characteristic Kβ radiation is often filtered out; this reduces the number of recorded reflections. A filter is chosen which has an absorption edge between the particular Kβ and Kα radiations which are being used and the thickness of the filter (3.4.1) is calculated so that it removes virtually all the Kβ radiation (fig. 3.6). The filter will at the same time reduce by absorption that portion of the white radiation spectrum with wavelengths just shorter than that of the absorption edge. Absorption of the Kβ and the adjacent continuous radiation will inevitably result in increased fluorescent radiation; if the wavelength of the radiation is longer than that of the Kα radiation then a screen placed between the specimen and the intensity recorder will decrease the background effect.

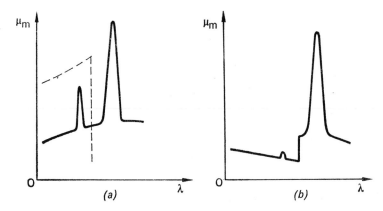

Fig. 3.6   Effect of a filter on the relative intensities of Kα and Kβ radiations:
(a) without filter; (b) with filter

3.5 Crystal-reflected Radiation

3.5.1 Introduction

A very effective way of cutting out both white and unwanted characteristic radiations is to use crystal-reflected radiation. For this purpose a large crystal with a set of planes that strongly reflect the characteristic lines is required. After reflection by these planes the various characteristic radiations are angularly dispersed, and it is possible to isolate the required line, say the Kα, with a suitable collimating system. Crystals used under these conditions are called **monochromators** and, after selection, the reflected radiation is called **monochromatic**. Substances suitable for use as monochromators are listed in the table below and plate I shows two powder

Crystal	Reflection	Spacing Å	Properties of reflection		Properties of crystal			Special uses
			Peak intensity	Breadth	Crystal imperfection	Stability	Mechanical properties	
β Alumina	0002 0004	11·24 5·62	Weak Weak–medium	Moderate	Great	Perfect	Hard, brittle	For long wavelengths, but usable crystals hard to obtain
Mica	001 004	10·1 2·53	Weak	Small	Negligible for selected specimens	Fair	Flexible, easily cleaved	For point-focusing devices; exhibits irradiation effects
Gypsum	020	7·60	Medium–strong	Very small	Good specimens hard to find	Poor	Soft, flexible	For small-angle scattering; focusing long wavelengths
Pentaerythritol	002	4·40	Very strong	Moderate	Great	Poor	Soft, easily deformed	General purposes; exhibits irradiation effects
Quartz	10Ī1	3·35	Weak–medium	Very small	Negligible	Perfect	Can be elastically bent	For small-angle scattering; focusing
Potassium bromide	200	3·29	Medium–strong	Moderate	Negligible	Slightly deliquescent	—	
Fluorite	111 220	3·16 1·94	Medium–strong Very strong	Moderate	Small	Perfect	Moderately hard	For eliminating harmonics; general purposes; short wavelengths
Urea nitrate	002	3·14	Strong	Very large	Very great	Very poor	Very easily deformed	For large specimens; soon decays
Calcite	200	3·04	Medium	Small	Negligible	Perfect	Moderately soft	For small-angle scattering; isolation of α_1 or α_2
Rock salt	200	2·82	Medium–strong	Large	Great	Slightly deliquescent	Can be plastically bent in warm supersaturated saline	For focusing
Aluminium	111	2·33	Very strong	Moderate to large	—	Good	Soft, can be seeded and grown to shape, then plastically shaped at room temperature	For focusing; diffuse scattering
Diamond	111	2·05	Weak	Very small	Negligible	Perfect	Very hard	For eliminating harmonics
Lithium fluoride	200	2·01	Very strong	Small–moderate	Negligible	Perfect	Hard, can be plastically bent at high temperature	For focusing; diffuse scattering; general purposes

photographs of Fe_3Si taken (a) without and (b) with $CoK\alpha$ radiation reflected in the 200 order of diffraction from lithium fluoride. Lithium fluoride monochromators are very good for general work, but if a particularly fine beam is required, as for the study of small-angle scattering (9.5.2), crystals such as those of quartz or calcite should be used.

In the beam obtained from a monochromator may be included wavelengths which are sub-multiples of that of the reflected characteristic radiation. The first-order reflection of the characteristic radiation, wavelength λ, the second-order reflection of radiation of wavelength $\lambda/2$, the third-order reflection of radiation of wavelength $\lambda/3$ and so on are all reflected at the same angle θ. These harmonics will be picked out from the white-radiation spectrum, and the more intense the white radiation, the stronger will be the harmonics. Because they are of shorter wavelength they will not be so easily absorbed and will not therefore, for the same intensity, blacken a photographic film to the same extent as will the longer-wavelength characteristic radiation; nevertheless, if accurate calculation of intensities is required, particularly with instruments like diffractometers which do not detect by photographic means, the harmonics must be eliminated. Harmonics will also be present in non-crystal-reflected radiation, but in investigations for which such radiation is adequate their presence will presumably not seriously affect the accuracy.

Harmonics can be eliminated from the diffracted beam by running the X-ray tube at such a low voltage that the sub-multiple wavelengths are not even produced, but to eliminate the radiation of wavelength $\lambda/2$ in this way the tube must for example be run under a potential difference of less than 18 kV for $CuK\alpha$ radiation. Since the excitation voltage is 9 kV, with a total applied potential difference of only 18 kV the radiation would be very weak indeed. Alternatively a reflection for which the harmonics are of negligible intensity could be used as a source. This is not possible however for all sub-multiples of the characteristic wavelength, and in practice the best solution is to choose a reflection in which the second-order reflection of the $\lambda/2$ component is insignificant and to run the X-ray tube so that $\lambda/3$ is not emitted. For $CuK\alpha$ radiation the tube voltage would still be of the order of only 24 kV and the intensity of the characteristic would still be only a small fraction of the intensity of the total radiation emitted.

The 222 reflection from fluorite is very much weaker than the 111 reflection and therefore the second-order reflection of radiation whose wavelength is half that of the characteristic $K\alpha$ wavelength will be very much weaker than the first-order reflection of the characteristic radiation. Fluorite is thus a suitable crystal for the suppression of harmonics. Even better from the point of view of purity of radiation are crystals of diamond, silicon or germanium from each of which the 222 reflection is identically zero.

For determinations of structure, relative intensities must be measured
and the partial polarization by the monochromator of the crystal-reflected
radiation taken into account (Azároff, 1955); this polarization will be
further modified by polarization of the diffracted beam by the specimen
itself (7.43).

3.5.2   Curved-crystal monochromators

The beam from an X-ray tube is divergent so that if parallel radiation is
required the beam must be suitably restricted. A plane monochromator
set to reflect parallel radiation uses so little of the original intensity of the
X-ray source that the monochromatic beam is not nearly so strong as it
could be. In investigations involving single crystals, a parallel beam is
essential and so, although the beam is weak, a plane monochromator must
be used. However, in many other types of investigation, it is possible to
use a bent monochromator which is capable of focusing the divergent
radiation and in the process producing a much more intense beam (4.8.1).
Crystals to be shaped as curved monochromators must be able to withstand bending and either cutting or grinding; quartz, which is elastically
bent, is the most popular, but both rock salt and aluminium, which are
plastically bent, are also used.

The principle of the curved-crystal monochromator, which has been
described by Johannson (1933), is illustrated in fig. 3.7.

P is the source of X-rays and we must find the condition that all rays of

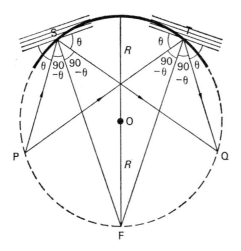

Fig. 3.7   The curved-crystal monochromator

a given wavelength leaving P should pass through another point Q after reflection by a set of planes of the crystal monochromator. All rays reflected by this set of planes must be incident at the same glancing angle θ in order to satisfy Bragg's relation $\lambda = 2d \sin \theta$ since both λ and d are fixed by the conditions. Thus all rays reflected by these planes must have turned through the same angle 2θ; that is, the angle between each pair of incident and reflected rays must be the same for all rays incident on the crystal and it must be equal to $(\pi - 2\theta)$. This can only be so if P, Q and all the points at which the rays are incident on the crystal lie on the same circle.

Let the radius of this circle be R (fig. 3.7). At S, PS and SQ must make equal angles, $\frac{1}{2}\pi - \theta$, with the normal to the reflecting planes at S. At T, PT and TQ must make the same equal angles with the normal to the planes at T and the two normals must meet at F on the circumference of the circle, radius R, because only then can the angle PSF be equal to the angle PTF ($= \frac{1}{2}\pi - \theta$).

Similarly the angles FSQ and FTQ are equal, and therefore PF = FQ.

Hence all the normals to the reflecting planes pass through F, and the reflecting planes must be bent into an arc of a circle of radius $2R$, centre F, while the surface of the crystal must lie on a circle of radius R, centre O (fig. 3.8). If the crystal surface is cylindrical then the X-ray source at P (fig. 3.7) must be linear and parallel to the axis of the cylinder; the rays will be brought to a line focus at Q, again parallel to the axis of the cylinder.

The way the monochromator is shaped depends on whether the crystal is elastically or plastically bent. Elastically bent quartz monochromators (Guinier, 1937, 1939, 1945a, 1952) are prepared by first grinding the oppo-

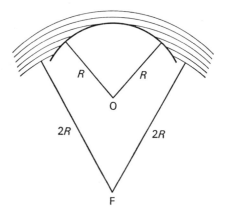

Fig. 3.8   Relation between the curvature of the reflecting planes and that of the reflecting surface

site surfaces of the crystal parallel to the ($10\bar{1}1$) planes, and then grinding these surfaces to produce a cylindrical surface of radius $2R$. The crystal itself is then bent elastically and held in supports so that the radius of curvature of the planes is $2R$; this will automatically make the radius of curvature of the ground cylindrical surface of the crystal R.

Plastically bent rock-salt monochromators (Bozorth and Haworth, 1938) are obtained by first preparing a rock-salt crystal with opposite faces parallel to the (100) planes and then bending the crystal plastically so that these planes are cylindrical with a radius of curvature $2R$; the cylindrical surface of the crystal is then ground until it has a curvature of radius R. Monochromators made from aluminium crystals are also plastically bent (Cauchois, Tiedema and Burgers, 1950; Hägg and Karlsson, 1952). The aluminium monochromator described by Hägg and Karlsson was designed for point focusing and necessarily has double curvature. Point focusing has also been achieved by using two cylindrical quartz mono-chromators arranged with their axes at right angles; the geometry of this system has been given by DuMond (1950). One of the advantages of a point focus is that it allows much shorter exposure times than does the line focus produced by the single-curvature monochromator; a point focus, too, makes the study of the texture of surfaces possible.

Crystal monochromators are normally between 2 and 5 mm thick The radius R is of the order of 200–400 mm but its value must be chosen to suit the wavelength and the working distance of the apparatus.

3.6 The production of X-rays

3.6.1 Types of tube

X-rays for diffraction work are produced by bombarding a suitable metal-lic surface with electrons which have been accelerated towards the target under a potential difference which may lie between 20 and 90 kV (3.1.2, 3.1.3). The target, which is earthed, forms the anode and relative to it the cathode is maintained at a negative potential of 20 to 90 kV. In the gas tube the electrons are produced by ionization of the residual gas and not by electrical heating of the cathode; in the hot-cathode tube the gas pressure is too low for ionization to occur, and the electrons are produced by a tung-sten filament which is heated electrically, and which acts as the cathode.

3.6.2 The gas tube

In early work on X-ray diffraction the home-made gas tube was favoured but this has been largely replaced by the hot-cathode tube which is now readily available commercially in a variety of designs. Gas tubes were, and remain, very temperamental; their outstanding virtue is that the radiation from them remains pure and does not become contaminated with radiation from anode deposits, notably tungsten, which arise in time from heated filaments. Because they are now seldom used we shall not describe them further.

3.6.3   The hot-cathode tube

The pressure inside a hot-cathode tube should be less than 0·0001 mmHg (0·01 N m^{-2}) in order to prevent any discharge in the residual gas, and for this purpose a diffusion pump backed by a rotary pump is required. The essential details of the tube, together with the electrical circuit which also acts as a half-wave rectifier, are shown in fig. 3.9.

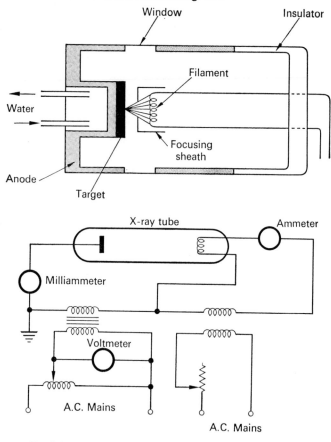

Fig. 3.9   The hot-cathode tube and the electrical circuit

A heated tungsten filament, which acts as the cathode, is aligned so as to be between 5 and 10 mm from the water-cooled copper anode. If a metal other than copper is to be the target then the required metal is either electroplated on to the anode or is fitted into it as an insert. The X-rays are focused by sheathing the filament in a metal hood; the end of the hood is slotted in a direction parallel to that of the filament so as to produce a line focus on the target. In some tubes the hood and the filament are at the same potential but it is difficult to see how this arrangement can provide for focusing; it seems rather to be straightforward collimation of the beam. Some designs do provide for an elementary focusing system by having the hood electrically insulated from the cathode assembly and maintained at 200 or 300 V negative to the filament.

The windows through which the X-rays pass from the target to the diffraction instrument can be made of aluminium, of beryllium or of beryllium and mica. The dimensions of the focal area on the target are roughly 10 mm by 1 mm and the focus is 'viewed' at an angle of 3° to 6° with the target face. End-on, the focus appears to be a spot of dimensions 1 mm by 1 mm, broadside on it appears to be a line 10 mm long by 0·1 mm wide. It is usual to focus the electron beam in the form of a line rather than of a spot because this results in more efficient heat dissipation at the anode, thus making it possible to run the tube at a higher power. Viewing the target at a narrow angle further increases the effective load per unit area of the focus; this increases the intensity of the X-ray beam, and to obtain a beam of maximum intensity the aperture of the collimating system should be greater than the apparent size of the focus.

There are two types of hot-cathode tube in general use; one is permanently sealed off and the other is continuously evacuated. With copper targets, both operate at a power of 1 to 2 kW and the demountable tube is perhaps the more powerful of the two on balance. When targets other than copper or silver or gold are in use, the power output of both types of tube is markedly reduced because heat is dissipated less efficiently at the anode. Sealed-off tubes are easy to operate and modern commercial equipment is compact and self-contained; the electrical supplies, including voltage stabilization, are built-in and the characteristic radiation can be changed in minutes by replacing what is called 'the insert'. This is an evacuated tube containing target, filament and focusing device.

Available commercially are 1 kW tubes with a focus of 10 mm by 1 mm and 2 kW tubes with one of 10 mm by 1·6 mm; the corresponding loadings are therefore 0·1 and 0·125 kW mm^{-2}.

The disadvantage of the sealed-off tube is that tungsten from the filament is deposited on both the target and the windows. In the course of time, tungsten L radiation will be produced (3.1.3) and the tungsten layer on the target and on the windows will reduce, by absorption, the intensity of the X-ray beam; thus the longer the tube has been in use, the less efficient it becomes as a source of X-rays.

Continuously evacuated tubes are fully demountable. The filament, the target and the windows can all be replaced without undue difficulty and there is no reason why this type of tube should not function efficiently for many years. With all demountable tubes occasional vacuum troubles occur but an elementary knowledge of vacuum techniques on the part of the operator should be sufficient to keep a tube in service.

3.6.4   The rotating-anode tube

The rate at which heat can be dissipated at a fixed anode sets an upper limit to the intensity of the X-ray-tube focus. By rotating the anode, so that no one part is subject to continuous electron bombardment, it is possible to increase the tube power (and the load per unit area of the focus) to 5 or 10 kW. There are two basic designs of rotating anode, both of which are illustrated in fig. 3.10. The shaft rotates in a vacuum-tight seal, and in the type

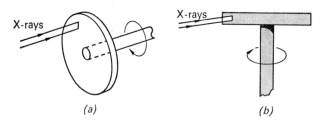

(a)   *(b)*

Fig. 3.10   The two basic designs of rotating anode

(a) the axis of rotation is parallel to the filament-anode axis of the tube but is slightly offset so that the electron beam is focused near the outer rim of the circular anode. In the type (b) the shaft is at right angles to the electron beam and the circular anode has a thick edge on which the electrons are focused.

3.6.5   The fine-focus tube

Another way in which the power load per unit area of focus can be increased is to operate the tube with a small focus; such tubes are called fine-focus tubes. The smaller the focal spot the greater the load per unit area that the target can withstand; this is presumably because the lines of flow of heat conduction are radially outwards, thus promoting more efficient target cooling than is possible with a line focus. Hence, with the same power, a fine-focus tube provides a much more intense source of X-radiation; indeed with relatively low power, less than 100 W for example, the intensity is still far greater than that obtainable from a conventional line-focus tube. This is easy to understand when one realizes that foci of about 50 μm and less in diameter are not now unusual.

The diffraction apparatus used with a fine-focus tube must be suitably designed (4.8.3). A fine focus is essential for a study of surface texture and for this purpose the collimator aperture must be very narrow. Also, the distances between the focus and the specimen, and the film, must be short. On the other hand in single-crystal investigations, where the relative intensities of the diffraction spots must be estimated by eye the use of a fine focus will be something of a drawback because the spots will be too small.

The production and control of a fine focus presents a problem in electron optics. In early tubes, such as that described by Goldsztaub (1947), a combination of an electrostatic lens followed by an electromagnetic lens was utilized for finely controlled focusing of the electron beam. Ehrenberg and Spear (1951) made a radical change in the design of fine-focus tubes simply by earthing the cathode and running the tube with the anode at a high positive potential relative to the earthed cathode. This arrangement allows the electrical supplies necessary for electrostatic focusing to be at, or near, earth potential so that the electromagnetic lens becomes unnecessary and there is no need for a long cathode-anode path and highly stable electrical supplies. A highly stable high-tension supply is not necessary for electrostatic focusing and interference from stray fields can be eliminated by making the filament-anode distance short, of the order of 20 to 30 mm. The anode is cooled by a stream of either carbon tetrachloride or oil pumped from a reservoir through plastic tubing. When the anode is not earthed, direct water cooling cannot be utilized.

Fine-focus demountable X-ray units which are based upon the Ehrenberg and Spear design are available commercially. The focal spot is 40 μm in diameter and the tube current can be as high as 400 μA at 50 kV. This gives a tube power of 20 W and a focus loading of $0 \cdot 2$ kW mm^{-2}.

3.6.6 Electroplating of targets

Separate sealed-off tubes are necessary for each type of radiation. With demountable tubes it is usual to have different targets where the required target material forms an insert in copper. Operators of demountable tubes may however wish to form a target other than copper by the method of electrodeposition and some practical details of this process are given in appendix I.

3.7 Safety precautions

When the X-ray camera is in position in front of the X-ray tube there is, unless special precautions are taken, a leakage of X-rays between the tube face and the beam collimator. There is also the chance that when the operator makes adjustments he could be irradiated by the direct beam when the camera is in position or if the tube shutter has been left open accidentally after removal of the camera. The most effective device for reducing these risks is that designed by Kennard, Martin and Woodget (1957). The device is a radiation trap consisting of two steel blocks in which annular grooves, 6 mm deep, have been machined so that when they are fitted together they form a labyrinth. One block is fitted to the X-ray tube and the other to the camera so that, when the camera is in position during an exposure, no scatter is possible.

Hughes and Taylor (1961) have developed a universal mounting and alignment system which allows all types of cameras, single crystal or powder, to be positioned rapidly and simply on an X-ray tube. A dummy tube where the source of X-radiation is replaced by a light source is used to align the camera safely and then the camera is transferred to the actual X-ray unit. Scattered radiation is eliminated by a modified version of the labyrinth trap devised by Kennard *et al.*

The International Committee for Radiological Protection has made recommendations which would give greater protection against accidental exposure to direct, rather than to scattered, X-radiation. These recommendations have been embodied in a system designed by Hughes (1962a, 1962b) which incorporates both the labyrinth of Kennard *et al.* and the alignment system of Hughes and Taylor. X-rays cannot be produced unless either the X-ray tube ports are closed by a shutter or a camera is in position with the labyrinth operational.

CHAPTER 4

Recording and Measuring Powder Patterns

4.1 Introduction

Many materials can be obtained only as a mass of randomly oriented crystals and, provided that the grain size is sufficiently small, the diffraction patterns from such materials consist of lines and are called powder patterns. For a given wavelength, incident rays reflected by a particular set of planes are deviated by 2θ (fig. 4.1), where θ is the Bragg angle which satisfies the relationship $\lambda = 2d \sin \theta$. For a system of randomly oriented crystals, reflected rays from corresponding sets of planes will all be deviated by 2θ from the direction of the primary beam. The directions of the reflected rays will therefore lie on the surface of a cone with its apex at the specimen, its axis in the direction of the X-ray beam and its semi-vertical angle equal to 2θ. The intersection of such a cone with the film (fig. 4.2) represents a line of the powder pattern. In general there will be many such lines, all concentric and each representing a set of reflections with the same value of θ, resulting in a powder pattern which is often extremely complicated. Rather than use a flat film it is generally preferable to place a narrow strip of film cylindrically around the specimen, as shown in fig. 4.3, because this intercepts a greater number of cones of reflection than a flat film does. Only a small part of each cone is recorded, but all reflections except those with θ just greater than 0° or just less than 90° can reach the film.

In this statement of the theory of the production of powder lines, it is assumed that in part of the specimen irradiated there are sufficient crystals in all orientations to give rise to smooth lines. This may be so if the grain-size is less than about 1 μm but even in finely ground material it usually exceeds this. When the grains are too large there is not enough room within the irradiated volume for a sufficient number of crystals to lie

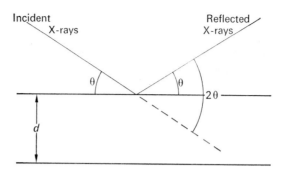

Fig. 4.1   Deviation of 2θ caused by reflection of X-ray beam

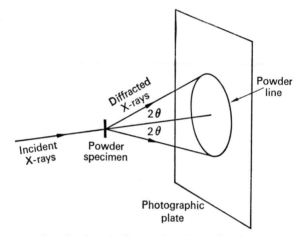

Fig. 4.2   Powder line produced on a flat plate

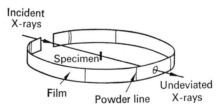

Fig. 4.3   Powder photograph produced on a cylindrical film

in all possible orientations. The reflections will still lie on a curve of constant θ but the resulting powder reflection will no longer be a smooth line, and with a very coarse-grained specimen the line may even be a series of spots. By rotating the powder specimen during exposure the number of crystals contributing to the formation of each powder line is considerably increased, and the line becomes smoother.

We have already seen (3.3.2) that characteristic radiation is not strictly monochromatic and that there is a spread of wavelengths about that at peak intensity. Now from Bragg's law, $\lambda = 2d \sin \theta$, it can be seen by differentiation that for a given value of d

$$\delta\lambda = d \cos \theta \, \delta\theta$$

or

$$\delta\theta = \delta\lambda/(d \cos \theta)$$

Thus for a given displacement $\delta\lambda$ from the peak of the characteristic radiation the corresponding change $\delta\theta$ in θ will increase as $\cos \theta$ decreases, that is, as θ increases. Hence $\delta\theta$ and therefore the resolution tend to a maximum as θ tends to 90°. In powder specimens there will almost inevitably be some crystals in a position to reflect the wavelengths which are displaced from the peak (Bragg and Lipson, 1938) and in coarse-grained material these give rise to spots which, because of the great resolution at high angles of reflection, will be visibly displaced from the centre of the line. Rotation of the specimen will again produce a powder line of the normal intensity distribution. This effect can be seen with the lines of the $K\alpha$ doublet.

So far we have been considering how characteristic radiation is diffracted. With ordinary filtered radiation and a stationary specimen however, each crystal will give rise to what is called a Laue pattern of spots by selection of some of the wavelengths of the white radiation to satisfy $\lambda = 2d \sin \theta$, where θ depends on the orientation of the crystal relative to the primary X-ray beam. The patterns from all the crystals will merge and contribute to the background of the film, but with a coarse-grained sample this background will be uneven. This effect is minimized by rotating the specimen, since each Laue spot is then drawn out into a streak as each successive wavelength of the continuous radiation is reflected; these streaks will merge with one another producing a more even background.

Powder photographs can be taken with very simple apparatus, but the importance of the method is such that a great deal of thought has been given to the construction of powder cameras which give the maximum amount of information about the pattern. The first essential is that θ should be accurately determined as this is the only characteristic by which the line can be identified. The lines must therefore be sharp and their

positions must be accurately recorded on the film. Powder lines are however weak compared with single-crystal reflections from the same mass of
material because the diffracted X-rays are no longer concentrated into a
spot; if a small pin hole is used to reduce the divergence of the beam, exposures will be unduly long. This difficulty may be overcome by the use of a
single slit parallel to the length of the specimen to admit the X-rays. Each
line will now be composed of a set of curves displaced along a line parallel

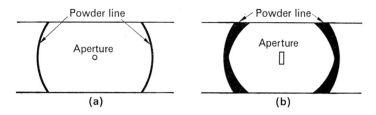

Fig. 4.4   Powder lines with X-rays admitted through (a) a small circular
aperture, (b) a vertical slit

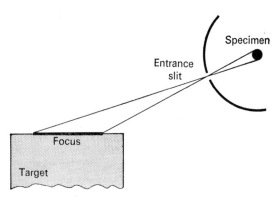

Fig. 4.5   Foreshortening effect with focus viewed at a small angle

to the length of the slit; as shown in fig. 4.4 this will not appreciably affect
the width of the line at the centre, which is the only part usually measured
(Alexander, 1950; Klug and Alexander, 1954).

Experience has shown that satisfactory photographs result if the length
of the slit is about 1/50 of the diameter of the camera. This dimension was
originally adopted in the cameras described by Bradley, Lipson and Petch
(1941). If such single-slit collimation is used in conjunction with line
focusing then in order to limit the length of specimen irradiated, the axis of

the specimen should be parallel to the target face of the X-ray tube and the line focus should be oriented with its length perpendicular to the direction of the specimen axis. Further, the orientation of the camera relative to the target face should be such that the rays entering it are those that come off the target at a small angle. The focus will then appear foreshortened (3.6.3) as shown in fig. 4.5. The greater the foreshortening the less the divergence of the beam perpendicular to the specimen axis and the sharper the powder lines will be. The dependence of the width of the diffraction lines on the tube focal area can be eliminated by introducing a two-slit collimating system (fig. 4.6). Thus, provided the width of the focal area exceeds the width of the aperture nearest the target, the line width of the reflections depends only upon the geometry of the collimating system.

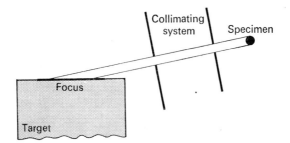

Fig. 4.6   A two-slit collimating system

With both single- and double-slit collimation, but particularly with single slit, the X-ray beam may not traverse the camera accurately along the axis of the collimator. The inclination of the beam to this axis may have both a horizontal and a vertical component. If the axis of the collimator is horizontal, the horizontal component will not matter very much, as it will merely shift the whole pattern by a constant amount, but the vertical component will lead to an asymmetric recording of the lines as shown in fig. 4.7. To avoid this the height of the camera can be adjusted with respect to the

Fig. 4.7   Asymmetric powder lines produced when the X-ray beam is not collinear with axis of the collimater

X-ray tube so that the beam emerges accurately along the axis of the collimator as shown in fig. 4.8.

Collimation with a single slit can be used only if the position of the
X-ray tube focus does not vary. This will be so in general in a hot-cathode

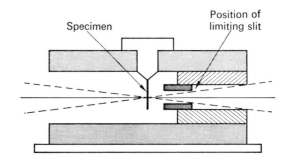

Fig. 4.8   Correct (————) and incorrect (— — —) paths for the X-ray beam
in a powder camera

tube, but may not be so in a gas tube. If a gas tube is used it is necessary to
have a double-slit system with the slit nearest the window of the tube having a narrow aperture.

4.2 Preparation of Powder Specimens

The specimens should be in the form of a long narrow cylinder of diameter
about 0·3 to 0·5 mm. Many ways of making such specimens, to suit particular requirements, have been described. For ordinary work the powder
may be coated on a gummed glass fibre, made into a paste and extruded, or
mixed with Canada balsam and coated on a hair kept taut with a lead
weight. Canada balsam gives a strong halo that may obscure some of the
low-angle lines but gum tragacanth does not and may be used as follows.
Mix the powder, which must be fine, with a fifth to a tenth of its volume of
powdered gum and make this into a paste with a small quantity of either
water or saliva. After a short time the paste will set sufficiently to be rolled
between the fingers into the form of a thin rod which can be made more
perfect by rolling between glass plates; microscope slides are ideal for
this purpose. Continue rolling until the paste finally hardens to counteract
any tendency of the rolled cylinder to warp. When it has set quite hard the
specimen can be set either horizontally or vertically in a camera. This
method cannot be used, of course, if the substance is soluble in water.

If the substance is deliquescent or efflorescent, or otherwise unstable under ordinary atmospheric conditions, it must be enclosed in a tube which is transparent to X-rays. Such a tube may be made of borosilicate glass, of collodion or of cellophane. If the substance is to be raised to high temperatures the enclosing tube may sometimes be made of silica. Silica glass softens at about 1100°C, but even at temperatures below this it may react with the specimen. A comprehensive list of methods of preparation and mounting of specimens is given in International Tables Vol. III (1962).

Metals and alloys may be examined directly in the form of wires or of thin rods, but it must be remembered that such samples may exhibit preferred orientation (Henry, Lipson and Wooster, 1960). Further, if a metal or an alloy is either filed or ground into a powder then it becomes work-hardened; before these powders can be used for X-ray diffraction purposes they must first be annealed under vacuum in order to relieve any strains which may have been induced.

The methods of preparation of specimens for diffractometer operation are described in section 4.5.3.

4.3 Measurement of Bragg Angles

4.3.1 Geometry of cameras

In all forms of powder cameras the usual way of deriving θ is to measure the distance S between two opposite parts of the diffraction cone that is recorded on the film. This overcomes any errors arising from an uncertainty in the position on the film which corresponds to zero θ.

If the photograph is taken on a flat film, the distances S are the diameters of the rings; θ is calculated from

$$\tan 2\theta = S/2D$$

where D is the distance of the specimen from the film (fig. 4.9).

Films taken in cylindrical or Debye-Scherrer cameras are also usually measured flat and θ is then proportional to S. It can be seen from fig. 4.10 that

$$\theta = S/4R$$

where R is the radius of the camera.

Although this is accurate enough for many purposes, for precision work the finite thickness of the film and its shrinkage during processing must be taken into account. The shrinkage can be allowed for by constructing the camera so that it casts on the film two shadows at a known angular separation; the angular separation of any two lines can then be found by proportion. The pieces of metal that cast these shadows should have sharp straight edges, which should be parallel to the axis of the camera and in close contact with the film; they are known as **knife edges**.

Bradley and Jay (1932a) first suggested measuring angles in this way. In their camera the X-ray beam enters between the ends of the film and leaves through a hole in the centre, as shown in fig. 4.11a. When the film is laid out flat it looks as shown. If ϕ_k is the value of the Bragg angle for a

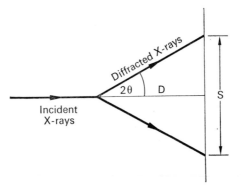

Fig. 4.9   The relation between θ, D and S

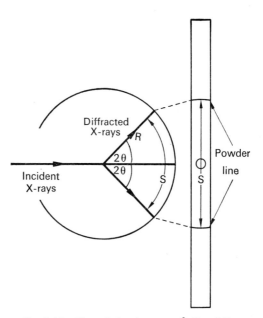

Fig. 4.10   The relation between θ, S and R

line which would fall at the knife edges, then θ for any pair of corresponding lines is connected with the distance S between them by the relation

$$\theta = \phi_k S/S_k,$$

where S_k is the distance between the shadows of the knife edges. The assumption is made that the film changes uniformly after processing; this has been found to be true within the accuracy of measurement for one specimen of film measured at intervals over a period of months (Lipson, 1942; D'Eye and Wait, 1960).

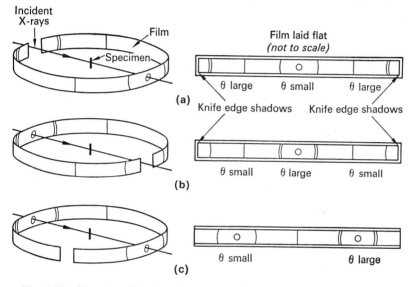

Fig. 4.11   Mounting films in 90 mm diameter cameras: (a) Bradley–Jay, (b) van Arkel, (c) Leviņš–Straumanis (asymmetric) method

The Bradley–Jay method of mounting films in the camera is useful for obtaining measurements of the low-order spectra, that is, the reflections for which θ is small. With this arrangement, corresponding low-angle lines are close together on the film so that S/S_k is very much less than unity for these reflections and the errors in the measured values of θ are appreciably less than are the errors in ϕ_k. For high angles, however, S approaches the value of S_k and for these angles the error in θ will almost equal that in ϕ_k. Although ϕ_k can be determined with an accuracy of about 0·01° from direct measurement of the camera diameter and the distance apart of the knife edges it would be better if the error were eliminated completely. Van Arkel (1926) had previously suggested that this could be done by mounting the film in the camera with the X-rays entering through a hole in the film and leaving between the ends (fig. 4.11b). If shadows of knife edges are cast at these ends, the value of θ for any pair of lines is connected with the distance S between them by the relation

$$\theta = 90° - \psi_\mathrm{k} S / S_\mathrm{k}$$

The angle ψ_k is again determined from direct measurement of the camera diameter and the distance between the knife edges and is the complement of the Bragg angle of a reflection that would fall at the positions of the knife-edge shadows. Thus for high-angle reflections, where S is now small, the error in ψ_k will have only a slight effect on the value of θ, if θ were 90° then S would be zero and the effect of an error in ψ would be eliminated.

A third method of mounting the film has been suggested by Ieviņš and Straumanis (1936); they call it the **asymmetric** method and it is illustrated in fig. 4.11c. The X-rays both enter and leave through holes in the film so that no calibration of the camera is involved because on the same film the positions corresponding to $\theta = 0°$ and $\theta = 90°$ can be located from the positions of the lines themselves. These positions are midway between corresponding low-angle ($2\theta < 90°$) and high-angle ($2\theta > 90°$) lines. Thus to make use of an asymmetric film it is necessary to measure several pairs of low-angle lines in addition to as many pairs of high-angle α-doublets as is possible. The average of the mid-positions of the low-order and high-order lines is then determined, and the distance T between these averages corresponds to a value of $2\theta = 180°$ or of $\theta = 90°$. Thus the value of θ for any pair of low-order lines which are separated by a distance S is $\theta = (45S/T)°$ and the value of θ for any pair of high-order lines which are separated by distance S is

$$\theta = (90 - 45S/T)°$$

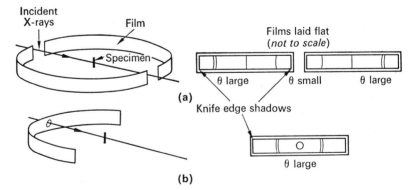

Fig. 4.12   Mounting films in 190 mm diameter cameras: (a) for the complete pattern, (b) for the high orders only

This advantage of self-calibration is not so important as might appear (Cohen, 1936); the chief advantage is, in fact, that the camera combines the properties of both the Bradley–Jay and the van Arkel mounting (Straumanis and Ieviņš, 1940). One pays for this advantage by having a rather more complicated camera and correspondingly more complicated devices for loading the film in it; but these factors have not detracted from the popularity of the camera.

These methods can be used with cameras of any size, but in practice it is inconvenient to have to handle films longer than about 250 mm. For cameras of 90 mm diameter (fig. 4.11) only one piece of film is required, but for cameras of 190 mm diameter two pieces of film are used, as shown in fig. 4.12, and four knife-edge shadows are cast. If accurate lattice parameters are required it is better to use the van Arkel mounting of the film, even if it involves recording only the high orders (Lipson and Rogers, 1944; Peiser, Rooksby and Wilson, 1955).

4.3.2 Measurement of powder photographs

The accuracy to which the Bragg angles of the reflections are required will determine how the distances between corresponding lines are measured. If interplanar spacings calculated from $\lambda = 2d \sin \theta$ are not required to an accuracy of better than 1 in 1000, then a steel rule can be used; with practice it is possible to estimate distances to 0·1 mm in this way. For accuracies greater than 1 in 1000 some form of travelling vernier microscope must be used. The graticule must have cross wires and the microscope should be capable of independent movement both along, and at right angles to, the length of the film; the film should be viewed in transmitted light from an opal-glass screen which is illuminated from the side remote from the film. The film must not be overheated during measurement and the instrument must be long enough to accommodate the full length of the film. It should also be made as wide as possible for greater versatility in application.

In determining line positions with a travelling microscope the magnification should be not more than two diameters, otherwise the lines may, because of the large grain size of the film, appear indistinguishable from the background. The positions of the lines should be read by placing the cross-wires on the middle of the arc as shown in fig. 4.13, but if the film contains some rather weak lines it may be better to have diagonal cross wires (fig. 4.14). The visibility of weak lines is also improved by putting opaque screens over that part of the film which has not received any radiation so that only the exposed part is visible (fig. 4.14). It is then necessary for the eye to accommodate itself to only a small range of brightness.

4+

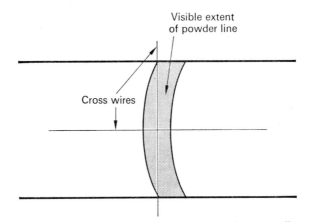

Fig. 4.13   Orientation of cross-wires for measuring strong lines

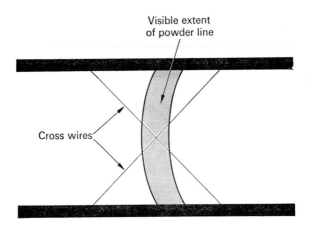

Fig. 4.14   Orientation of cross-wires for measuring weak lines

4.3.3   Correction for absorption

In deriving Bragg angles from powder photographs, one source of sys-
temic error is the absorption of X-rays in the specimen; because of this
absorption the distribution of intensity across a powder line may be irregu-
lar and the maximum may not be in the centre. Such errors will decrease
both as θ increases and as the linear absorption coefficient decreases. This
may be seen by considering a low-angle reflection from a specimen of
medium absorption.

Suppose that a specimen has a circular section and is irradiated by a parallel beam of X-rays, as shown in fig. 4.15. A reflected ray such as B_2C_2, coming from the centre of the specimen, will be greatly weakened by absorption, whereas a ray such as B_3C_3 is weakened less and a ray such as B_1C_1 less still. On the other hand there will be more reflected rays from the middle of the specimen because it is thicker there. If the absorption is low, this latter effect will be the more important, and the recorded line, which covers the range $C_1C_2C_3$ will have a maximum near the centre. If, however, the absorption effect is predominant, the total reflection from the middle of the specimen will be weaker than that from the edges and the distribution of intensity across the powder line will have the general form shown in fig. 4.15; that is, to the eye the line will appear doubled.

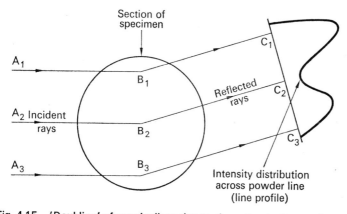

Fig. 4.15   'Doubling' of powder lines due to absorption in the specimen

Taylor and Sinclair (1945a) have considered the problem in detail and have drawn line profiles for different values of μr, where μ is the linear absorption coefficient (3.4.1) and r is the radius of the specimen. They have shown that doubling of the lines due to reflection from the edges of the specimen occurs only for low values of θ, and that, as θ increases, the weaker component becomes smaller and finally disappears. Only the stronger component will then be measured, and since this is at the higher angle (fig. 4.15), sin θ will be higher than its true value and the spacing derived from it will therefore be low. If the value of μr for the specimen is known, the error can be calculated. A more satisfactory way of making the corrections is to mix with the specimen a substance which gives reflections at known Bragg angles and to obtain the unknown ones by interpolation. For example, a curve of spacing, d, against arc, S, may be plotted for the lines of known spacing; then, from the measured values of S for the unknown spacings, the required value of d can be read from the curve.

4.4 Cylindrical Powder Cameras

4.4.1 Diameter of camera

Many different cameras are described in the literature. The smaller cameras are satisfactory for simple patterns and also for the general outlines of the more complicated ones. Larger cameras give better resolution (for a given angular separation $\delta\omega$ at the specimen, the distance between two lines on a cylindrical film of radius r is $r\delta\omega$) which is useful for measuring cell dimensions and detecting faint lines because the intensity of a line does not decrease with distance so rapidly as the background does. Lines arising from characteristic radiation are confined, to a close approximation, to particular values of θ, and since the length of any one complete powder line (fig. 4.2) is therefore proportional to the distance of that line from the specimen, the intensity at any one point of the line must, to a close approximation, be inversely proportional to that distance. On the other hand white radiation scattered by the specimen is largely responsible for the background intensity well away from an absorption edge; from a stationary specimen white radiation reflected from each grain of the polycrystalline material (as Laue spots) will also be confined to a particular angle θ but the reflections will be so numerous that they will effectively cover the whole of the exposed film to give a continuous background, especially if the specimen is rotated so that the individual spots are drawn out into streaks. Under these conditions the scattered white radiation behaves like incoherent radiation and since it emanates from a small source the intensity in any one direction will be inversely proportional to the square of the distance from the source. Some fluorescent radiation and some radiation which has been modified by the Compton effect (3.41) will also be present; these are truly incoherent and their contribution to the background intensity will also follow an inverse-square law. The fall-off in intensity of the background with distance will clearly be greater than that of the lines produced by the incident characteristic radiation, and so the lines will be relatively stronger with respect to the background if the radius of the camera is increased.

If, however, the characteristic reflections are not confined accurately to particular values of θ, as they will not be if the crystals are very small or are deformed by internal strains (9.3.2), then there will be no increase in contrast of the powder lines relative to the background when the camera radius is increased.

Air in a camera has two important effects. First, it scatters the radiation and so produces a general fogging of the film, particularly near the position of zero θ; and secondly, it absorbs the incident and the diffracted radiation.

To observe faint lines or measure large inter-planar spacings, it is advisable either to evacuate the camera or to fill it with hydrogen, which scatters X-rays much less than air does. Absorption effects are not usually important for cameras of 90 mm diameter or less, but for larger cameras they may be considerable, especially for the longer wavelengths. For example in a path of 350 mm in air about 70 per cent of chromium radiation is absorbed.

Exposures are, of course, longer with larger cameras, and the early history of the subject was a constant attempt to make powder cameras as large as possible consistent with reasonable exposure times; thus as more powerful tubes (3.6.3) and faster film became available, larger cameras came into use. It is interesting that the original powder camera of Hull (1917) had a radius of 200 mm; the lines turned out to be extremely broad because a large specimen, close to the X-ray-tube window, was used. Subsequent development was carried out with cameras of 25 mm and 50 mm diameter, and, when more complicated patterns were dealt with, a camera of 90 mm diameter was introduced. This size of camera is still extensively used and cameras of 190 mm diameter (Bradley, Lipson and Petch, 1941) are also used for special purposes but it would seem that this is a reasonable limit.

Camera diameters used in Great Britain seem to have been chosen fortuitously but in the United States attempts have been made to employ more logical considerations, and diameters of 57·3 and 114·6 mm (Parrish and Cisney, 1948) have been introduced since they lead to simple relationships between the lines—they give separations of 2 and 4 mm respectively per degree of Bragg angle. This property is of course only of value for approximate work: for precision work errors, such as those caused by film shrinkage, have to be eliminated.

4.5 The X-ray Powder Diffractometer

4.5.1 Recording diffracted radiation

The most efficient and accurate way of interpreting X-ray-diffraction patterns, of the single-crystal or powder type, is to use a counter diffractometer. In the powder camera the intensity of the diffracted X-radiation is measured by the degree of blackening the radiation produces on a photographic film, but in a counter diffractometer the intensity is measured directly in terms of the rate at which the diffracted photons trigger the counting device. As its name implies, the powder diffractometer measures the intensity of the radiation diffracted by polycrystalline material, and the instrument originates directly from the Debye-Scherrer powder camera. The counting device, which may be either of the Geiger, the proportional or the scintillation type (7.3), depends on the ionization produced by the incident photons and scans the circumference of the circle which would have carried the photographic film in a powder camera. A proportional or scintillation counter used to record X-ray intensities has the advantage that a pulse-height analyser can be incorporated in the associated electronic circuits. The analyser can be adjusted to eliminate most of the white radiation although at the same time it allows through a large proportion of the higher intensity characteristic radiation; this is another way of reducing the background radiation.

4.5.2 Diffractometer geometry

The circle scanned by the counter may be in either a vertical or a horizontal plane; in recent designs the tendency has been to build the horizontal-circle type with a vertical common axis of rotation of both specimen and detector. In operation the counter is driven at twice the angular velocity of the specimen. The specimen itself is flat, and the line focus of the X-ray tube is, as usual, viewed at an angle of about 6° with the face of the target. The focus however is parallel to the axis of oscillation of the specimen so that as presented to the specimen it appears a very fine line (fig. 4.16a).

From fig. 4.16b, where O is the centre of the scanning circle, it can be seen that the combination of line source and flat reflecting plate, P, produces a focusing effect. A beam of X-rays diverging from the focus, F, is incident on the flat specimen and is diffracted through the Bragg angle by all the crystal grains which happen to be correctly oriented. By the ordinary laws of reflection, those rays reflected from the surface will be brought to an approximate focus at C (FO = OC) where the entrance slit to the detector is situated; the entrance slit is parallel to the line focus, to the surface of the specimen and to the axis of rotation of the specimen. The rays would be perfectly focused if the specimen were curved so as to lie on the circumference of the circle, called the focusing circle, which passes through F, O and C. The radius of this circle is $FO/2 \sin \theta$ and hence to maintain the true-focusing condition the radius of curvature of the specimen would have to vary as the diffraction pattern was scanned.

A device which will do this has been described by Ogilvie (1963). The specimen holder is so designed that the surface of the specimen always contains the axis of rotation while at the same time the specimen is bent automatically to a uniform radius of curvature which is at the correct value for focusing at each diffraction angle. Bending is provided by the three-point action of two forces producing oppositely directed turning moments about a fulcrum placed between them, and the magnitude of the bending moment is varied by a cam and a cam follower. In order to maintain successive uniform radii of curvature of the specimen with this system the holder must be carefully designed and machined. If the specimen itself is in the form of a packed powder then some disintegration must be expected at values of 2θ which are greater than about 120°; this difficulty can

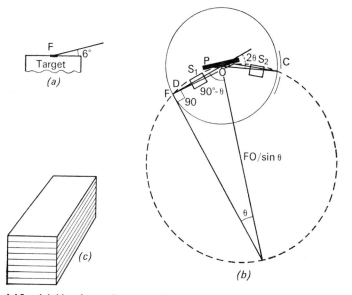

Fig. 4.16 (a) Line focus, F, perpendicular to plane of paper and 'viewed' at an angle of 6°; (b) Focusing effect produced by the combination of a line source and a flat specimen; (c) Soller slits

be overcome if the specimen is made from a mixture of powder, amyl acetate and a drop of collodion.

Generally however some defocusing is tolerated and a flat specimen is so oriented that it is always tangential at O to the focusing circle; it is for this reason that the rate of rotation of the specimen is half that of the scanner. To reduce the extent of the defocusing the divergence of the incident beam is limited in the plane at right angles to the axis by the slit D. Further, focusing of the radiation from a line source requires that the radiation in the plane containing the line should be parallel, and consequently the divergence of the beam in the plane parallel to the axis is limited by the Soller slits S_1 and S_2. Soller slits are assemblies of thin slits made with thin metal plates (fig. 4.16c), and to limit the axial-plane divergence the slits are placed with the planes of the metal plates at right angles to the rotation axis.

4.5.3   Practical considerations

With the photographic-film technique all the reflections are recorded simultaneously and the relative intensities of the lines are therefore unaffected by changes in the intensity of the primary radiation. On the other hand, a diffractometer records the line intensities successively and so for accurate work the intensity of the incident beam must remain constant; this is generally effected by stabilizing the high-voltage supply and the tube current to better than 1 per cent.

The scanning unit moves continuously or step-by-step over a circular scale which, if required, permits direct readings of the angle 2θ. In this, and indeed in all methods of recording, the continuous method is the quicker but the step-by-step method is the more accurate. For continuous recording, scanning rates may be varied for rates of change of 2θ which may range from $0 \cdot 1°$ up to $2°$ or $3°$ per minute, but the higher the scanning speed the lower the accuracy; the output is registered on a strip-chart recorder, the linear speed of which is geared to the angular scanning speed. For step-by-step recording the intensities are plotted by hand on a chart; the stepwise movement can be adjusted to advance the detector, that is to increase 2θ, by steps which may be as small as $0 \cdot 01°$.

Whether the camera operates continuously or step-by-step the maximum value of 2θ at which measurements can be made will depend upon the nature of the source at F (fig. 4.16b). If it is the actual line focus of an X-ray tube then scanning at values of 2θ greater than about $160°$ is not possible because the counter then fouls the casing of the tube. Readings at values of 2θ greater than about $160°$ can be obtained either by replacing the tube line focus at F by a slit or by placing F at the focus of a focusing monochromator (3.5.2).

Several factors should be considered when preparing the specimens. The flat surface should be large enough to intercept the whole of the primary X-ray beam at all angles of diffraction and to do this the area may have to be as large as 10 mm × 20 mm. Since the area irradiated depends on θ the sample must be homogeneous, otherwise the intensity of the diffracted beam will depend upon which section of the specimen happens to be exposed to the radiation at a given value of θ. Specimens which have been prepared from fine powders mixed with some binding material may be non-homogeneous and the possibility of preferred orientation occurring in pressed powders or in bulk material must be considered.

If possible, the specimen should have high absorption (Parrish and Wilson, 1959; Kaelble, 1967), which is in contrast to the requirements of a Debye–Scherrer camera. As was shown in 4.3.3 high absorption in a cylindrical specimen in a powder camera will produce a shift in the maximum intensity of a line; in a diffractometer with a low-absorption specimen the radiation will penetrate into the body of the material and this will result in a shift in the maximum intensity of the reflection.

For good resolution the line focus should not appear to be wider than 0·1 mm, the divergence of the X-ray beam, both parallel and at right angles to the specimen surface, should not exceed 4°, and the receiving slit for the detector should cover a range of the order of 0·05° in 2θ. It is possible,

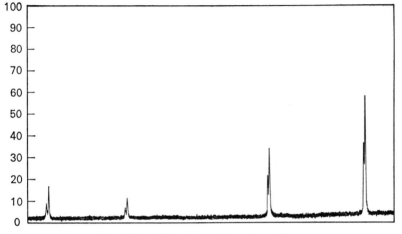

Fig. 4.17   Diffractometer trace from silicon

under these conditions and with a scanning circle of diameter of the order of 300 mm, to resolve the CuKα doublet at a value of 2θ between 30° and 40°. The trace in fig. 4.17 is an example of a typical output from a diffractometer. Detailed accounts of X-ray diffractometers have been given by Peiser, Rooksby and Wilson (1955), by Klug and Alexander (1954) and by Parrish (1962); Arndt and Willis (1966) have given a detailed treatment of single-crystal diffractometry.

4.6 High-temperature and Low-temperature Cameras

4.6.1 High-temperature cameras

X-ray diffraction data at high temperatures are required for the investigation of phase transitions and for the determination of coefficients of expansion. The long list of adaptations to powder cameras which have been made for high-temperature operation is reviewed in an excellent biblio-

4*

graphy by Goldschmidt (1964) and of these that of Goldschmidt and Cunningham (1950) probably has the widest application.

If the specimen oxidizes at high temperatures the camera must be evacuated, or filled with an inert gas or the specimen placed in a sealed enclosure. If the first method is chosen it may be difficult to obtain a uniform temperature around the specimen (a temperature gradient of 2–3°C per mm may exist along it) as convection and gaseous conduction help to produce thermal equilibrium inside the camera. Moreover if the specimen is an alloy there may be differential evaporation of the constituents leading to a change in composition. If an inert gas is used, thermal equilibrium will be established sooner and differential evaporation will occur on a greatly reduced scale. If the sealed enclosure is chosen, differential evaporation can be prevented by maintaining the enclosure at a uniform temperature; distillation cannot then take place from the hotter parts to the colder. The material of the enclosure must be amorphous and transparent to X-rays. Fused silica satisfies these conditions well but it reacts with many materials at temperatures well below that of its softening point and so cannot always be used even below 1000°C.

If accurate determination of temperature is of overriding importance then internal calibration should be used and diffraction patterns of the specimen and a standard substance obtained simultaneously. If the variation of lattice spacing of the standard with temperature is known, the temperature can be determined to better than ± 1 percent.

X-ray diffractometers are adapted for high-temperature measurements by incorporating high-temperature attachments. There are furnaces of various designs and a review of the devices has been published by Campbell, Stecura and Grain (1962). Houska and Keplin (1964) have described a diffractometer furnace capable of attaining temperatures of 2000°C, and Cornish and Burke (1965) have designed an attachment with which lattice parameters can be determined to an accuracy of 1 in 20,000 at temperatures of up to 1000°C.

4.6.2   Low-temperature cameras

X-ray data at reduced temperatures are also necessary for the investigation of phase changes and for the determination of expansion coefficients at temperatures which are outside the normal range. Operation at low temperatures also enables the structures of solid phases of substances which are either liquid or gaseous at normal temperatures to be examined.

Cameras may be cooled by liquid streaming, by gas streaming, or by conduction through a solid. Typical of those liquid-streaming methods which are applicable to Debye–Scherrer cameras is that described by Lonsdale and Smith (1941) in which liquid refrigerant contained in a Dewar vessel streams out through a fine capillary at the lower end of the Dewar (fig. 4.18) continuously bathing the specimen, which must be supported from the base of the camera. The photographic film can be protected from direct contact with the coolant by surrounding the specimen with

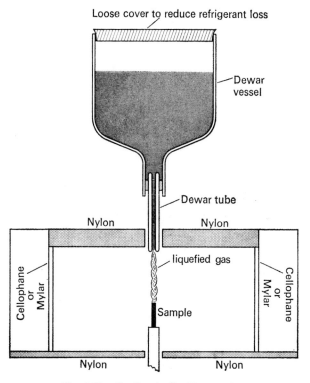

Fig. 4.18   Cooling by liquid streaming

either cellophane or Mylar or any other substance which is transparent to X-radiation (fig. 4.18); nylon strips can be inserted to prevent cooling of the film by metallic conduction. It is not essential, however, to protect photographic film from the effects of cold since the properties of the emulsion are not drastically affected by exposure to low temperatures. To prevent the formation of frost and ice the camera can be kept in an atmosphere of dry air or, better still, enclosed in a cellophane or polythene bag if the refrigerant is liquid nitrogen; the evaporating refrigerant will then generate an excess pressure which will exclude the outer air. Since the temperature of the specimen is that of the coolant the method is not convenient if varying temperatures are required. Temperatures higher than that of liquid nitrogen can be obtained by streaming with a mixture of methyl and ethyl alcohols (Tombs, 1952) which may be cooled by immersing a

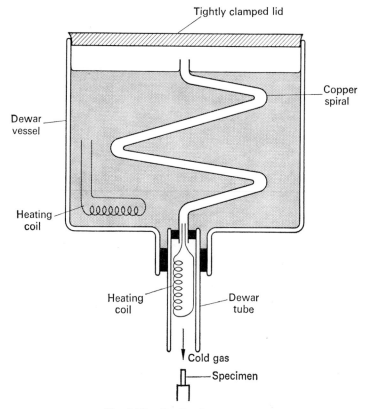

Fig. 4.19   Cooling by gas streaming

tube containing liquid nitrogen in the mixture. Copper–constantan thermocouples provide a satisfactory means of measuring temperature.

The liquid-streaming method involves the use of large quantities of refrigerant and, in general, is not recommended for temperatures below that of liquid nitrogen. Gas streaming is more economical and it also provides a better means of temperature control. Robertson (1960) has cooled a single crystal to 23°K by streaming evaporated liquid hydrogen over it and powders can be cooled in a similar way. The method is usually confined to liquid-nitrogen temperatures however. The basic principle is illustrated in fig. 4.19. Liquid nitrogen contained in a Dewar vessel produces nitrogen gas which is forced through the spiral copper tube immersed in the liquid nitrogen. A Dewar tube of glass, stainless steel or any low thermal-conductivity alloy is connected to the lower end of the copper tube and directs the nitrogen gas from the Dewar vessel on to the specimen in the powder camera. By including an electric heater to promote rapid boiling of the liquid nitrogen, specimen temperatures as low as 85K can be obtained. For higher temperatures the rate of boiling can be reduced and if a spiral heater is inserted in the Dewar tube virtually any value of the temperature up to room temperature can be maintained. A copper–constantan thermocouple can be used as a thermometer. Once more the most effective way to stop frosting is to place the camera in a bag so that it is in an atmosphere of dry nitrogen gas.

If the level of the liquid nitrogen in the Dewar vessel falls during evaporation the temperature of the effluent gas rises at a rate of approximately 1K for each centimetre drop in the liquid level, so that the nitrogen level should be maintained constant during an exposure, preferably by an automatic filling device. Many systems have been designed which can maintain the level to within a fraction of a centimetre. A simple, but effective type has as sensing element a thermistor which undergoes a large change in resistance as it passes between liquid and gaseous nitrogen. These resistance changes activate a solenoid valve which controls a supply of liquid nitrogen to the Dewar vessel.

The third method of cooling—by conduction—has not found general favour in its application to cylindrical powder cameras, presumably because of the constructional difficulties, but an instrument which utilizes conduction cooling, and has liquid nitrogen as the refrigerant, has been developed by Keeling, Frazer and Pepinsky (1953). Conduction from the liquid nitrogen is through a goniometer head and full temperature adjustment is provided by a heating coil located near the top of the head. Extension of the range down to helium temperatures presents a difficult exercise in low-temperature instrumentation but one such device suitable for both single crystal and powder specimens is now in operation (Hughes and Steeple, 1970).

A metal cryostat, the lower end of which is shown in fig. 4.20, has been designed so as to reduce the liquid-helium consumption to less than 150 cm³ per hour when the specimen is at a temperature of 9 or 10K. When the electric heater is switched on, the specimen temperature can be raised to any value up to 80K at which point the helium in the inner dewar can be replaced by nitrogen to allow the temperature to be raised even higher. The goniometer head is of copper and screws directly into the bottom of the helium Dewar, and both the Dewar and the goniometer head can be continuously oscillated through an angle of 200°. Since the compartment containing the goniometer head, the specimen and the radiation shields is continuously evacuated, the Mylar cover is very close to room temperature so that no frosting occurs.

At liquid-helium temperatures the thermoelectric power of a copper-constantan thermocouple is very low and this makes it unsuitable for use

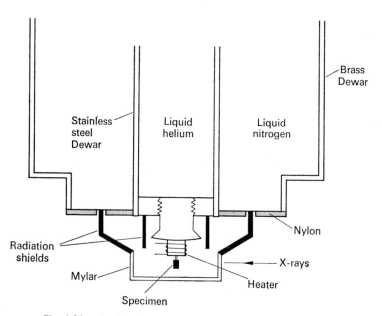

Fig. 4.20   Cooling from liquid helium by metallic conduction

in this temperature range. However, a thermocouple of chromel-gold + 0·03 at% iron which has been developed by Berman, Brock and Huntley (1964) has, at helium temperatures, a thermo-electric power of 15 μVK^{-1}, and this is ample for temperature measurement.

Low-temperature techniques are more readily applicable to the powder diffractometer, in which the specimen is easily accessible, than to the powder camera. Little use, however, seems to have been made of either the liquid-streaming or the gas-streaming methods. In one of the few applications, Weltman (1962) covered the range from 77K to 273K by streaming with a gas which was dried and then cooled by passage through liquid nitrogen. On the other hand there are numerous attachments for conduction cooling, all of which are Dewar systems which are either bolted or clamped to the goniometer head so that they rotate with the specimen holder. Examples of attachments which can be cooled to helium temperatures by conduction are those of Barrett (1956), of Black, Boltz, Brooks, Mauer and Peiser (1958) and of Lytle (1964).

4.7 High-pressure cameras

Crystals can be structurally changed by the application of high pressures, but the design of the necessary equipment calls for specialized techniques. The recording method may be either that of the powder camera or of the diffractometer. The techniques were reviewed by Jamieson and Lawson

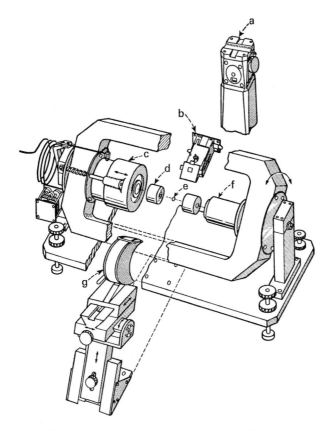

Fig. 4.21 (a)   High-pressure focusing camera: (a) fine focus Mo X-ray tube; (b) bent quartz monochromator, 600 mm radius; (c) 50-ton hydraulic ram; (d) Bridgman anvil; (e) sample disc; (f) load cell; (g) cassette

(1962) but since then cameras for operation at pressures of hundreds of kilobars have been designed. Using a focusing camera (4.8.1) and a fine-focus X-ray tube (4.8.3), McWhan (1967) obtained X-ray-diffraction photographs from polycrystalline material held at pressures of up to 200 kbar. The apparatus (fig. 4.21) has a pressure vessel which consists of a disc 2·38 mm in diameter and 0·64 mm high made from a casting of a mixture of 1·25 g amorphous boron per cm³ of epoxy resin; a hole in the centre of this disc contains the sample. Pressure is applied to the flat surfaces of the disc by Bridgman anvils and a 50-ton hydraulic ram. The anvils, which are in the shape of cylindrical rods, form the axis of the powder camera and are seated in a small press frame; during an exposure the complete press frame and anvil assembly is oscillated in the camera with the anvils acting as the axle.

4.8 Other Types of Cameras

4.8.1 Focusing cameras

Focusing cameras give improved resolution over ordinary powder cameras and enable shorter exposure times to be used (plate I(c)).

The principle of the focusing camera is illustrated in fig. 4.22. The specimen, AB, the whole of which is irradiated, is in the form of a thin strip of powder, 30 or 40 millimeters long and up to 10 millimeters wide, lying on the surface of a cylinder. From the discussion in section 3.52 on focusing monochromators it will be clear that the source of X-rays and the photographic film on which the reflections are recorded must also lie

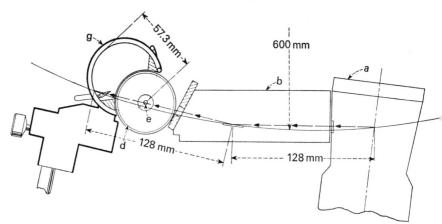

Fig. 4.21 (b)   The principle of the focusing camera

on this cylinder, and from the figure it can be seen that all reflected rays with the same Bragg angle will be recorded as a single line on the film CD. Since the specimen is cylindrical, the source at E must be a line parallel to the axis of this cylinder and to take advantage of the focusing effect the slit must be very fine; the width of the slit must be limited so as to reduce the divergence of the X-ray beam and hence to reduce the defocusing effect.

Because the specimen is large, exposures will be correspondingly short. The fact that the specimen has to be stationary is not a great disadvantage because, compared with the ordinary powder specimen, so many more crystals are irradiated that a larger number should be correctly oriented to produce any particular reflection; smoother lines should therefore result. One disadvantage of focusing cameras of this type is that the low-angle reflections, which generally are more important than are those of high

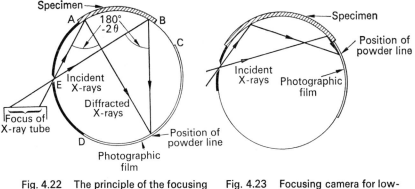

Fig. 4.22   The principle of the focusing camera   Fig. 4.23   Focusing camera for low-angle powder lines

order, are incident obliquely on the film (fig. 4.23) and are thus always rather broad; high-angle reflections are always sharp because the diffracted rays producing them are incident on the film at almost normal incidence. A further disadvantage is that, as can be seen from figs. 4.22 and 4.23, the full angular range of reflections cannot be recorded in a single camera and separate cameras have to be used for high, medium and low angles.

The symmetrical focusing camera (Gayler and Preston, 1929) is useful for accurate measurement of cell dimensions; the principle is illustrated in fig. 4.24. Cohen (1935) claimed, on theoretical grounds, that it is inherently the most accurate instrument for this purpose, but this is not borne out by results. In effect all cameras are focusing cameras for the high orders; the ordinary focusing camera gives sharp lines because, like a good camera lens, it has been correctly designed, whereas the ordinary powder camera gives sharp lines because the aperture is stopped down. Consequently the latter requires longer exposures, but it has the advantage over the focusing camera that *all* the lines on the film are reasonably sharp. The focusing camera, however, has the advantage that the linear separation it gives for the lines is twice that given by an ordinary camera of the same diameter; but, if the latter camera is made larger to allow for this, the results it gives seem to be quite as accurate as are those obtained with the focusing camera.

If experimental circumstances require an intense beam of crystal-reflected radiation then such a beam will have to be produced by a focusing monochromator (3.5.2). The radiation from the focused source will necessarily be widely diverging and as such is unsuitable for use with a Debye–Scherrer camera, which ideally requires a parallel beam. On the other hand,

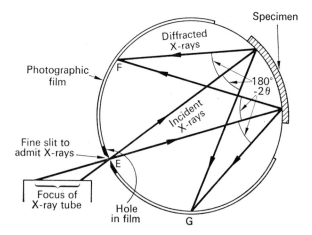

Fig. 4.24   The principle of the symmetrical focusing camera. F and G are the two positions of the powder lines with Bragg angles θ, and are at equal distances from the entrance slit E

the focusing camera should be used in conjunction with radiation which is diverging in the plane at right angles to the axis of the camera and is therefore eminently suitable for use with a focusing monochromator. There are two possible combinations of monochromator and camera; fig. 4.25a shows the arrangement necessary for surface-reflected radiation from the specimen and fig. 4.25b shows that for transmitted radiation. In both arrangements the focus F, the specimen and the photographic film on which line spectra such as A and B are recorded, all lie on the surface of the same cylinder. The surface-reflected radiation still suffers from the drawback that the low-angle reflections are broadened, but with the camera designed to record transmitted radiation (Guinier, 1945b), the low-angle lines strike the film almost normally and are very sharply focused.

The adjustment of a system with a focusing camera and a focusing monochromator will depend upon the wavelength of the crystal-reflected radiation which is incident upon the specimen. When a different exciting

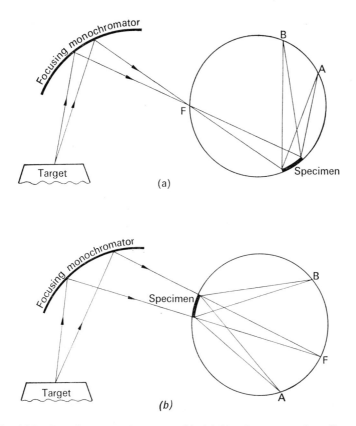

Fig. 4.25   Focusing monochromator with (a) focusing camera for reflected radiation, (b) focusing camera for transmitted radiation

wavelength is required either the curvature of the monochromator or the linear and angular relationships between the source and the camera must be changed. Guinier (1939) made the necessary adjustments by keeping the curvature of the monochromator constant and resetting the camera; de Wolff's solution (1948) was to devise a system whereby the curvature of the collimator could be modified as required, and Hofmann and Jagodzinski

(1955) adopted the expedient of changing the whole of the monochromator unit in order to ensure that the distance between source and camera could be kept constant. Fischmeister (1961) has described a system comprising a Guinier camera with exchangeable spacers that are adjusted once and for all; these spacers define all critical angles and distances and so enable the camera to be adjusted quickly for radiation of different wavelengths.

4.8.2   Cameras for block specimens

The powder method can be used for any polycrystalline specimen, even if it is not in the form of a powder. X-ray photographs of materials such as metals and alloys are often required, and, if the crystals are small, quite good 'powder' lines may be given.

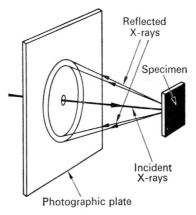

Fig. 4.26   Arrangement of photographic plate and specimen in the back-reflection method

The most common type of photograph taken from a block specimen is of the **back-reflection** type; the incident X-rays pass through a hole in the film, impinge on a reasonably flat part of the specimen, and are reflected back on to the film (fig. 4.26). Some of the high-angle powder lines will be recorded, and from these valuable information can be obtained about the state of the specimen.

If the crystal size is large, the photograph will consist mainly of one or more Laue photographs, and so may not give the information desired. In order to give characteristic reflections, the specimen must be oscillated through a small angle, say 5°; if it is also translated (Wainwright, 1941; Thewlis and Pollock, 1950) or rotated (Lipson, Shoenberg and Stupart, 1941) in the plane of the surface irradiated, then still more crystals are brought into the beam and so a more truly representative photograph is obtained.

Back-reflection methods will, of course, give only the high-angle reflections; to record the lower orders the X-rays must strike the surfaces at a small glancing angle. At these angles good photographs can be obtained only from specimens with reasonably flat surfaces, since, if the surface is rough, only discrete regions can contribute to the reflected beams. Even if the surface is smooth, the medium-angle reflections will be broad, as shown in fig. 4.27. To remedy this, the area of the beam should be reduced by

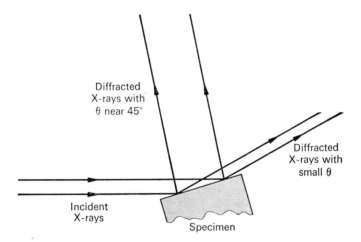

Diffracted X-rays with θ near 45°

Diffracted X-rays with small θ

Incident X-rays

Specimen

Fig. 4.27   X-ray beam incident at a low angle on a large specimen showing why some reflections are broadened

using a narrow slit rather than a circular hole to limit the X-ray beam falling on the specimen. As was shown in 4.1 this should still result in reasonably sharp lines.

Sometimes it is necessary to correlate the structure as revealed by X-rays with other physical properties such as microstructure or hardness. In these cases small rods of square cross-section are used and if they are fine grained, they give quite good X-ray photographs (Lipson and Parker, 1944).

4.8.3 Microbeam cameras

With the advent of microfocus tubes with a high power output per unit area of focus (3.6.5), it is now possible to irradiate specimens with a very fine beam of X-rays and at the same time to obtain photographs without using

110   X-Ray Powder Diffraction

overlong exposures. As has been indicated (3.6.5) the cameras, called micro-beam cameras, must in general be of much smaller dimensions than are general purpose types so that source-to-specimen and specimen-to-film distances are both short; the flat-film method of recording is normally used with either forward transmission or back reflection.

The beam is less than 0·1 mm in diameter, and Otte and Cahn (1959) have used beams of diameter as small as 10 to 50 μm. With this finely collimated radiation it is possible to select and study the orientation of single grains in aggregates and to determine the preferred orientation of different parts of a single fibre.

4.8.4 Factors affecting choice of camera

There is no one design of camera which is suitable for all types of work. For general routine purposes an ordinary camera with a diameter of about 90 mm is suitable and the choice of film-mounting, the Bradley–Jay, the van Arkel or the Ieviņš–Straumanis mounting, will depend on the particular investigation. If the identification of crystalline material is the main aim then either the Bradley–Jay or the Ieviņš–Straumanis mounting is preferable because they are the more convenient for the rapid calculation of the Bragg angles of the low-θ lines which are most important in this connection (10.2.1). The Bragg angles of these lines can be read directly on a scale if values uncorrected for film shrinkage are sufficiently accurate. Some workers prefer a scale which is graduated in lattice spacings, d (appendix 2), but this has the disadvantage that interpolation on a rapidly varying scale has to be made between the units.

The van Arkel mounting is less convenient for the rapid calculation of low values of θ from a number of different films. Also, because of the large separation of corresponding low-θ lines, film shrinkage makes the values obtained from a scale much less accurate than with either the Bradley–Jay or the Ieviņš–Straumanis mounting. It should be noted that photographs taken with the van Arkel mounting can be compared directly with those taken with either of the other two mountings, provided that the same radiation and the same diameter of camera are used.

If the camera is to be used to measure cell dimensions accurately, then either the van Arkel or the Ieviņš–Straumanis method is the best. Although the latter method has become very popular, it should be realized that it involves much more measurement than the former; in effect each photograph has to be calibrated by measuring low-order as well as high-order lines. For very accurate measurements, however, a diameter of 90 mm is probably too small. For this purpose it is better to use an ordinary camera of 190 mm diameter (which is standard in Great Britain) or to use a symmetrical focusing camera of about half this diameter which will give equal accuracy (4.8.1). The ordinary camera has the advantage that it can also be used for general purposes where separation of the lines is desirable.

Where short exposure times are required, either with or without crystal-reflected radiation, then the focusing camera is the most suitable; when the specimen to be examined is in block form then either the back-reflection or the glancing-angle technique must be adopted. For complex studies of preferred orientation or of the orientation of single grains the microbeam camera is essential.

The counter diffractometer is an expensive instrument but it is to be preferred under circumstances where there is a recurring requirement for the rapid determination of X-ray intensities. At the other extreme, when the step-by-step method of operation is adopted, the diffractometer is the best source of accurate intensities (provided, of course, that the specimen is not unstable) and is therefore suitable when structures have to be determined to the highest degree of accuracy.

4.9 Neutron Diffraction

A beam of neutrons exhibits wave properties corresponding to a wavelength λ given by h/mv where h is Planck's constant and mv is the momentum of the neutron. For thermal neutrons with a Maxwellian distribution of velocities appropriate to room temperature, the most probable value of the velocity gives a value of λ between 1Å and 2Å and therefore, with wavelengths of this order, it is possible to obtain diffraction effects from crystalline matter. A nuclear reactor is required as a high-intensity source of neutrons but even with beams of the highest available intensity specimens are required of about ten times the linear dimensions of those normally employed in X-ray work.

X-radiation is scattered by the electrons in an atom and the diffracted amplitude increases as the number of electrons in the atom increases; thus the hydrogen atom scatters very weakly compared with uranium or, indeed, compared with almost any other atom of which crystalline matter is composed. Neutrons, on the other hand, are scattered by nuclei and there is no systematic increase in the diffracted amplitude from successively heavier atoms, as there is with X-radiation. Thus the scattering power of hydrogen and of the light elements is not much less, numerically, than is the average for other elements; this makes the location of hydrogen atoms, and of light atoms in a structure which also contains heavy atoms, much easier with neutron diffraction than with X-ray diffraction.

There can, too, be quite a spectacular difference between the amplitudes of neutron radiation scattered by two atoms of nearly equal atomic number or even by isotopes of the same element; under such conditions the atoms can be readily distinguished from one another.

A further example of the use of neutron diffraction lies in its application to certain magnetic systems. Atoms that have a resultant magnetic moment scatter neutrons with an amplitude which is modified by both the orientation and the magnitude of such moments, and this leads to information on an atomic, or microscopic, scale about magnitization on a macroscopic scale (Bacon, 1962).

A beam of thermal neutrons from a reactor is white radiation and to obtain monochromatic radiation the beam of neutrons is reflected from a large single crystal and a beam of appropriate wavelength selected in accordance with Bragg's law, $\lambda = 2d \sin \theta$. As with X-ray monochromators, there remains the problem of harmonics (3.5.1); these are reduced to negligible proportions by selecting a suitable wavelength from the range which comprises the white-radiation spectrum.

Neutron-diffraction data are obtained with a spectrometer which is, in effect, a very large diffractometer, with a proportional counter filled with boron trifluoride gas as the detector. Intensities of diffracted beams are particularly easy to determine (Bacon, 1962) as they are very little affected by absorption but they do depend upon thermal effects and decrease as the temperature of the sample rises.

CHAPTER 5

Interpretation of Powder Photographs

5.1 Introduction

By measuring the position of a line on a powder photograph only one parameter can be directly determined, namely, the Bragg angle, θ, for the line. Once θ is known for every line on the photograph then, in principle, the whole pattern can be indexed. For the indexing to be unambiguous, the Bragg angle must be determined to the necessary degree of accuracy and the limits of error in both the measured and in the derived quantities must be known. The actual measurements made are those of the distances between corresponding powder lines (4.3.1); from these distances the experimental value of either $\sin^2 \theta$ or $1/d^2$ for each line is determined, and indexing is, in general, carried out with one or other of these sets of values. With complex powder patterns the values of $\sin^2 \theta$ (or of $1/d^2$) for the high-angle reflections may be required to an accuracy of ± 0.02 per cent so that both the random and systematic errors of measurement must be minimized. As described in detail in chapter 7 the random errors can be reduced by measuring several films and the systematic errors can be eliminated by using an internal standard.

When the unit cell of the substance is known, indexing the lines is a straightforward matter, although the calculations are rather lengthy for systems of low symmetry. In this chapter ways of systematizing these calculations in order to reduce the amount of work involved are discussed. A semi-graphical method using the reciprocal lattice (5.2.4; 5.3.4) is often useful for calculating spacings in the systems of low symmetry. When the unit cell is unknown, the problem of indexing is considerably more difficult, and trial-and-error methods have to be used. With a simple cubic pattern, such as given by sodium chloride, it is easy to index all the lines and to derive the length of the cell side. As the size of the unit cell increases the complexity of the pattern increases but cubic patterns are always fairly easy to index. Patterns from substances belonging to the uniaxial systems (tetragonal, trigonal and hexagonal) are more difficult but can usually be indexed. The trouble involved increases with increase of cell size. An orthorhombic pattern may prove to be very troublesome, especially if the unit cell is large. Because of the larger number of variables involved, the monoclinic, and even more the triclinic, system, presents the greatest difficulty, but with new methods of computation the problem is by no means insoluble, even when nothing is known about the unit cell. Completely graphical methods are often helpful in indexing, both when the unit cell is known and when it is unknown, and these are considered in a separate section.

5.2 Interpretation when the unit cell is known

When the unit cell is known, three stages in the interpretation are necessary. These are:

(a) derivation of Bragg angles from the diffraction pattern;
(b) calculation of Bragg angles for all possible combinations of indices hkl;
(c) the comparison of the two sets of results.

The problems which are considered here are those that arise in the second stage. Instead of calculating values of θ directly, it is better to calculate $\sin^2 \theta$, since the expression for $\sin^2 \theta_{hkl}$ is usually a fairly simple function of h, k and l. In order to simplify the third stage, a table of expressions for of $\sin^2 \theta$ as a function of θ is given in table 4.

5.2.1   The cubic system

For the cubic system the expression for $\sin^2 \theta$ can be obtained by combining Bragg's equation

$$\lambda = 2d \sin \theta$$

with the expression (2.1.2)

$$d = a/\sqrt{(h^2 + k^2 + l^2)}$$

for the spacing, d, of the (hkl) planes in terms of the unit-cell edge, a, and the indices h, k and l.

This gives

$$\sin^2 \theta_{hkl} = \frac{\lambda^2}{4a^2} (h^2 + k^2 + l^2) \qquad \ldots 5.1$$

The quantity $h^2 + k^2 + l^2$ is an integer so that all that is required is to calculate the value of $\lambda^2/4a^2$ and to multiply it by the possible values of $h^2 + k^2 + l^2$ which are given in table 5. Because $\sin^2 \theta$ can never be greater than unity it is not necessary to include values of $(\lambda^2/4a^2)(h^2 + k^2 + l^2)$ which are greater than this. Values of $\sin^2 \theta$ for the observed angles can be obtained from table 4 and the two sets then compared in order to determine which lines are actually present.

5.2.2 Tetragonal, hexagonal and trigonal systems

The formula for the tetragonal system is

$$\sin^2 \theta_{hkl} = \frac{\lambda^2}{4a^2} (h^2 + k^2) + \frac{\lambda^2}{4c^2} l^2 \qquad \ldots 5.2$$

where a and c are the unit-cell edges. Here again the values of $\sin^2 \theta$ are calculated for all possible values of h, k and l and then compared directly with the values of $\sin^2 \theta$ determined from the observed angles.

For convenience let $\lambda^2/4a^2 = A$ and $\lambda^2/4c^2 = C$. Then we require values of A multiplied by all possible values $h^2 + k^2$ (table 5) and values of C multiplied by l^2. For example, if $A = 0.10$ and $C = 0.07$ we can draw up a table as follows:

h, k	1, 0	1, 1	2, 0	2, 1	2, 2	3, 0
$A(h^2 + k^2)$	0·10	0·20	0·40	0·50	0·80	0·90

l	1	2	3
Cl^2	0·07	0·28	0·63

Possible values of $\sin^2 \theta$ are then obtained by adding the appropriate terms and are best set out in tabular form:

Possible Values of $\sin^2 \theta$

hk	0, 0	1, 0	1, 1	2, 0	2, 1	2, 2	3, 0
l 0	0·00	0·10	0·20	0·40	0·50	0·80	0·90
1	0·07	0·17	0·27	0·47	0·57	0·87	0·97
2	0·28	0·38	0·48	0·68	0·78		
3	0·63	0·73	0·83				

For example, the value of $\sin^2 \theta_{112}$ is 0·48.

The procedure is similar for crystals based on a hexagonal lattice; $\sin^2 \theta$ is in this case given by

$$\sin^2 \theta_{hkl} = \frac{\lambda^2}{3a^2}(h^2 + hk + k^2) + \frac{\lambda^2}{4c^2}l^2. \qquad \ldots 5.3$$

If the crystal belongs to the trigonal system and is based on a rhombohedral lattice then the reflections can be indexed on the assumption either of a primitive rhombohedral cell containing one lattice point or of a primitive hexagonal cell containing three lattice points (Henry, Lipson and Wooster, 1960). When the plane indices p, q, r are referred to the rhombohedral unit cell the expression for $\sin^2 \theta$ becomes

$$\sin^2 \theta_{pqr} = \frac{\lambda^2}{4} \left[\frac{\cos^2 (\alpha/2)}{a_R^2 \sin (\alpha/2) \sin (3\alpha/2)} \right]$$
$$\times \{(p^2 + q^2 + r^2) - (1 - \tan^2 \alpha/2)(pq + qr + rp)\} \qquad \ldots 5.4$$

where a_R is the side and α is the angle between the axes of the rhombohedral unit cell. This relationship can be rewritten as

$$\sin^2 \theta_{pqr} = \frac{\lambda^2}{4} \left\{ \frac{(p^2 + q^2 + r^2) \sin^2 \alpha + 2(pq + qr + rp)(\cos^2 \alpha - \cos \alpha)}{a_R^2(1 - 3\cos^2 \alpha + 2\cos^3 \alpha)} \right\}$$

If, however, the planes are to be indexed in terms of the triply primitive hexagonal cell, equation 5.3 is used and the rhombohedral indices pqr must be transformed to the indices hkl which are appropriate to the hexagonal unit cell.

The relationships between the two sets of indices and the two sets of axes are

$$h = p - q; \qquad k = q - r; \qquad l = p + q + r$$

and

$$9a_R^2 = 3a^2 + c^2$$

The transformation must be carefully carried out and the procedure is discussed in detail in Henry, Lipson and Wooster (1960) and, in rather less detail, by Azároff (1968).

5.2.3 Orthorhombic system

The formula for the orthorhombic system is

$$\sin^2 \theta_{hkl} = \frac{\lambda^2}{4a^2} h^2 + \frac{\lambda^2}{4b^2} k^2 + \frac{\lambda^2}{4c^2} l^2 \qquad \ldots 5.5$$

which we may write as

$$\sin^2 \theta_{hkl} = Ah^2 + Bk^2 + Cl^2 \qquad \ldots 5.6$$

A, B and C can be calculated and lists made of the quantities Ah^2, Bk^2 and Cl^2; sums of these quantities will then give possible values of $\sin^2 \theta$.

For example, $NiAl_3$ is orthorhombic with $a = 6 \cdot 61$, $b = 7 \cdot 36$ and $c = 4 \cdot 81$ Å. For cobalt Kα radiation, $\lambda = 1 \cdot 790$ Å. Therefore $A = 0 \cdot 01833$, $B = 0 \cdot 01478$, $C = 0 \cdot 0346$. Values of Ah^2, Bk^2 and Cl^2 are:

h, k or l	Ah^2	Bk^2	Cl^2
0	0·0000	0·0000	0·0000
1	0·0183	0·0148	0·0346
2	0·0733	0·0591	0·1386
3	0·1650	0·1330	0·3118
4	0·2933	0·2365	0·5544
5	0·4582	0·3695	0·8662
6	0·6599	0·5321	—
7	0·8982	0·7242	—
8	—	0·9459	—

Values of $\sin^2 \theta_{hkl}$ can now be obtained by adding the quantities in the appropriate columns. For example,

$$\sin^2 \theta_{200} = 0 \cdot 0733$$
$$\sin^2 \theta_{210} = 0 \cdot 0733 + 0 \cdot 0148 = 0 \cdot 0881$$
$$\sin^2 \theta_{213} = 0 \cdot 0733 + 0 \cdot 0148 + 0 \cdot 3118 = 0 \cdot 3999$$
$$\sin^2 \theta_{543} = 0 \cdot 4582 + 0 \cdot 2365 + 0 \cdot 3118 = 1 \cdot 0065$$

The last reflection obviously will not occur. Again the results should be set out in tabular form, but the tables will be more extensive than that in the previous section since three different indices are involved. It is necessary to calculate *all* the values of $\sin^2 \theta$ since a given line may be composed of more than one set of reflections. For example, a line with $\sin^2 \theta = 0 \cdot 1325$ might be 030; actually 030 is absent and the reflection is 220 ($\sin^2 \theta = 0 \cdot 1324$). It is therefore not enough to find possible indices for all the lines; all the calculated values must be compared with the observed ones.

5.2.4 The monoclinic and triclinic systems

For the monoclinic system, and still more so for the triclinic system, the expression for $\sin^2 \theta$ in terms of the direct-lattice parameters is rather cumbersome and is inconvenient for indexing purposes. In the monoclinic system

$$\sin^2 \theta_{hkl} = \frac{\lambda^2}{4} \left\{ \frac{h^2/a^2 + l^2/c^2 - 2hl \cos \beta/ac + k^2/b^2}{\sin^2 \beta} \right\} \qquad \ldots 5.7$$

where b is perpendicular to the plane of a and c, and β is the angle between a and c.

For a triclinic crystal

$$\sin^2 \theta_{hkl} =$$

$$\frac{\lambda^2}{4} \left\{ \frac{h^2 \sin^2 \alpha/a^2 + k^2 \sin^2 \beta/b^2 + l^2 \sin^2 \gamma/c^2 + 2hk(\cos \alpha \cos \beta - \cos \gamma)/ab}{1 - \cos^2 \alpha - \cos^2 \beta - \cos^2 \gamma + 2 \cos \alpha \cos \beta \cos \gamma} \right.$$

$$\left. + \frac{2kl(\cos \beta \cos \gamma - \cos \alpha)/bc + 2lh(\cos \gamma \cos \alpha - \cos \beta)/ca}{1 - \cos^2 \alpha - \cos^2 \beta - \cos^2 \gamma + 2 \cos \alpha \cos \beta \cos \gamma} \right\} \quad \ldots 5.8$$

where α, β, γ are the angles between b and c, c and a, a and b respectively. It is clearly possible, from these expressions, to compare the observed and calculated values of $\sin^2 \theta$ by the methods already described for tetragonal, hexagonal and orthorhombic systems, but the procedure is laborious. The tedium can be reduced using the simpler relations obtained when $\sin^2 \theta$ is expressed in terms of the parameters of the reciprocal unit cell. With these parameters, namely $a^* = 1/a$, $b^* = 1/b$ and $c^* = 1/c$,

$$\sin^2 \theta_{hkl} = \frac{\lambda^2}{4} (h^2 a^{*2} + k^2 b^{*2} + l^2 c^{*2} + 2lh c^* a^* \cos \beta^*) \quad \ldots 5.9$$

for the monoclinic system, and for the triclinic system

$$\sin^2 \theta_{hkl} = \frac{\lambda^2}{4} (h^2 a^{*2} + k^2 b^{*2} + l^2 c^{*2} + 2kl b^* c^* \cos \alpha^*$$

$$+ 2lh c^* a^* \cos \beta^* + 2hk a^* b^* \cos \gamma^*) \quad \ldots 5.10$$

If great accuracy is not necessary then for either of these two systems the labour can be further reduced with the help of a graphical plot of the reciprocal lattice. As was shown in section 2.2.1, in a three-dimensional reciprocal lattice the distance from the origin of a lattice point representing a set of parallel planes (hkl) of spacing d in real space is K/d where K is an arbitrary constant. If this constant is made equal to λ, the wavelength of the incident radiation, then the distance of the reciprocal-lattice point from the origin is λ/d; that is, it is equal to $2 \sin \theta$ where θ is the Bragg angle for the planes (hkl) of spacing d. The position, P, of the reciprocal point (fig. 5.1) may be defined in terms of cylindrical coordinates ξ and ζ where ξ is the radius of a cylinder whose axis OA passes through the origin of the reciprocal lattice and is parallel to the axis of oscillation of the crystal; the coordinate ζ is the distance from the origin of a plane perpendicular to OA and passing through P. Thus, since OP has been made equal to λ/d it follows that

$$\lambda^2/d^2 = \xi^2 + \zeta^2$$

or

$$\sin^2 \theta_{hkl} = \tfrac{1}{4}(\xi^2 + \zeta^2) \quad \ldots 5.11$$

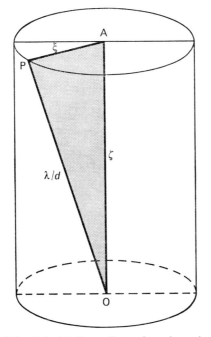

Fig. 5.1   Cylindrical coordinates in reciprocal space

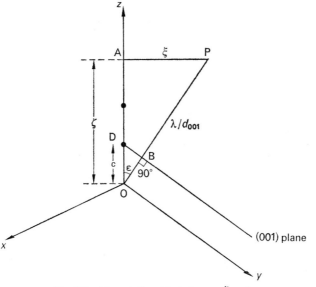

Fig. 5.2   The relationship between ζ and c

If the axis of oscillation OA of the crystal is parallel to the z axis, say, then the plane containing AP will be parallel to x^* and y^*, and all the reciprocal points in that plane will have the same value of l; these points can be plotted and ξ measured for each point. As shown below, $\zeta = l\lambda/c$, so that $\sin^2 \theta_{hkl}$ can then be determined.

To show that $\zeta = l\lambda/c$, consider fig. 5.2

$$OA = \zeta = OP \cos \epsilon = (\lambda/d_{001}) \cos \epsilon$$

where $OB = d_{001}$, the perpendicular distance between the (001) planes. But $OB = OD \cos \epsilon$, i.e.

$$d_{001} = c \cos \epsilon \quad \text{or} \quad 1/c = \cos \epsilon/d_{001}$$

Therefore $\qquad\qquad\qquad \zeta = \lambda/c$

For the lth layer the interplanar spacing will be $1/l$ times that for $l = 1$, the first layer, and hence $\zeta = l\lambda/c$.

5.3 Interpretation when the unit cell is unknown

When the unit cell is unknown, assigning indices to the lines on a powder photograph is much more complicated. In general the substance should first be assumed cubic and if this does not work then successively more complicated systems, through to triclinic if necessary, should be tried. In the following sections, which are intended to act as a guide to possible methods of solution, the characteristics of the various systems are described.

5.3.1 The cubic system

The characteristic of the cubic system is that the values of $\sin^2 \theta$ have a common factor. For example, the first seven lines on a powder photograph of an oxide of cobalt and iron of the spinel type have the following values of $\sin^2 \theta$:

0·0343, 0·0919, 0·1258, 0·1370, 0·1839, 0·2752, 0·3097.

The first two have a common factor of about 0·0115, and if we divide all the values of $\sin^2 \theta$ by 0·0115, we obtain the numbers 2·98, 7·99, 10·94, 11·91, 15·99, 23·93 and 26·93. These closely approximate to the values 3, 8, 11, 12, 16, 24 and 27, and the oxide is therefore cubic. The photograph was taken with CoKα radiation ($\lambda = 1·790$ Å) and therefore

$$a = \lambda/2\sqrt{0·0115} = 8·34 \text{ Å}$$

Plate 1   (a) (*above*) Photograph of Fe₃Si taken with non-crystal-reflected CoK$_\alpha$ radiation

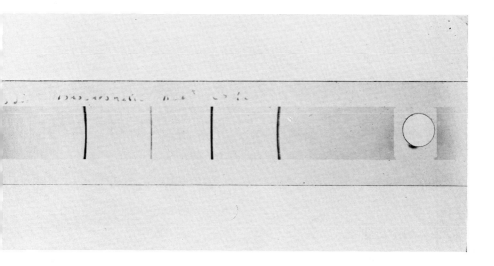

Plate 1   (b) (*above*) Photograph of Fe₃Si taken with crystal-reflected CoK$_\alpha$ radiation

Plate 1   (c) (*below*) Photograph of mineral anorthite obtained with a focusing camera and crystal-reflected CoK$_\alpha$ radiation. The reflection extends over an angular range of 20–90°. (Courtesy of United Steel Companies Ltd.)

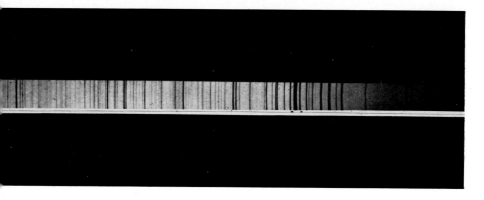

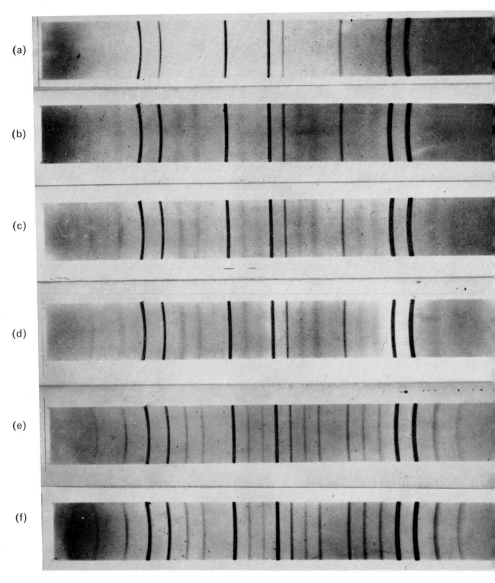

Plate 2 Powder photographs of AuCu$_3$ showing (a) main lattice lines and (b) to (f) main lattice lines and progressively sharper superlattice lines

The fact that the values of $\sin^2 \theta / 0.0115$ are all slightly less than integers means that a slightly smaller value of the common factor would have given better results. This lower value could not have been reliably deduced from the first two lines since their accuracy is low; on the other hand, the evidence from the first two lines gave the main indication that the oxide was cubic. This illustrates a general rule: *the lower orders are essential for the correct but approximate determination of the cell dimensions; the higher orders are used for accurate determination.* Therefore, in indexing a powder photograph, one must always be prepared to modify the common factor deduced from the lower orders in order to produce better agreement with the higher orders.

After a sufficient number of lines have been indexed on a powder photograph, it is a simple matter to determine the type of lattice, and this is particularly easy for cubic substances. The general rules for determining the lattice type are given in 8.2.1, and in the following paragraph cubic lattices are considered in detail. From the indexed lines on a powder photograph of a cubic substance, it is very easy to derive the length of the unit-cell side. This is done by inserting in equation (5.1) values of $\sin^2 \theta$ and hkl for one or more of the high-θ lines.

It is shown in 8.2.1 that reflections are possible from a primitive (P) lattice for all values h, k and l. Consequently the factor $(h^2 + k^2 + l^2)$, which we shall call N, can have as possible values every integer which can be expressed as the sum of three squares. It is obvious that the numbers 7 and 15, for example, cannot be so expressed, and we can generalize these 'forbidden numbers' in the form (appendix 3)

$$m^2(8n - 1)$$

where m and n are integers. The five numbers below 32 which satisfy this expression are given in the table below. The extension of this table up to $N = 100$ is given in table 5, and data up to $N = 1000$ can be found in the International Tables, Vol. 2 (1959).

For the all-face-centred (F) and body-centred (I) lattices there are restrictions on the possible values of h, k and l. For F, h, k and l must be either all even or all odd. For I, $h + k + l$ must be even. Some reflections may, of course, be absent for other reasons (7.5.5), but here we are considering the reflections which are always absent with these types of cubic lattice.

As shown above, the values of N that occur on the photograph of the iron-cobalt oxide are 3, 8, 11, 12, 16, 24 and 27; that is, very few of the lines corresponding to the primitive lattice P are present. But if we compare the numbers with those possible for the face-centred lattice F, we find that they all fit in and that only three, 4, 19 and 20, are missing in the range

5+

Possible Values of *N* for Cubic Lattices

Forbidden numbers	P	F	I	Forbidden numbers	P	F	I
	1				17		
	2		2		18		18
	3	3			19	19	
	4	4	4		20	20	20
	5				21		
	6		6		22		22
7				23			
	8	8	8		24	24	24
	9				25		
	10		10		26		26
	11	11			27	27	
	12	12	12	28			
	13				29		
	14		14		30		30
15				31			
	16	16	16		32	32	32

covered. These facts are sufficient to prove that the lattice is F; the few missing lines are absent for other reasons (7.5.5).

Recognizing the lattice I is not so simple, as there is a risk of error unless a sufficient number of lines is indexed. For the I lattice, N must be even so that the first six possible lines for I have values of N equal to 2, 4, 6, 8, 10 12, and these of course have the same relative values of $\sin^2\theta$ as the first six lines for the P lattice, 1, 2, 3, 4, 5, 6. The next line for I is 14, and this cannot be mistaken for the next line for P which is 8 and occurs after the gap corresponding to the forbidden number 7. If therefore we had taken the lattice to be P, the apparent existence of a line corresponding to $N = 7$ would have shown the mistake. This, of course, is only one example; several other lines should also be used, if possible, to differentiate between I and P. It is important to use several lines because some can be absent for structural reasons, and it might be that line 14 for I, in a particular photo-graph, was missing for this reason. If indexing had continued on the assumption of a P lattice, the line for which $N = 30$ in the true I lattice would have been called 15, which is a forbidden number. Just before this line was reached, however, it would have been noticed that there was no line corresponding to $N = 14$ on the assumption of the P lattice; this number, 14, is not a forbidden one, but the true number, 28, of the I lattice, is a forbidden one.

It is not always possible to observe a sufficient number of lines to make a definite distinction between P and I, on the basis of the Bragg angles of the lines only. For example, there are many body-centred alloys of iron that give only four or five lines with cobalt Kα radiation which is most suitable for such alloys (3.4.4). Evidence from the intensities (chapter 8) has then to be used, but structures that give so few lines are usually so simple that the calculations are not lengthy, and give quite definite results. Photographs can also be taken with radiations of shorter wavelengths to provide more lines for indexing.

With experience it is often possible to recognize at a glance whether a powder photograph is that of a cubic substance or not. If it is cubic, some regularity in the sequence of lines can usually be seen, with occasional gaps corresponding to vanishingly weak reflections and to the missing numbers given above. If the substance has a primitive lattice there must be a gap at least at every eighth line, because of the missing numbers, 7, 15, 23, 31, 39 etc.; these gaps, as can be seen from the solution to problem 6, form a useful check on the indexing. If the lattice is body-centred, the sequence of lines 2, 4, 6, 8, . . . , will still be regular, but since the only even numbers in the list of forbidden values of N below 100 are 28, 60, and 92, the number of lines between the gaps can be greater than 8. The number of lines with N even between 28 and 60, and between 60 and 92, is fifteen in each case.

Interpretation based on these principles should not be attempted without experience, and conclusions should always be checked by measurement and calculation. In particular, the recognition of a face-centred cubic structure is not always easy. For example, the powder photograph of silicon (Plate III) seems not to be cubic because the combination of absences due to the lattice and to the structure produces a grouping of the lines which is not quite regular; nevertheless it *is* cubic.

5.3.2   The tetragonal system

If the system is not cubic, a more complicated pattern will be formed, and no rough survey of the photograph will give any information. The simplest procedure is to see if any relationships beteeen the values of sin^2 θ will give an indication of the symmetry.

Suppose the substance is tetragonal. Then

$$\sin^2 \theta_{hkl} = A(h^2 + k^2) + Cl^2$$

The problem is therefore to find values of A and C which give the observed values of sin^2 θ with integral values of h, k and l. This may seem difficult in general, but some special cases provide a certain amount of assistance. Consider the spectra with l zero;

$$\sin^2 \theta_{100} = A, \quad \sin^2 \theta_{110} = 2A, \quad \sin^2 \theta_{200} = 4A,$$
$$\sin^2 \theta_{210} = 5A, \quad \sin^2 \theta_{220} = 8A \qquad \qquad \ldots 5.12$$

The ratio 2 occurs frequently. Except by chance, the only other system in which this ratio is found is the cubic one. If, therefore, the substance is not cubic and the $\sin^2 \theta$'s of two low-angle lines are in the ratio of 2, it is probable that the substance is tetragonal and that the two lines are 100 and 110, or 110 and 200. The quantity A can be derived and the presence of other $hk0$ lines can be tested. If other lines do fit in, the original guess is probably correct.

The remaining problem is to find the value of C. A method for doing this is illustrated by the following example. The first nine lines on a powder photograph have the following values of $\sin^2 \theta$:

a	0·0445	d	0·1767	g	0·2245
b	0·0888	e	0·1811	i	0·3117
c	0·1449	f	0·2204	j	0·3554

$\sin^2 \theta_b$ and $\sin^2\theta_a$ have a ratio of 2. Suppose then that the substance is tetragonal and that line a is 100 and line b 110. Take $A = 0.0444$. Then $2A = 0.0888$, $4A = 0.1776$, $5A = 0.2220$, $8A = 0.3552$. These values suggest that d is 200, f is perhaps 210 and j is 220. Now subtract the multiples of A from all the values of $\sin^2 \theta$, even those that are thought to be indexed correctly.

Line	$\sin^2 \theta$	$\sin^2 \theta - A$	$\sin^2 \theta - 2A$	$\sin^2 \theta - 4A$	$\sin^2 \theta - 5A$
a	0·0445	0·0001			
b	0·0888	0·0444	0·0000		
c	0·1449	0·1005	0·0561		
d	0·1767	0·1323	0·0879		
e	0·1811	0·1367	0·0823	0·0035	
f	0·2204	0·1760	0·1316	0·0418	
g	0·2245	0·1801	0·1357	0·0469	
i	0·3117	0·2673	0·2229	0·1341	0·0897
j	0·3554	0·3110	0·2668	0·1777	0·1334

If the original guess is correct, each horizontal line in this table of differences should contain a quantity Cl^2, and so we must look for quantities that occur frequently. The value 0.1329 ± 0.0012 occurs in the differences for lines d, f, i and j. But it does not occur for lines c, e and g. Suppose then that 0.1329 corresponds with $l = 2$. Then $C = 0.0322$. This does not occur at all. The value 0.1354 ± 0.0013, which occurs for lines e, g and i is no better. We must therefore suppose that the original guess is wrong, although there has been enough success to show that it has some relation to the truth.

Suppose that line a is 110 and line b 200. Then $A = 0.0222$, and the possible multiples are 0.0444, 0.0888, 0.1110, 0.1776, 0.1998. These values may now be subtracted from the observed values of sin² θ; the whole table should be completed, but for brevity we shall take only the first five lines.

Line	$\sin^2\theta$ $-A$	$\sin^2\theta$ $-2A$	$\sin^2\theta$ $-4A$	$\sin^2\theta$ $+5A$	$\sin^2\theta$ $-8A$	$\sin^2\theta$ $\perp 9A$
a	0·0223	0·0001				
b	0·0666	0·0444	0·0000			
c	0·1227	0·1005	0·0561	0·0339		
d	0·1545	0·1323	0·0879	0·0657	0·0035	
e	0·1589	0·1367	0·0923	0·0701	0·0418	0·0206

We now see that the value 0·0339 occurs for line c; this is near the value for which we were looking. Four times it is 0·1356, which is near the second value considered before. The values of the constants so derived are $A = 0·0222$, $C = 0·0339$, and it is possible to index all the lines on this basis. More accurate values are $A = 0·0220$ and $C = 0·0340$. From these values it will be found that j is 312, not 400; there are usually chance coincidences like this to trap the unwary.

The substance is $CuAl_2$.

5.3.3   The hexagonal and trigonal systems

The equation 5.3 for hexagonal axes is:

$$\sin^2\theta_{hkl} = A(h^2 + hk + k^2) + Cl^2$$

where $A = \lambda^2/3a^2$ and $C = \lambda^2/4c^2$.
For the $hk0$ spectra,

$$\sin^2\theta_{100} = A; \quad \sin^2\theta_{110} = 3A; \quad \sin^2\theta_{200} = 4A;$$
$$\sin^2\theta_{210} = 7A; \quad \sin^2\theta_{300} = 9A; \quad \sin^2\theta_{220} = 12A \ldots 5.13$$

It will be seen that the most frequently occurring ratio is now 3; this cannot occur in the tetragonal system except by chance.

If the unknown system has trigonal symmetry then it may be based on either a hexagonal or a rhombohedral lattice. If the lattice is hexagonal then the procedure is that just described for a hexagonal system, but if the underlying lattice is rhombohedral then the hexagonal cell determined by this method will be triply primitive (5.2.2) and certain lines which would be present if the lattice were truly hexagonal will be absent. The particular absences depend on the relative orientation of the true rhombohedral cell and the selected hexagonal cell and fall into one of two categories (Henry, Lipson and Wooster, 1960). For one possible orientation the only lines present on hexagonal indexing are those for which

$$-h + k + l = 3n, \quad n = 0, 1, 2, \ldots$$

For the other orientation, the condition for a reflection to occur is

$$h - k + l = 3n, \quad n = 0, 1, 2, \ldots$$

5.3.4   The orthorhombic system

If the substance is orthorhombic, the problem is much more difficult (Lipson, 1949); three constants have to be found and these are related to the $\sin^2 \theta$'s by the equation

$$\sin^2 \theta_{hkl} = Ah^2 + Bk^2 + Cl^2$$

Some idea of the probable magnitudes of A, B and C may be obtained from the number of lines on the photograph; the smaller these constants are, the greater the number of lines that can appear.

In what follows it is convenient to use dimensionless reciprocal units (2.2.1). Since the distance from the origin of any point hkl in reciprocal space is $2 \sin \theta_{hkl}$ points with θ equal to θ_m will lie on the surface of a sphere of radius $2 \sin \theta_m$. The volume of this sphere is $4\pi (2 \sin \theta_m)^3/3$, and within it lie all the points with θ less than θ_m. The number of such points is given approximately by the ratio of this volume to the volume of the unit cell of the reciprocal lattice,

$$32\pi \sin^3 \theta_m/3a^*b^*c^*$$

This is not equal to the number of lines with θ less than θ_m on the powder photograph for several reasons. First, groups of reciprocal points represent planes of the same spacing; for example, the reflections hkl, $\bar{h}kl$, $h\bar{k}l$, $hk\bar{l}$, $h\bar{k}\bar{l}$, $\bar{h}k\bar{l}$, $\bar{h}\bar{k}l$ and $\bar{h}\bar{k}\bar{l}$ all give lines with the same value of θ (table 8). Thus the number of possible different values of θ is reduced to be of the order of $4\pi \sin^3 \theta_m/3a^*b^*c^*$. Further, some of the reflections will be of vanishingly small intensity (7.5.5) and some will be absent for space-group reasons (8.2.1); others will overlap because their Bragg angles are almost identical. These effects are difficult to assess quantitatively, but it is probably fair to assume that, in all, only about half the theoretically possible number of lines will be observed. If M is the number of observed lines, then

$$M \simeq 2\pi \sin^3 \theta_m/3a^*b^*c^* \qquad \ldots 5.14$$

Now $A = \lambda^2/4a^2 = \frac{1}{4}a^{*2}$ and therefore $a^* = 2\sqrt{A}$. Similarly

$$b^* = 2\sqrt{B} \quad \text{and} \quad c^* = 2\sqrt{C}$$

Therefore

$$M \simeq 2\pi \sin^3 \theta_m/24\sqrt{(ABC)} \simeq \sin^3 \theta_m/4\sqrt{(ABC)},$$

or

$$\sqrt{(ABC)} \simeq \sin^3 \theta_m/4M \qquad \ldots 5.15$$

If A, B and C are of the same order of magnitude, $\sqrt{(ABC)}$ may be replaced by $A^{3/2}$ (or $B^{3/2}$ or $C^{3/2}$), and the approximate value of A is given by $(\sin^3 \theta_m/4M)^{2/3}$. That is,

$$A \simeq (\tfrac{1}{4})^{2/3} \sin^2 \theta/M^{2/3} \simeq 0{\cdot}4 \sin^2 \theta/M^{2/3} \qquad \ldots 5.16$$

For example, on a powder photograph of $NiAl_3$, twenty lines occur with values of $\sin \theta$ less than 0·5 when cobalt Kα radiation is used. Thus we should expect that A, B and C would be about $0{\cdot}4 \times 0{\cdot}25/20^{2/3} = 0{\cdot}014$. Actually,

$$A = 0{\cdot}0148, \quad B = 0{\cdot}0346, \quad C = 0{\cdot}0183$$

This method thus gives the correct order of magnitude for A, B and C.

The problem is still formidable but not necessarily insoluble, as we can see by considering some simple cases.

$$\sin^2 \theta_{100} = A, \quad \sin^2 \theta_{010} = B, \quad \sin^2 \theta_{001} = C \qquad \ldots 5.17$$

$$\sin^2 \theta_{011} = B + C = \sin^2 \theta_{010} + \sin^2 \theta_{001}$$
$$\sin^2 \theta_{101} = A + C = \sin^2 \theta_{100} + \sin^2 \theta_{001}$$
$$\sin^2 \theta_{110} = A + B = \sin^2 \theta_{100} + \sin^2 \theta_{010}$$
$$\sin^2 \theta_{111} = A + B + C = \sin^2 \theta_{100} + \sin^2 \theta_{010} + \sin^2 \theta_{001} \qquad \ldots 5.18$$

It might be possible to find lines whose values of $\sin^2 \theta$ had this type of relation. This, however, is unlikely because the chances that 100, 010 and 001 will be absent for one or more reasons are high. We therefore rewrite the equations in another way:

$$\begin{aligned} C = \sin^2 \theta_{001} &= \sin^2 \theta_{101} - \sin^2 \theta_{100} \\ &= \sin^2 \theta_{011} - \sin^2 \theta_{010} \\ &= \sin^2 \theta_{hkl} - \sin^2 \theta_{hk0}, \text{ in general} \qquad \ldots 5.19 \end{aligned}$$

That is, C should often occur as the difference between the $\sin^2 \theta$'s for two lines, and so conversely, any difference that occurs often may be tried as A, B or C.

A way of trying to find such differences is illustrated by the following analysis of $NiAl_3$ (p. 129). The values of $\sin^2 \theta$ for the first twenty lines are given in the first column of the table below. In the succeeding columns are given the differences, up to 0·1, between the lines (lettered from a to w) shown at the heads of the columns and the lines immediately following. This table of differences may then be examined to see if any value occurs particularly frequently. One way of doing this is to plot the differences shown in the table along horizontal lines, one for each powder line, as shown in fig. 5.3. The points should be elongated to take possible errors into account; in this example the error has been taken as $\pm 0{\cdot}0005$, but in general it must be found from experience with known structures.

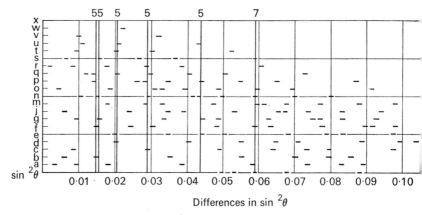

Fig. 5.3   Differences in sin² θ for lines on a powder photograph of NiAl$_3$

When the diagram is complete, a vertical line is run across it and the points where several values are cut simultaneously are noted. In fig. 5.3, the outstanding figure is 0·0591, which occurs seven times; 0·0437, 0·0205, 0·0153 and 0·0146 occur five times, and many values occur four times. The immediate deduction is that one of our constants, say A, is 0·0148 as this agrees both with 0·0146 and $\frac{1}{4}$ of 0·0591, and it is of the order of magnitude that we expect. We cannot yet make use of the other observations, and in any event it is better to take only one step at a time.

First we deduce that line c, sin² θ = 0·0591, is 200. We next note that sin² θ_b and sin² θ_d differ by 0·0148. This suggests that line b and line d have the same k and l indices and that h = 0 and 1 respectively. Similarly, sin² θ_f differs from sin² θ_g by 0·0148 and from sin² θ_n by 0·0590. This suggests that lines f, g and n have the same k and l indices and h = 0, 1 and 2 respectively.

It is wise not to consider more than six or eight lines to begin with; and we find that, of the first seven lines, we have indexed one completely, have some information about four, and know nothing positive about two, a and c. We may guess that line a has a non-zero h index since the differences 0·0146 and 0·0591 do not occur in connection with it. It is true that 0·0586 (sin² θ_f − sin²θ_a) may possibly be 0·0591, but line f itself is supposed to have h = 0. If line a does not have h = 0, then the simplest indices to try for it are 110; it cannot of course be 100; this would make B = sin² θ_a − A = 0·0496 − 0·0148 = 0·0348. This is encouraging, because 0·0350 occurs four times in the table. Also 4 × 0·0348 = 0·1392, which suggests that line m is 020.

We also note that sin² θ_a − sin² θ_b = 0·0352. This would suggest that these lines have the same h and l indices and k = 1 and 0 respectively. But

Differences of sin² θ Values ($\times 10^4$) for $NiAl_3$

Line symbols

sin² θ	a	b	c	d	e	f	g	j	m	n	o	p	q	r	s	t	u	v	w
a 0496																			
b 0530	034																		
c 0591	095	061																	
d 0678	182	148	087																
e 0882	386	352	291	204															
f 1082	586	552	491	404	200														
g 1230	734	700	639	552	348	148													
j 1325	829	795	734	647	443	243	095												
m 1386	890	856	795	708	504	304	156	061											
n 1565			974	887	683	483	335	240	179										
o 1672				994	790	590	442	347	286	107									
p 1716					834	634	486	391	330	151	044								
q 1858					976	776	628	533	472	293	186	142							
r 1980						898	750	655	594	415	308	264	122						
s 1999						917	769	674	613	434	327	283	141	019					
t 2063						981	833	738	677	498	391	347	205	083	064				
u 2156							926	831	770	591	484	440	298	176	157	093			
v 2266								941	880	701	594	550	408	286	267	203	110	100	
w 2366									980	801	694	650	508	386	367	303	210	100	
x 2587											915	871	729	607	588	524	431	321	221

line b has $h = 0$ and so must be 001 or 002. This fixes C as 0·0530 or 0·0132, but both these occur only once in the table. Thus we have probably been led astray by a chance coincidence.

The next simplest indices to try for line b are 011. C then equals $\sin^2 \theta_b$ − 0·0348 = 0·0182. This is much better, since 0·0182 occurs four times.

We are now in a position to see if our values of A, B and C explain all the lines. The first eight give:

Line	hkl	Calculated	Observed
a	110	0·0496	0·0496
b	011	0·0530	0·0530
c	200	0·0590	0·0591
d	111	0·0678	0·0678
e	102	0·0876	0·0882
f	012	0·1076	0·1082
g	112	0·1224	0·1230
j	{300 / 202	{0·1328 / 0·1318	0·1325

This is the correct solution, although the choice of axes is not the same as that used earlier in this chapter. If the axes used here are x', y' and z', their relation to the x, y and z axes in 5.2.3 is $x' = y$, $y' = z$, $z' = x$.

The values of A, B and C have to be adjusted as higher orders are taken into account. It is obvious, for example, that C is too low; 0·0183 gives better agreement for the lines with $l = 2$ and will not appreciably worsen the agreement for the lines with $l = 1$.

It is possible that these methods would not work for a structure much more complicated than that of $NiAl_3$; even for this, only one of the constants is readily found, probably because it is the smallest. Nevertheless, the method does provide a way of starting on the problem and is worth trying on a photograph that does not appear to be that of a cubic, tetragonal or hexagonal substance.

5.3.5 The monoclinic system

In the monoclinic system where b is perpendicular to the plane of a and c, $\sin^2 \theta$ is given by equation 5.7. The equation has four unknowns, a, b, c, and β, and this presents a problem in indexing which is much more difficult than any so far considered. The method described by Lipson (1949) for indexing the $0kl$, $h0l$ and $hk0$ reflections of an orthorhombic crystal depends on the fact that the $[h00]$, $[0k0]$ and $[00l]$ directions are mutually perpendicular; that is, that the unit-cell edges a, b and c are at right angles to each other. But in the monoclinic system b is perpendicular to both a and c and therefore Lipson's method for orthorhombic crystals can be applied to the $hk0$ and $0kl$ reflections from a crystal with monoclinic symmetry.

To illustrate this, consider the reflections from a monoclinic crystal for which $l = 0$. $\sin^2 \theta$ for this system then becomes

$$\sin^2 \theta_{hk0} = Ah^2 + Bk^2 \qquad \qquad \dots 5.20$$

where

$$A = \lambda^2/4a^2 \sin^2 \beta = \sin^2 \theta_{100} \quad \text{and} \quad B = \lambda^2/4b^2 = \sin^2 \theta_{010}$$

Similarly

$$\sin^2 \theta_{0kl} = Bk^2 + Cl^2 \qquad \qquad \dots 5.21$$

where

$$C = \lambda^2/4c^2 \sin^2 \beta = \sin^2 \theta_{001}$$

It follows from 5·20 that

$$\sin^2 \theta_{0k0} = Bk^2 \quad \text{and} \quad \sin^2 \theta_{h00} = Ah^2$$

and therefore

$$\sin^2 \theta_{hk0} = \sin^2 \theta_{h00} + \sin^2 \theta_{0k0}$$

Also, from 5.21 we have

$$\sin^2 \theta_{0kl} = \sin^2 \theta_{0k0} + \sin^2 \theta_{001}$$

These relations can be rewritten in the form

$$\sin^2 \theta_{0k0} = \sin^2 \theta_{hk0} - \sin^2 \theta_{h00}$$
$$\sin^2 \theta_{0k0} = \sin^2 \theta_{0kl} - \sin^2 \theta_{001}$$
$$\sin^2 \theta_{h00} = \sin^2 \theta_{hk0} - \sin^2 \theta_{0k0}$$
$$\sin^2 \theta_{001} = \sin^2 \theta_{0kl} - \sin^2 \theta_{0k0}$$

Thus by following the procedure for orthorhombic crystals (5.3.4), the values of $\sin^2 \theta_{100}$, $\sin^2 \theta_{010}$ and $\sin^2 \theta_{001}$ can be determined, and then from the relations 5.20 and 5.21 the $hk0$ and $0kl$ reflections can be indexed by comparison of the calculated and observed values of $\sin^2 \theta$. The parameter b can be obtained directly from the relation $\sin^2 \theta_{010} = \lambda^2/4b^2$.

The values of a, c and β have still to be found, and since only two independent equations involving the three unknowns can be formed it is necessary to resort to trial and error. It is convenient to express equation 5.7 as

$$\sin^2 \theta_{hkl} = h^2 \sin^2 \theta_{100} + l^2 \sin^2 \theta_{001} - 2hl \sin \theta_{100} \sin \theta_{001} \cos \beta + k^2 \sin^2 \theta_{010}$$

$\sin^2 \theta_{100}$, $\sin^2 \theta_{010}$ and $\sin^2 \theta_{001}$ are, of course, all known and hence $\cos \beta$ can be found by comparison of the observed and calculated values of $\sin^2 \theta$

for the, as yet, unindexed hkl reflections. Once β has been determined, the values of a and c can be calculated.

As with an orthorhombic crystal, the success of the method applied to monoclinic systems will depend both upon the nature of the substance and the ingenuity with which the problem is tackled. The application of computer techniques will increase the chances of success but even without these aids the method has proved useful (de Wolff, 1963).

5.3.6   The triclinic system

In triclinic substances three quantities have to be determined to define the edges of the unit cell, and three more to define the angles between these edges, and these substances would present overwhelming problems if it were not for the simplifying factor that the axes are to some extent arbitrary so that there is a good chance of finding a possible set. Ito (1949, 1950) was the first to develop a method for indexing any powder pattern irrespective of the crystal symmetry but which was particularly useful for systems having triclinic symmetry.

Any calculation of $\sin^2 \theta$ values for a triclinic substance using equation 5.8, which refers to the direct lattice, must be prohibitively long, and consequently Ito based his method on equation 5.10 which expresses $\sin^2 \theta$ in terms of the reciprocal-lattice parameters.

The very fact that this relation is used to determine the unit-cell parameters means that the unknown cell so derived will be triclinic. The actual lattice may or may not be triclinic, but if it is then the unit cell given by the calculated values of a^*, b^*, c^*, α^*, β^* and γ^* will display the full symmetry of the lattice. If the cell does not display the full symmetry of the lattice then there are procedures to be described later (5.3.7) by means of which the cell can be transformed into one that does.

As is implicit in the application of Lipson's method to orthorhombic and monoclinic systems, the first step in Ito's procedure is to establish one of the zones $hk0$, $0kl$ or $h0l$. For the $hk0$ zone the relation for $\sin^2 \theta$ reduces to

$$\sin^2 \theta_{hk0} = \frac{\lambda^2}{4} \left(h^2 a^{*2} + k^2 b^{*2} + 2hka^*b^* \cos \gamma^* \right) \qquad \ldots 5.22$$

When both k and l are zero

$$\sin^2 \theta_{h00} = \lambda^2 h^2 a^{*2}/4 = h^2 \sin^2 \theta_{100}$$

Similarly, $\sin^2 \theta_{0k0} = k^2 \sin^2 \theta_{010}$ and hence

$$\sin^2 \theta_{hk0} = h^2 \sin^2 \theta_{100}$$
$$+ k^2 \sin^2 \theta_{010} + 2hk \sin \theta_{100} \sin \theta_{010} \cos \gamma^* \qquad \ldots 5.23$$

Now the quantity $h^2a^{*2} + k^2b^{*2} + 2hka^*b^*\cos\gamma^*$ (equation 5.22) is d^{*2}_{hk0} where d^*_{hk0} is the reciprocal of the spacing d_{hk0} and it can be seen from fig. 5.4 and the cosine law that the reversal of the direction of kb^* results in

$$d^*_{h\bar{k}0} = h^2a^{*2} + k^2b^{*2} - 2hka^*b^*\cos\gamma^*$$

Hence

$$\sin^2\theta_{h\bar{k}0} = \lambda^2(h^2a^{*2} + k^2b^{*2} - 2hka^*b^*\cos\gamma^*)/4$$

or

$$\sin^2\theta_{h\bar{k}0} = h^2\sin^2\theta_{100}$$
$$+ k^2\sin^2\theta_{010} - 2hk\sin\theta_{100}\sin\theta_{010}\cos\gamma^* \quad \ldots 5.24$$

Adding (5.23) and (5.24)

$$\sin^2\theta_{hk0} + \sin^2\theta_{h\bar{k}0} = 2h^2\sin^2\theta_{100} + 2k^2\sin^2\theta_{010},$$

and thus the two powder lines whose indices are 110 and 1$\bar{1}$0 must have values of $\sin^2\theta$ whose mean is equal to $\sin^2\theta_{100} + \sin^2\theta_{010}$. Cos γ^* can now be found from equation (5.23) and by repeating the operation for another zone, say $h0l$, $\sin^2\theta_{001}$, $\cos\alpha^*$ and $\cos\beta^*$ can also be determined.

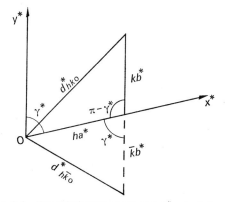

Fig. 5.4   The relationship between d^*_{hk0} and $d_{h\bar{k}0}$

Before giving an actual example it is convenient at this stage to draw attention to two possible drawbacks to the application of Ito's method. First, one or more of the values of $\sin^2\theta_{100}$, $\sin^2\theta_{010}$ and $\sin^2\theta_{001}$ might be missing from the list of observed values; the reflections may be too weak to be measured or systematically absent. In either event the corresponding values of $\sin^2\theta_{100}$, $\sin^2\theta_{010}$ and $\sin^2\theta_{001}$ would be incorrectly assigned and the powder pattern could not be completely indexed. Secondly the accuracy of the observed values of $\sin^2 \theta$ may be so low that the $100, 010, \ldots$ reflections have been incorrectly assigned to $\sin^2 \theta$ values which happen to fall within the limits set by the experimental errors (5.1). The effects of these two factors can be reduced by an extension of Ito's method which will be explained later, but first an application of the method itself will be outlined.

The values of $\sin^2 \theta$ given below were obtained from four superimposed single-crystal rotation photographs with the crystal set in four unrelated arbitrary orientations and the reflections represent the first fifteen lines of a powder pattern.

a	0·0100	f	0·0310	n	0·0425
b	0·0165	g	0·0338	o	0·0437
c	0·0195	i	0·0384	p	0·0469
d	0·0223	j	0·0399	q	0·0500
e	0·0262	m	0·0420	r	0·0606

Choose a and b as 100 and 010; then $\sin^2 \theta_{100} + \sin^2 \theta_{010}$ is 0·0265, and we must therefore look for two other $\sin^2 \theta$'s with a sum of 0·0530; there are two such pairs—c and g (0·0533) and d and f (0·0533)—both of which possibilities should be explored. In fact it is found that the former gives a consistent scheme more naturally, as we shall now show.

Since this scheme starts by accounting for the lines a, b, c and g, let us assume that line d is 001. Then we must look for pairs of lines with sums

		sin² θ	
Line	Indices	Calc.	Obs.
a	100	0·0100	0·0100
b	010	0·0165	0·0165
c	$\bar{1}10$	0·0193	0·0195
d	001	0·0223	0·0223
e	$10\bar{1}$	0·0262	0·0262
f	$01\bar{1}$	0·0308	0·0310
g	110	0·0337	0·0338
i	101 —	0·0384	0·0384
j	$\begin{cases} \bar{1}\bar{1}1 \\ 200 \end{cases}$	$\begin{cases} 0·0397 \\ 0·0400 \end{cases}$	0·0399
m	$11\bar{1}$	0·0419	0·0420
n	$\bar{2}10$	0·0421	0·0425
o	$\bar{1}11$	0·0435	0·0437
p	011	0·0468	0·0469
q	$20\bar{1}$	0·0501	0·0500
r	$\bar{2}\bar{1}1$	0·0602	0·0606

equal to $2(0.0100 + 0.0223) = 0.0646$, and $2(0.0165 + 0.0223) = 0.0776$. Pairs of lines which satisfy this condition are e and i (0.0646) and f and p (0.0779). Thus, if these results are correct, we have indexed the nine lines a, b, c, d, e, f, g, i and p, and have found the values of the six constants as follows:

$$A = 0.0100 \quad D = \tfrac{1}{2}(\sin^2 \theta_p - \sin^2 \theta_r) = 0.0080$$
$$B = 0.0165 \quad E = \tfrac{1}{2}(\sin^2 \theta_i - \sin^2 \theta_e) = 0.0061$$
$$C = 0.0223 \quad F = \tfrac{1}{2}(\sin^2 \theta_g - \sin^2 \theta_c) = 0.0072$$

These values must be tested on the remaining lines; the indices in the table on p. 134 are found by trial. (Line $1\bar{1}0$ is given a lower value of $\sin^2 \theta$ than line 110 to correspond to an obtuse angle γ.)
This indexing gives a unit cell with the following dimensions:

$$a = 8.10, \quad b = 6.33, \quad c = 5.34 \text{ Å}, \quad \alpha = 99° 17', \quad \beta = 98° 57',$$
$$\gamma = 104° 18'$$

The substance is hexamethylbenzene, of which the usual unit-cell dimensions are:

$$a = 8.92, \quad b = 8.86, \quad c = 5.30 \text{ Å}, \quad \alpha = 44° 27', \quad \beta = 116° 43',$$
$$\gamma = 119° 34'$$

The latter axes are related to the former by algebraic methods (Henry, Lipson and Wooster, 1960).

5.3.7 Ito's procedure for transformation of axes

In the last section it was pointed out that, in principle, every powder photograph could be treated as that of a triclinic substance and that, if the symmetry were higher than triclinic, relationships should appear when certain transformations of axes were made. Ito (1949) has described a procedure which is based upon a reduction process described by Delaunay (1933); this should lead automatically from an arbitrary unit cell to the simplest possible. The theory is described in Ito's book and in the International Tables, Vol. 1 (1952), Azároff and Buerger (1958) and Kaelble (1967), and only an outline of the method will be described here. It depends simply on the principle that a unit cell should have nearly equal sides with obtuse angles as near as possible to 90°.
We start by evaluating the following six quantities based upon the arbitrary unit cell found:

$$b_0 c_0 \cos \alpha_0 = P_0 \qquad a_0 d_0 \cos \phi_a = S_0$$
$$c_0 a_0 \cos \beta_0 = Q_0 \qquad b_0 d_0 \cos \phi_b = T_0$$
$$a_0 b_0 \cos \gamma_0 = R_0 \qquad c_0 d_0 \cos \phi_c = U_0$$

where, in addition to the usual quantities, d_0 is the length of the $[\bar{1}\bar{1}\bar{1}]$ diagonal of the unit cell, and ϕ_a, ϕ_b and ϕ_c are the angles between this diagonal and the a_0, b_0 and c_0 axes respectively. Formulae for S_0, T_0 and U_0 are as follows:

$$S_0 = -(a_0{}^2 + Q_0 + R_0)$$
$$T_0 = -(P_0 + b_0{}^2 + R_0)$$
$$U_0 = -(P_0 + Q_0 + c_0{}^2)$$

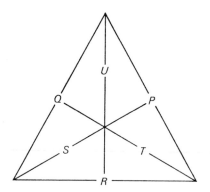

Fig. 5.5   *P, Q, R, S, T,* and *U* in tetrahedral formation

The transformation of the arbitrary unit cell to an orthogonal one would correspond to the reduction of the quantities P, Q and R to zero. The procedure for finding whether this reduction is possible can be most simply stated by arranging the six quantities in tetrahedral formation as shown in fig. 5.5. If β, for example, is acute, Q will be positive; the first transformation is made by changing the sign of Q, subtracting Q from the quantity, T, at the opposite edge of the tetrahedron, adding Q to the other four quantities, and fully interchanging either pair of quantities, S and R *or* P and U, which meet at one end or the other of the edge representing Q. The procedure is repeated until quantities P_n, Q_n, R_n, S_n, T_n and U_n are produced which are all either zero or negative.

An example will make this procedure clear. Suppose we have a unit cell with dimensions:

$$a_0 = 14 \cdot 76, \ b_0 = 9 \cdot 33, \ c_0 = 11 \cdot 43 \ \text{Å}, \ \alpha = 144 \cdot 7°, \ \beta = 140 \cdot 7°,$$
$$\gamma = 18 \cdot 45°$$

(This is a rather artificial example devised solely for the purpose of testing the method.) Then the values of P_0, etc., are those shown in the first line of the table below, in which each column is given a number, m. (The equality of the magnitudes of Q, R and T immediately suggests that there are some relationships between the unit-cell dimensions.)

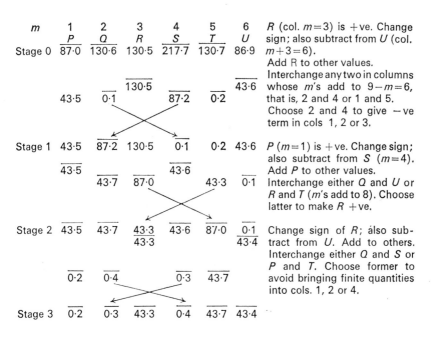

All the quantities are now negative and three are approximately zero. The unit cell however, is still not orthogonal, because R is not zero. We may, however, continue the procedure starting with P, assuming it to be exactly zero; the only change will then be to interchange R and S, and we arrive at the final results that, to the necessary accuracy,

$$P = Q = R = 0; \quad S = T = U = \overline{43 \cdot 5}$$

These values correspond to a cubic unit cell of side $\sqrt{43 \cdot 5}$ Å $= 6 \cdot 6$ Å.

Thewlis and Hutchison (1953) have suggested another method for finding whether the initial unit cell can be transformed to a simpler one; their method may be compared to that used in finding the symmetry of the external form of a crystal from measurement of the angles between the various faces. The poles of the various reflecting planes are plotted on a stereographic projection, and attempts are made to find, by inspection, whether any symmetry is present; any suspected symmetry is tested by rotating the projection through the angle required to bring the symmetry elements on to the vertical or horizontal axis. The authors claim that the method will work even if a few of the lines on the powder photograph cannot be indexed at first, and they used the method to check the unit cell of α-uranium.

5.3.8 Extension of Ito's method for the triclinic system

The uncertainties inherent to Ito's method (5.3.6) can be reduced by establishing a zone of reflections using values of h and k which are greater than unity (Novak, 1954; de Wolff, 1958, 1963). Consider for example the $hk0$

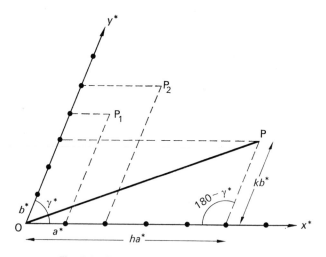

Fig. 5.6   The $hk0$ zone in reciprocal space

zone in reciprocal space. In fig. 5.6 the distance OP is the reciprocal of the spacing, d, in real space of the ($hk0$) planes, and the value of OP is given by

$$OP^2 = h^2 a^{*2} + k^2 b^{*2} + 2hk a^* b^* \cos \gamma^* \qquad \ldots 5.25$$

or

$$\sin^2 \theta_{hk0} = h^2 \sin^2 \theta_{100} + k^2 \sin^2 \theta_{010} + 2hk \sin \theta_{100} \sin \theta_{010} \cos \gamma^*$$

from which

$$\cos \gamma^* = \frac{\sin^2 \theta_{hk0} - h^2 \sin^2 \theta_{100} - k^2 \sin^2 \theta_{010}}{2hk \sin \theta_{100} \sin \theta_{010}} \qquad \ldots 5.26$$

Thus if $\sin^2 \theta_{100}$ and $\sin^2 \theta_{010}$ are chosen correctly from the list of observed values then for given h and k the correct selection of $\sin^2 \theta_{hk0}$ from the remaining observed values will give $\cos \gamma^*$. This same value of $\cos \gamma^*$ will result from all the correct combinations of h, k and $\sin^2 \theta_{hk0}$ which are possible; these combinations of h, k and $\sin^2 \theta_{hk0}$ correspond to the points P, P_1, P_2, ... shown in fig. 5.6. If the observed values of $\sin^2 \theta$ were known without error, then, apart from sheer accident, only the correct selections of h, k, $\sin^2 \theta_{hk0}$, $\sin^2 \theta_{100}$ and $\sin^2 \theta_{010}$ would give identical values for $\cos \gamma^*$. The observed values of $\sin^2 \theta$ are however subject to error so that various combinations of incorrect choices might well produce the same (incorrect) value of $\cos \gamma^*$, but one may still expect the correct value to occur with the highest frequency.

Novak (1954) found γ^* graphically. For all the observed reflections, d^* (OP in equation 5.25) is calculated and with these values of d^* as radii a set of concentric circles, centre O, one circle to each value of d^* (fig. 5.7) is

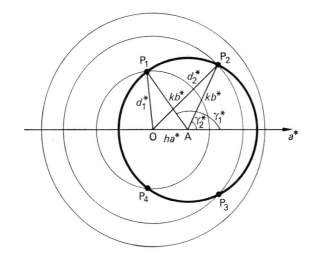

Fig. 5.7   Graphical construction for γ^*, two possible values of which are γ_1^* and γ_2^*

drawn. Possible values of a^* and b^* are then determined from two of the low-angle lines and with these values fixed, a distance $OA = ha^*$ can be marked off along the x^* axis. With centre A an arc of radius kb^* can now be drawn which will cut some of the circles in points like P_1, P_2, P_3, P_4 and the angles between a^* and b^* such as γ_1^*, can then be measured. If this construction is repeated for all combinations of h and k, within predetermined limits, then, for the chosen values of a^* and b^*, γ^* can be obtained for all possible combinations of d^*, h and k; the value of γ^* which occurs with the highest frequency is assumed to be correct, as are the associated values of a^* and b^*. The operation is then repeated for the $h0l$ and $0kl$ zones. If it is then not possible to index all the lines on the pattern with the resulting values of the six lattice parameters the process must be repeated with different values for one or more of a^*, b^* and c^*.

de Wolff (1958, 1963) applied computer techniques to reduce the labour of the method as described by Novak. Essentially de Wolff uses equation (5.26) and first selects pairs of values of $\sin^2 \theta_{100}$ and $\sin^2 \theta_{010}$ from the first few of the low-angle lines. Each pair is then substituted in turn in (5.26) and $\cos \gamma^*$ is calculated for all possible combinations of h and k (up to pre-determined maximum values) and the remaining observed values of $\sin^2 \theta$. From the values of $\cos \gamma^*$ which occur with the highest fre-quencies a suitable choice of γ^* is made and this angle, combined with a^* and b^* as calculated from the appropriate pair of values of $\sin^2 \theta_{100}$ and $\sin^2 \theta_{010}$, defines the $hk0$ zone. Now the low-angle reflections might equally well include $\sin^2 \theta_{001}$ so that it is equally likely that the low-angle pairs of values of $\sin^2 \theta$ will, in addition to the $(\sin^2 \theta_{100}, \sin^2 \theta_{010})$ pair, include also the $(\sin^2 \theta_{010}, \sin^2 \theta_{001})$ pair and the $(\sin^2 \theta_{100}, \sin^2 \theta_{001})$ pair. In this event, it may be that the angles occurring with the highest frequencies will also include the values of α^* and β^*. The three zones, $hk0$, $h0l$ and $0kl$ will then have been determined and the pattern can be indexed. The upper limit on the indices used in the calculation will depend upon the powder pattern, but in general suitable values are $h = k = l = 4$.

It might happen that this analysis leads to the identification of only two interplanar angles, that is that only two zones have been identified. The parameters determined will include a^*, b^*, c^* and, say, α^* and β^*; there still remains the problem of finding γ^*. It will now be necessary to refer to expression 5.10 for $\sin^2 \theta_{hkl}$ in the triclinic system. If we write, as usual,

$$A = \lambda^2 a^{*2}/4 \, (= \sin^2 \theta_{100})$$
$$B = \lambda^2 b^{*2}/4 \, (= \sin^2 \theta_{010})$$
$$C = \lambda^2 c^{*2}/4 \, (= \sin^2 \theta_{001})$$

this expression becomes

$$\sin^2 \theta_{hkl} = Ah^2 + Bk^2 + Cl^2 + 2\sqrt{(BC)}\, hl \cos \alpha^*$$
$$+ 2\sqrt{(CA)}\, lh \cos \beta^* + 2\sqrt{(AB)}\, hk \cos \gamma^* \quad \ldots 5.27$$

Thus, if the $0kl$ and $h0l$ zones are already known for a given reciprocal-lattice point hkl, $\cos \gamma^*$ remains the only unknown on the right-hand side of this expression. For the $hk0$ zone, (5.10) reduces to

$$\sin^2 \theta_{hk0} = Ah^2 + Bk^2 + 2\sqrt{(AB)}\, hk \cos \gamma^*$$

where A and B are already known. A computer programme will calculate $\cos \gamma^*$ for all combinations of h, k and the $\sin^2 \theta$ values of the lines not yet indexed; the angle calculated most frequently should be γ^*. In practice, once this stage has been reached, γ^* can be found by trial and error without the aid of computer techniques.

5.3.9   The method of Zsoldos for indexing triclinic systems

Zsoldos (1958) has developed a method for triclinic systems which is a generalized version of that described by Vand (1948) for indexing powder patterns of crystals in which the unit cells are known to have one large edge. If there is one large edge the patterns will have identifiable long-spacing reflections because if that edge is c, say, then the diffraction lines will be grouped in bands, the lines in each band having the same h and k values but different l values. Vand's procedure is to consider first the expression for the distance from the origin, d_{hkl}^* of the reciprocal-lattice point hkl. If we combine (5.10) with Bragg's law in the form $2 \sin \theta = \lambda d^*$, then

$$d_{hkl}^{*2} = h^2 a^{*2} + k^2 b^{*2} + l^2 c^{*2} + 2klb^*c^* \cos \alpha^*$$
$$+ 2lhc^*a^* \cos \beta^* + 2hka^*b^* \cos \gamma^*$$

This can be rearranged in the form

$$d_{hkl}^{*2}/c^{*2} - l^2 = (1/c^{*2})(h^2 a^{*2} + k^2 b^{*2} + 2hka^*b^* \cos \gamma^*)$$
$$+ (l/c^*)(2kb^* \cos \alpha^* + 2ha^* \cos \beta^*)$$

i.e. in the form

$$d^{*2}/c^{*2} - l^2 = pl + q \qquad \qquad \ldots 5.28$$

where

$$p = (1/c^*)(2kb^* \cos \alpha^* + 2ha^* \cos \beta^*) \qquad \ldots 5.29$$

and

$$q = (1/c^{*2})(h^2 a^{*2} + k^2 b^{*2} + 2hka^*b^* \cos \gamma^*) \qquad \ldots 5.30$$

Since h and k are constant for a given band, p and q will be constant for that band; thus, within any one band the graph of $(d^*/c^*)^2 - l^2$ against l will be a straight line. In practice each observed value of d^* is fixed in turn and $(d^*/c^*)^2 - l^2$ is plotted against l for all values of l. The graph is then inspected to find points lying on a straight line; equation (5.27) is satisfied for such a group of points. The intercept made by the line on the $(d^*/c^*)^2 - l^2$ axis gives q and the slope of the line gives p; from this information reflection within the same band can be indexed.

In the generalized procedure developed by Zsoldos (1958) for indexing triclinic crystals, one of the measured low-angle reflections, usually the first, is chosen as (001) and thus a value is assigned to c^* in equation (5.28). As in Vand's method $(d^*/c^*)^2 - l^2$ is plotted against l for all values of d^* and l, and again all points belonging to identical h and k but different l will lie on a straight line, of slope p with intercept q. In principle the values of h and k for each straight line can now be determined by substituting the values of p and q for that line in equations (5.29) and (5.30). It is best to look initially for straight lines which contain large numbers of points and have low values of p and q; special relationships between the lines, or important features of individual lines may possibly then be uncovered. Two lines may have the same slope or one line may have $p = 0$. Should the latter occur then it is clear from (5.29) that at least either $h = 0$ and $\alpha^* = 90°$ or $k = 0$ and $\beta^* = 90°$; that is, the crystal is at least monoclinic.

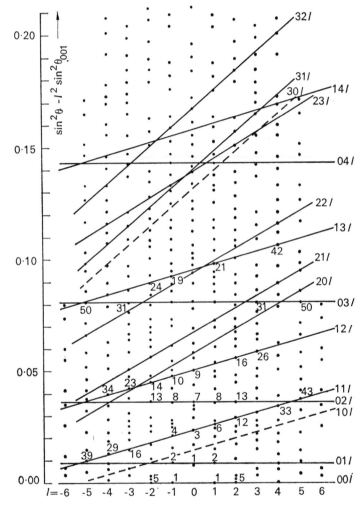

Fig. 5.8   Graph of $\sin^2\theta - l^2 \sin^2\theta_{001}$ against l from the powder pattern of $Na_2S_2O_3$

Further, if such a line passes through a point in the column $l = 0$ (fig. 5.8) then either d^*_{h00} or d^*_{0k0} (i.e. either a^* or b^*) can be determined. In fig. 5.8 values of $\sin^2 \theta - l^2 \sin^2 \theta_{001}$ are plotted as a function of l.

5.4 Interpretation by Graphical Methods

5.4.1 The principle of graphical methods

For certain crystal systems it is possible to draw curves representing the calculated spacings, or quantities related to them, as functions of the cell dimensions. To index a given powder photograph, the spacings, or the related quantity derived from the lines on the photograph, are plotted on a diagram and compared with the curves obtained theoretically. If the cell dimensions are known, the diagram of experimental results can be placed directly on the theoretical curves and the indices read off; if the cell dimensions are not known, the diagram has to be moved about over the curves until a fit is obtained.

In practice the method is not easily applied to systems of symmetry lower than tetragonal, hexagonal or trigonal. It is most conveniently

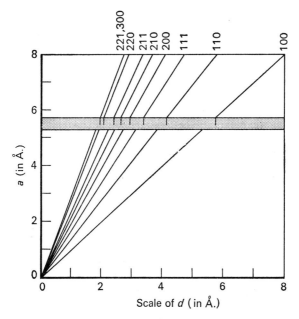

Fig. 5.9   Chart for indexing the powder photograph of a cubic material

illustrated with reference to the cubic system, in which the expression for the spacing d of the planes hkl is

$$d = a/\sqrt{(h^2 + k^2 + l^2)},$$

where a is the lattice parameter. It is, therefore, possible to calculate d as a function of a for given indices; obviously d is proportional to a, and so the curve of d against a is a straight line passing through the origin. For each set of indices a line of different slope will be produced, as shown in fig. 5.9. In order to index a photograph, all that is necessary is to mark on a strip of paper the values of the spacings ($\lambda/2 \sin \theta$) derived from the photograph and to find at which value of a the observed values agree with the set of theoretically derived lines.

This application is described as an introduction to the methods to be described in the following sections; the indexing of powder photographs of cubic specimens (5.3.1) is usually simple enough not to need such methods. For the tetragonal, hexagonal and trigonal systems, however, graphical methods are quite useful. In these systems there are two cell dimensions, a and c, on which the spacings depend, and thus the method as described for the cubic system cannot be used because another dimension is needed. The following methods avoid this difficulty.

5.4.2 Bjurström's method

The basic equation of Bjurström's method (Bjurström, 1931) for the tetragonal system is

$$1/d^2 = (h^2 + k^2)/a^2 + l^2/c^2 \qquad \ldots 5.31$$

We may consider two extreme forms of equation (5.31); with $a = \infty$ it becomes

$$1/d^2 = l^2/c^2 \qquad \ldots 5.32$$

and with $c = \infty$ it becomes

$$1/d^2 = (h^2 + k^2)/a^2 \qquad \ldots 5.33$$

If c is taken as unity, possible values of the right-hand side of equation (5.32) can be marked along a horizontal line, as shown at the top of fig. 5.10; these values are, of course, the squares of the natural numbers. Similarly, possible values of $h^2 + k^2$ can also be plotted, as shown at the bottom of fig. 5.10. These two horizontal lines may be labelled as $c/a = 0$ and $c/a = \infty$. Each point on the top line is now joined by a straight line to every point on the bottom line, and we obtain the complete diagram.

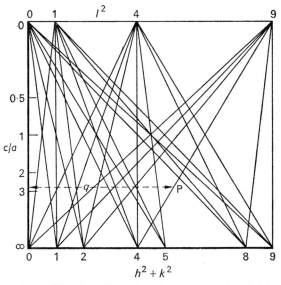

Fig. 5.10   Bjurström's chart for a tetragonal material

Consider a point, such as P, which lies on one of the cross lines, at a fraction p of the distance from the top. Then the distance q from the line joining the left-hand ends of the top and bottom lines is

$$q = p(h^2 + k^2) + (1 - p)l^2 \qquad \qquad \ldots 5.34$$

which may be re-written as

$$q = \frac{(h^2 + k^2)}{1/p} + \frac{l^2}{1/(1 - p)} \qquad \qquad \ldots 5.35$$

By comparing this equation with equation (5.31), we see that q is the value of $1/d^2$ for a tetragonal substance with $a^2 = 1/p$, $c^2 = 1/(1 - p)$, and $(c/a)^2 = p/(1 - p)$. Each horizontal line on fig. 5.10 therefore represents a set of values of $1/d^2$ for a crystal with a given axial ratio, which may thus be marked on the side.

A difficulty still remains. Not only is c/a fixed as $\sqrt{[p/(1 - p)]}$, but a is also fixed as $\sqrt{(1/p)}$ and c as $\sqrt{[1/(1 - p)]}$, and another variable is needed in order to express variation in either of these axes. This is a general rule for the application of graphical methods to the uniaxial systems; two variables are always needed. In Bjurström's method the variable is obtained as follows: the values of $1/d^2$ from the photograph to be indexed are marked along a horizontal line, and a vertical line is drawn from the left-hand end; the values of $1/d^2$ are then joined to a fixed point on this line, and, as seen in fig. 5.11, any horizontal section of the resulting diagram is a representation of the set of values of $1/d^2$ on a reduced scale.

To use the charts the diagram of the experimental results (fig. 5.11) is drawn on transparent paper, and placed on the Bjurström chart (fig. 5.10) with the left-hand edges of both coincident, but with their relative vertical positions arbitrary. A horizontal straight edge is then moved slowly down the composite diagram. If at a certain position of the straight edge it is found that all the observed values of $1/d^2$ coincide with lines on the chart, a fit will have been found; the indices may be read off and c and a calculated. If a fit is not found, the scale of $1/d^2$ may be displaced vertically and the operation repeated until it is successful.

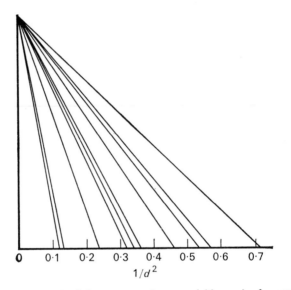

Fig. 5.11   Values of $1/d^2$ represented on a variable scale for use with Bjurström's chart

It is generally sensible to concentrate first on the lower orders as it is useless trying to index the higher orders if the lower ones do not fit satisfactorily.

Similar methods can be used for the hexagonal system and for trigonal crystals referred to a hexagonal lattice. The equation is

$$\frac{1}{d^2} = \frac{4}{3a^2}(h^2 + hk + k^2) + \frac{1}{c^2}l^2 \qquad \ldots 5.36$$

The same equation can be used for rhombohedral crystals if these are referred to hexagonal axes; in this case the lines which are absent because of the rhombohedral lattice type can be omitted from the chart.

The value of Bjurström's method is that the charts, being composed of straight lines only, are easily constructed. On the other hand, it may be found that the superposition of two sets of lines is rather confusing. Bjurström (1931) has suggested a device which overcomes this difficulty. It consists of two bars connected together so that they form two opposite sides of a parallelogram. Along these bars, points are marked at distances proportional to possible values of $h^2 + k^2$ (for the tetragonal system), and corresponding points are joined by straight wires, as shown in fig. 5.12;

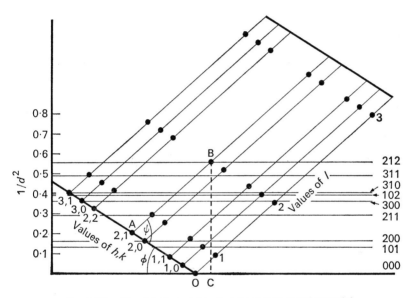

Fig. 5.12   Bjurström's construction for a tetragonal material

along each wire distances are marked proportional to the squares of the natural numbers.

To use this construction a set of parallel lines is drawn at distances from a fixed line representing the values of $1/d^2$ obtained from the photograph to be analysed. The corner corresponding to $h^2 + k^2 = 0$ and $l = 0$ is placed on the line $1/d^2 = 0$. Then the construction is moved round until a point on it lies over each $1/d^2$ line, as shown in fig. 5.12. If no such fit is found, the angle of the parallelogram is altered and the operation repeated until it is successful.

The theory is illustrated by the figure; $OA = h^2 + k^2$ and $AB = l^2$, and therefore

$$BC = (h^2 + k^2) \sin \phi + l^2 \sin (\psi - \phi) \qquad \ldots 5.37$$

Comparing this equation with equation 5.31, we see that BC represents $1/d^2$ if $1/a^2 = \sin \phi$ and $1/c^2 = \sin (\psi - \phi)$. Thus the different values of ϕ and ψ simulate different values of a and c.

5.4.3 Hull–Davey charts

If graphical methods without the superposition of two sets of lines are preferred the Hull–Davey (1921) charts may be used. In these charts the function $\log d$ is used. The usefulness of this function can be seen by applying it to the method described for the cubic system in 5.4.1.

From the equation $d = a/\sqrt{(h^2 + k^2 + l^2)}$, we have

$$\log d = \log a - \tfrac{1}{2} \log (h^2 + k^2 + l^2) \qquad \ldots 5.38$$

For any set of indices this represents a straight line, of slope 45°, making an intercept of $-\tfrac{1}{2} \log (h^2 + k^2 + l^2)$ on the axis of d; for several sets of indices we thus obtain the diagram shown in fig. 5.13. Any horizontal line

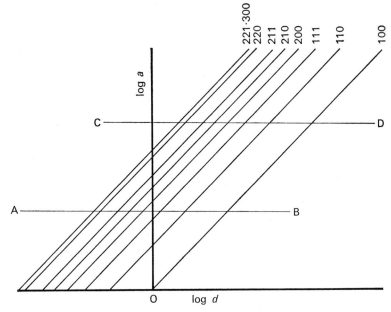

Fig. 5.13   Logarithmic form of chart shown in fig. 5.9

such as AB or CD will represent a set of values of log d for a particular value of a. It can be seen, however, that the relative positions of the intersections of the lines on the chart with the lines AB and CD are the same; they differ only in position. Thus we can represent the complete chart by the series of points given by the intersection of a line such as AB with the lines of the chart, and we can represent different values of a by translating the line of points along its own length. Having compressed the data for the cubic system into one dimension we can now plot, at right angles, the variation of log d produced when c/a changes from unity, and so obtain charts for the tetragonal system. In a similar way charts for the hexagonal system can also be produced.

In the Hull–Davey charts, the values of log d for different values of c/a are plotted against c/a; but Bunn (1961) has shown that if log d is plotted against log c/a then charts may be obtained which are much simpler to construct. Copies of these charts are available commercially.

The charts are used by plotting on a strip of paper the values of log d obtained from the photograph to be indexed. This strip of paper can be moved vertically (representing a change in a or c), or horizontally (representing a change in c/a) until a match is found as shown in fig. 5.14. The scale of log d values must, of course, always be kept vertical. As in Bjurström's method, it is important to concentrate on the low orders first.

Harrington (1938) has simplified the construction of the Hull–Davey type of chart by adopting a procedure which can, if required, be extended to apply to the orthorhombic system. He used the relationship

$$1/d_{hkl}^2 = 1/d_{hk0}^2 + 1/d_{00l}^2$$

which holds for the cubic, tetragonal and orthorhombic systems and for the hexagonal system where c is the unique axis. In particular, for the tetragonal system

$$1/d_{hkl}^2 = (h^2 + k^2)/a^2 + l^2/c^2,$$

$$1/d_{hk0}^2 = (h^2 + k^2)/a^2,$$

and $$1/d_{00l}^2 = l^2/c^2$$

and from these three relationships it can be seen that $1/d_{hkl}^2$ tends asymptotically to $1/d_{hk0}^2$ for large values of c, and to $1/d_{00l}^2$ for large values of a. By writing $1/d_{hk0}^2 = (h^2 + k^2)/a^2$ in the form

$$\frac{c^2}{d_{hk0}^2} = \left(\frac{h^2 + k^2}{a^2}\right)\left(\frac{c}{a}\right)^2$$

it becomes evident that the relationship between log d and log (c/a) is linear and that a series of parallel straight lines is obtained when all possible values of $h^2 + k^2$ are considered; similarly the graph of $\log d_{00l}$ against $\log (c/a)$ for all l is a series of parallel straight lines. The slopes of the two sets of lines are opposite in sign. Thus each of the curves of $\log d_{hkl}$ against $\log (c/a)$ will tend asymptotically to the corresponding lines for $\log d_{hk0}$ and $\log d_{00l}$, and they will all have the same shape since the relation

$$1/d_{hkl}^2 = 1/d_{hk0}^2 + 1/d_{00l}^2$$

is independent of the particular values of hkl.

By exactly the same procedure, Harrington's curves can be constructed for the hexagonal system, merely replacing $(h^2 + k^2)$ by $\frac{4}{3}(h^2 + hk + k^2)$.

Because there are three unknowns to be determined, the extension to

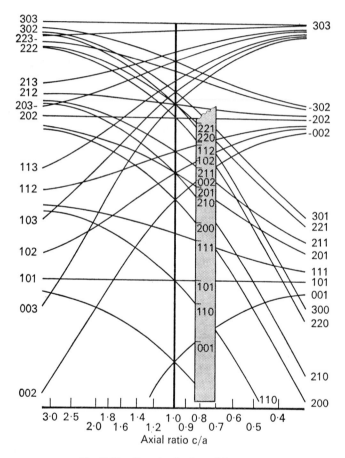

Fig. 5.14   Bunn's chart used for urea

cover the orthorhombic system is rather more complex, but the procedure is essentially the same. The expression

$$1/d_{hkl}^2 = h^2/a^2 + k^2/b^2 + l^2/c^2$$

is rewritten in the form

$$c^2/d_{hkl}^2 = c^2h^2/a^2 + c^2k^2/b^2 + l^2$$

If c/b is kept constant then log d_{hkl} can be plotted against log c/a for all sets of hkl values for this particular value of c/b. The ratio c/b can then be changed step by step and a whole range of log d_{hkl} − log (c/a) charts constructed.

5.5 Interpretation by Other Methods

Included in other methods for indexing patterns from crystals of symmetry at least as high as orthorhombic are those described by Schieltz (1964), and by Williams (1966). Schieltz's method is for powder photographs obtained with a single-crystal rotation cylindrical camera (described in most textbooks on diffraction by single crystals). The pattern will consist of almost complete rings and from these rings, with the aid of a Bernal chart (1926), it is possible to locate what would have been one or more higher-layer lines had a single-crystal pattern been obtained. From the distance between one of these lines and the zero-layer line, one unit-cell edge can be calculated. After measuring the powder film, the reciprocal vectors d_{hkl}^* are determined for all of the observed reflections and a set of concentric circles, whose radii are the d_{hkl}^* values, is drawn. A reciprocal-space net is now constructed, the origin is placed on the common centre of the circles, and the net is rotated; by trial and error a net is found whose rotating intersections fit the concentric circles and indexing is then completed. Fig. 5.15 shows the completed analysis of urea obtained by Schieltz.

The method of Williams (1966), a graphical one, applies to the indexing of complex superlattice structures and is based upon the properties of Fourier transforms (Lipson and Taylor, 1958). When a disordered structure of, say, an alloy, becomes ordered, a **superlattice** is formed, the unit cell of which is larger than that of the disordered structure. The ordering process is accompanied by the appearance of faint lines which are additional to the main lines which originate from the disordered structure (7.6.2). The lattice on which the disordered structure is based is sometimes called a sublattice of the lattice of the ordered alloy, and in reciprocal space the cell derived from this sublattice will, of course, be larger than that derived from the superlattice. In the construction by Williams, the reciprocal net for, say, the hk0 zone of the real-space sublattice is drawn, and with each reciprocal-lattice point as centre, circles of radius d^*_{hkl}, corresponding to every one of the observed reflections from the ordered structure, are drawn. Each point is thus the centre of a set of concentric circles, and from

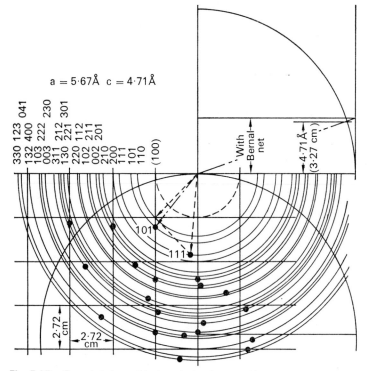

Fig. 5.15   Completed graphical analysis of urea with powder data indexed

the convolution property of Fourier transforms as applied to X-ray-diffraction data, the superlattice spots in reciprocal space are the points where many circles intersect. Circles with radii d^*_{hkl} where $l \neq 0$ tend to confuse the diagram and in fig. 5.16 these have been omitted. The figure shows the results produced by Williams for ordered Ni_4Mo and indicate clearly that a^* and b^* for the superlattice are each one fifth of the corresponding values for the main lattice; that is, the values of a and b for the superlattice are each five times bigger than are those for the main lattice. The third dimension c can be determined from the $0kl$ section in the same way.

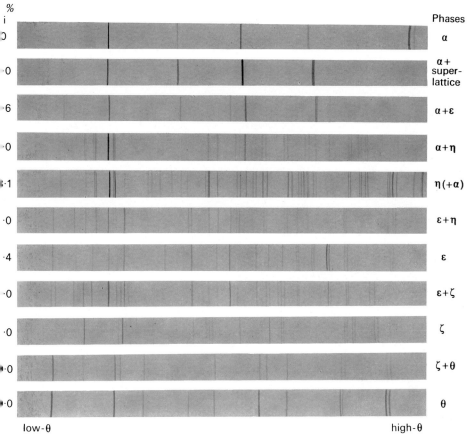

%
Si

0

0

6

0

1

0

4

0

0

0

0

low-θ   high-θ

Phases

α

α +
super-
lattice

α + ε

α + η

η (+ α)

ε + η

ε

ε + ζ

ζ

ζ + θ

θ

Plate 3   Powder photographs of iron–silicon alloys taken with CoK_α radiation

In general it is not possible to apply the method if there are only a few superlattice reflections present in the plane considered. If this is so for the $hk0$, $0kl$, or $h0l$ planes, then the $hk1$ plane can be tried provided the crystal symmetry is at least orthorhombic. The distance of this plane from the

000   020

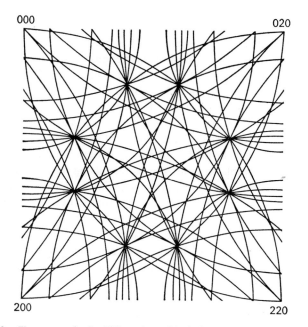

200   220

Fig. 5.16   First square in the $hk0$ section with circles corresponding to super-lattice spots in this section only included. The spots can clearly be located where several circles intersect

origin of the reciprocal lattice is d^*_{001}, which is known from the dimensions of the unit cell of the disordered structure. It is a simple matter to calculate for each line the value of $\sqrt{(d^{*2}_{hkl} - d^{*2}_{001})}$ for each spot, and with these distances as radii circles can now be drawn and the analysis carried out as for the $hk0$ plane.

5.6 Comparison of Analytical and Graphical Methods

The choice between graphical and analytical methods of indexing powder photographs depends to a great extent on personal preferences, but some general advice can be given to those who have not used either method.

6+

If the problem of indexing a powder photograph is an isolated one, and charts are not available, it is probably simpler to try analytical methods rather than to construct the charts. If, however, the problem is likely to arise frequently, charts should certainly be prepared; those found in the literature are usually too small for convenient use. Suitable sizes are of the order of 1 m by 0·5 m.

As a first step in the analysis of any powder photograph, lists of both $\sin^2 \theta$ and $1/d^2$ should be prepared for all lines; if graphical methods are to be used then values of the appropriate functions should also be calculated. Examine the values of $\sin^2 \theta$ to determine analytically if the substance is cubic (5.3.1). If it appears not to be cubic try one or other of the tetragonal or hexagonal graphical methods. If no fit can be obtained with these, check analytically the values of $\sin^2 \theta$ for any clues (5.3.2, 5.3.3), because the unit cell may be too large for the indices of the lines to be easily found graphically; this will be so in particular if many of the low orders are absent. If it is possible to index the pattern using hexagonal indexes the crystal may still exhibit trigonal symmetry (5.3.3); in this case transformation to rhombohedral indices should be carried out (5.2.2). If all these methods fail either the analytical method of Lipson (5.3.4) or the graphical method of Harrington (5.4.3) should be tried in order to find whether the substance is orthorhombic. Finally, the extension of Lipson's method (5.3.5) can be tried for monoclinic crystals, and for both monoclinic and triclinic crystals either Ito's method (5.3.6) or its extension (5.3.8) or the method of Zsoldos (5.3.9) may lead to a solution.

It should be emphasized, however, that powder methods should be used only when the material is so finely subdivided that single-crystal methods are impracticable. Using powder methods, especially with systems of low symmetry, difficulties will be enhanced by lack of precision in the data and by the omission from the data of some low-angle reflections, whereas with single-crystal methods the effects of these factors are not nearly so pronounced. Crystals with dimensions of the order of 0·01 mm can be set on an X-ray goniometer and this should always be tried before attempting to analyse a complicated powder photograph. Single-crystal photographs not only provide a simple means of carrying out what is a major task by powder methods, but they give definite results, independent of chance coincidences. For example, analytical methods tried on Co_2Al_9 gave some indication of an orthorhombic unit cell; single-crystal methods later showed the cell to be monoclinic (Douglas, 1950).

If a single crystal cannot be obtained, a coarse-grained powder may give some useful information. Lipson and Steeple (1951) have shown that it is sometimes possible to obtain information about the indices of a line from the number of spots on it; a line of relatively few spots must have simple indices such as 00*l* in the tetragonal system or 000*l* in the hexagonal system because the multiplicity factor (7.4.6) is small.

After completing the indexing a result should be carefully checked. The agreement between the observed and the calculated values of $\sin^2 \theta$ (or of $1/d^2$) for *every* line must be within the limits set by the experimental errors; these limits must be predetermined in order to reduce the inclination to widen them so as to include values which would otherwise have to be rejected. Particular attention must be paid to the possibility of accidental absences in the low-angle region and to the possibility that some low-angle lines may have too small a value of $\sin^2 \theta$ to be recorded on the film. The latter will present a serious problem if the unit cell is large and may result in indices such as 3, 4 . . . for the first line which is actually observed.

5.7 Check on Validity of a Result

In view of the degree of subjectivity involved in many of the operations described in this chapter, and the high probability that some sort of result will be obtained, it is worth stating the conditions that must be satisfied for a correct answer.

(*a*) The accuracy of the values of $\sin^2 \theta$ should be known and agreement *must* be obtained for each line within the known limits of experimental error. It is always tempting to stretch the limits somewhat for lines that do not fit well; this temptation must be avoided.

(*b*) Although not all the calculated lines will be observed, a result should be suspect if a large proportion—say, more than half—is absent. It is easy to obtain a good fit if a very large unit cell is postulated, but then only a few possible values of $\sin^2 \theta$ will be used, and even the low-angle lines will have high indices.

(*c*) For crystals other than triclinic, the absences should fit into a reasonable space-group scheme.

(*d*) The density of the material should be accurately measured, to 0·1 percent if possible, and this should lead to a precisely integral number of atoms or molecules in the unit cell.

If all these conditions are satisfied, the result may be accepted with a considerable degree of assurance, but there can never be complete certainty from powder methods only, unless a complete set of atomic positions is obtained and all the intensities are found to be correct.

A test for the reliability of the interpretation of an X-ray powder pattern has been suggested by de Wolff (1968). He points out that good agreement is the main criterion but that this must not be obtained simply by postulating a large unit cell; high accuracy is not significant if almost all the possible lines are absent.

De Wolff therefore suggests the adoption of a figure of merit M, which takes these two criteria into account:

$$M = \sin^2 \theta_{20}/\bar{\varepsilon}N_{20},$$

where $\sin^2 \theta_{20}$ is the value of $\sin^2 \theta$ for the twentieth line, $\bar{\varepsilon}$ is the mean discrepancy in $\sin^2 \theta$ and N_{20} is the number of theoretically possible lines up to the twentieth. The larger the value of M, the more likely it is that the interpretation is correct. M will be larger if $\bar{\varepsilon}$ is small and if N_{20} is as close as possible to 20.

De Wolff suggests that an interpretation that leads to a value of M less than about 10 should be looked at with suspicion. For $NiAl_3$ (5.2.3, 5.3.4) the agreement published by Bradley and Taylor (1937) was extremely good; $\bar{\varepsilon}$ is $1 \cdot 5 \times 10^{-7}$, $\sin^2 \theta_{20} = 0 \cdot 26$, and $N_{20} = 25$. M therefore has a value of about 35. Lester and Lipson (1970) report a plausible result that was later found to be wrong: it had $\bar{\varepsilon} = 6 \times 10^{-4}$, $\sin^2 \theta_{20} = 0 \cdot 22$, and $N_{20} = 60$; thus $M = 6$. De Wolff quotes correct results for which $\bar{\varepsilon}$ had about this value, and others for which N_{20} is also near to 60; but the combination of both values makes the result doubtful.

CHAPTER 6

Accurate Determination of Cell Dimensions

6.1 Introduction

The accurate determination of cell dimensions is important in the study of many materials, particularly metals and alloys. Very accurate results can be obtained if measurements are made on the spectra that are reflected almost back into the incident beam, since large Bragg angles are very sensitive to small changes in cell dimensions. This can be seen by differentiating the equation $d = (\lambda \operatorname{cosec} \theta)/2$, which gives

$$\delta d = -\lambda \operatorname{cosec} \theta \cot \theta \ \delta\theta/2$$
$$= -d \cot \theta \ \delta\theta \qquad\qquad \ldots 6.1$$

Since $\cot \theta$ tends to zero as θ tends to $90°$, an error $\delta\theta$ in θ will result in a relatively much smaller error δd in d. Consequently, at this angle, high accuracy can be obtained with simple apparatus and few precautions.

The most direct way of determining cell dimensions involves the measurement of several diffraction lines on a powder photograph, and therefore a circular camera is to be preferred to a flat plate. The ordinary powder camera (4.3.1) or the symmetrical focusing camera (4.8.1) may therefore be used.

As recommended in 4.3.1, the van Arkel method of mounting the film should be used if accurate results are required; the high-angle reflections are then close together on one film and no accurate knowledge of any camera constant is needed. An approximate value of the radius may be required, but the way the final result is derived eliminates any error due to it (Cohen, 1936). It is this that makes this method of mounting the film so valuable.

6.2 Elimination of Systematic Errors in Powder Photographs

There are several effects that may cause powder lines to be displaced from their true positions, so that angles measured according to the methods described in 4.3 will not be the true Bragg angles. But as we have seen in the previous section, a small error in θ should produce a vanishingly small error in d as θ approaches $90°$; if we could determine d from a reflection with $\theta = 90°$, all the systematic errors should be zero. Although this is impossible to achieve, since a Bragg angle of $90°$ corresponds with reflection directed back into the X-ray beam, it can be done in effect by determining d from several reflections with different θ's and extrapolating the result to $\theta = 90°$.

Although any function of θ will, in theory, do for plotting such an extrapolation curve, it is obviously better that the curve should be linear. Finding the correct function for this purpose involves a consideration of the various possible sources of error. These are listed below.

6.2.1 Non-coincidence of the axis of the camera and the rotation axis of the specimen

It can be seen from fig. 6.1 that any displacement of the specimen perpendicular to the axis of the slit system will cause no error in θ; on one side of

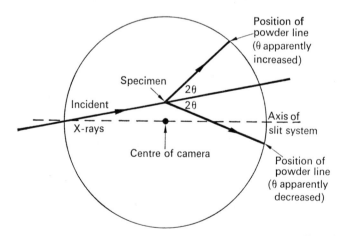

Fig. 6.1   Specimen displaced perpendicular to the direction of the X-ray beam

the photograph the line will be displaced to a higher value of θ and on the other side to a lower, and the average value will not be appreciably changed. On the other hand, fig. 6.2 shows that a displacement along the axis will produce an error since the two lines will move nearer or further apart. Bradley and Jay (1932a) have shown that the error in d produced in this way is proportional to $\cos^2 \theta$.

Beu and Scott (1962) have developed an exact analytical method for correcting errors in θ arising from eccentric mounting of the specimen. They measure with high accuracy the radius of the camera in three directions and these values are then used to calculate both the displacement of the specimen from the axis of rotation and the true camera radius, and hence the correction for the resulting error in θ. The displacement of the sample

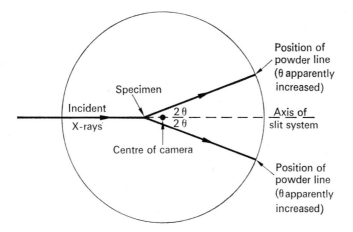

Fig. 6.2   Specimen displaced along the direction of the X-ray beam

can be calculated to 0·001 mm which may be compared with the value of 0·01 mm usual in precision measurements for lattice-parameter determinations.

6.2.2   Absorption and divergence of the X-ray beam

Although the effects of absorption in the specimen and of divergence of the X-ray beam are quite distinct from each other, it has been shown by Bradley and Jay (1932b) that they produce similar errors, and so are best treated together.

Only for a specimen in which absorption is negligible can the centres of the powder lines be taken as the correct positions. Usually the measured value of θ will be larger than the true value (4.3.3), and so the calculated spacing will appear too small. A similar effect is produced by divergence of the X-ray beam.

The exact form of the error has not yet been found. Bradley and Jay (1932a) showed that it will be approximately proportional to $\cos^2 \theta/\theta$; Jay (1945) suggested that it is nearer to $\cos^2 \theta/\sin \theta$; Taylor and Sinclair (1945b) proposed the mean of these two functions. Nelson and Riley (1945) showed experimentally that the mean does give better results for the specimens they have used; this function, $\frac{1}{2}(\cos^2 \theta/\sin \theta + \cos^2 \theta/\theta)$, is therefore recommended, and a table of its values is given (table 6).

6.2.3   Finite height of the specimen

In (4.1) it was pointed out that each line on a powder photograph is composed of a series of overlapping lines, each component coming from a different point of the specimen. Every component except that which comes from the centre will be displaced in the same direction, so the measurement of the composite line will involve an error which can be minimized by keeping the height of the specimen irradiated small. Bradley and Jay (1932b), however, showed that the effect is quite small, and that if the height of the irradiated part of the specimen is kept to less than 2 mm in a camera of 90 mm diameter, the error is negligible. Lipson and Wilson (1941) showed that the error depends also upon the shape of the focus of the X-ray tube, and that the error, for reflections with θ near 90°, is probably even smaller than Bradley and Jay claimed. In the methods described in this chapter the error is neglected.

6.2.4   Film shrinkage

A photographic film will, in general, shrink when processed so that values of θ determined from measurement of the film will require correction. The four methods for making this correction are:

(1) printing fiducial marks on the film before it is processed;
(2) using a specially designed camera;
(3) printing a scale on the film, again before it is processed;
(4) using an internal standard.

With the first procedure, which is that most widely used, the form of the film correction will depend upon the type of film mounting (4.3.1). Fiducial marks are obtained by constructing the camera in such a way that knife-edge shadows at a known angular separation are cast near the ends of the film (4.3.1). As we have already shown, with the Bradley–Jay mounting, the value of the Bragg angle θ for a given line is obtained from

$$\theta = \phi_{\mathrm{k}}(S/S_{\mathrm{k}})$$

where S and S_{k} are both measured on the processed film and are respectively the distance between the relevant pair of corresponding lines and the distance between the knife-edge shadows; ϕ_{k} is the Bragg angle for a line that would fall at the knife edges.

Calibration of the camera depends on ϕ_{k}, and with the Bradley–Jay mounting this angle must be known as accurately as possible. The best way to determine it is to measure the dimensions of the camera or to measure the angle directly with a spectrometer; the camera is centred on the table and turned through a measured angle so that first one knife edge and then the other is brought into coincidence with the cross-wires in the telescope.

If fiducial marks are to be used to correct for film shrinkage, the van Arkel mounting is more suitable than is that of Bradley and Jay since it avoids the necessity for accurate camera calibration (4.3.1). With the van Arkel type

$$\theta = 90° - \psi_{\mathrm{k}}(S/S_{\mathrm{k}})$$

where ψ_{k} is the complement of the Bragg angle of the reflection that would fall at the positions of the knife-edge shadows. As θ tends to 90°, S tends to zero and therefore at high angles an error in ψ_{k} will have only a small effect on θ; extrapolation to $\theta = 90°$ will eliminate completely the film-shrinkage error and an approximate value of ψ_{k} will be good enough (6.1).

The second method of correction for film shrinkage is that in which the self-calibrating camera of Ieviņš and Straumanis (1936) is used; the procedure was described in (4.3.1).

Both methods of correction so far considered depend on the assumption that the shrinkage itself is uniform. Hägg's method (1947) in which a fine scale is printed on the film before processing, depends on no such assumption. The distance, S, between corresponding lines can be read directly from the scale, and any non-uniformity in the length of the film will be compensated for automatically by the same degree of non-uniformity in the printed scale.

The fourth method of correction for film shrinkage—whether uniform or not—is the use of an internal standard (4.3.3); this corrects also for other systematic errors. A standard cubic substance, whose lattice parameter is accurately known, is mixed in the form of a powder with the material under investigation; the unknown material need not necessarily have a cubic structure. Diffraction patterns of both the known and the unknown materials are then obtained under identical experimental conditions. If

6*

the particle sizes are comparable (4.1) and if account is taken of the refraction correction the effect of systematic errors will be the same for both specimens. The diffraction patterns must not interfere with each other and a correction must be applied for any difference between the temperature of the mixture and the temperature at which the lattice parameter of the standard was determined.

One way of doing the comparison is to calculate the value of $\sin^2 \theta$ for each line of the standard pattern and to determine the observed value of $\sin^2 \theta$ for that line from measurement of the film; a graph of $\sin^2 \theta_{obs} - \sin^2 \theta_{calc}$ against the extrapolation function $\frac{1}{2}(\cos^2 \theta/\theta + \cos^2 \theta/\sin \theta)$ can then be drawn for the standard (D'Eye and Wait, 1960). This graph gives the correction to be applied to the observed $\sin^2 \theta$ value for any reflection from the unknown material; after making the correction the lattice parameters of the unknown can be determined.

6.2.5 Refraction

The process of extrapolation to $\theta = 90°$ does not eliminate the error due to refraction. Exact treatment of the effects of refraction is difficult except for a flat, single-crystal specimen. In practice, there is reason to suppose that the only effect of importance is the change in wavelength within the specimen. In order to correct for refraction, the calculated cell dimensions should be divided by the refractive index n of the material. This can be found from the equation given in 3.2. For a cubic crystal the following expression is sometimes simpler to apply.

$$(1 - n)a = 4\cdot47 \times 10^{-6}(\lambda/a)^2 \sum Z$$

where $\sum Z$ is the sum of the atomic numbers of the atoms in the unit cell. It will be seen that the correction is small and is rarely greater than the experimental error.

6.2.6 Limits of accuracy

The accuracy of measured cell dimensions or spacings depends greatly on the quality of the photographs obtained. The highest accuracy will be obtained from photographs consisting of sharp, well resolved lines. There are factors that suggest a limit of the order of one part in 100,000 on this accuracy. Errors in the wavelength used are probably of the order of 0·001 per cent, as indicated by the differences between the values obtained by Siegbahn's school using photographic methods and those obtained by Bearden and Shaw using a double-crystal spectrometer. But it is not certain that either of these wavelengths should be used for the calculations from powder photographs. The emission lines are not symmetrical (3.3.2) and the differences between the peak positions and the centres of gravity are of the order of 0·003 per cent. For calculations of spacings from powder lines, wavelengths corresponding to the centres of gravity may be more significant than those corresponding to the peaks. Bearden (1967) suggests that the wavelength of the tungsten $K\alpha_1$ line should be adopted as the wavelength standard (3.3.1). He has determined the wavelength of the line to within $\pm 0·005$ per cent, but although this line itself is symmetrical the differences between the centres of gravity and the peaks of the asymmetrical lines from other radiations will still pose a problem.

Subjective errors, and, according to Parrish (1960), incomplete elimination of systematic errors by the conventional extrapolation procedures, are among the main factors limiting the accuracy to which lattice parameters can be determined. Even with sharp, well resolved doublets the precision with which the position of a single line can be measured will limit the accuracy to something of the order of 0·003 per cent; to confirm this, the average of several photographs taken, if possible, with several radiations, is necessary. For most photographs a much lower accuracy must be expected and the resolution of the $K\alpha$-doublet gives a rough idea of what can be attained. If the high-angle doublets are resolved, but are not very sharp, the accuracy should be better than 0·2 per cent and may approach 0·02 per cent. If they are not resolved one cannot expect better than 0·1 per cent accuracy. If the high orders are so blurred that they are almost invisible and the low orders have to be used, an accuracy of about 1 per cent is all that can be expected.

The high accuracy obtainable from good powder photographs is not merely of academic interest; for many applications (6.9), small changes in lattice parameter are involved, and the highest possible precision is needed to measure these changes with reasonable accuracy.

6.3 Procedure for Powder Photographs of Cubic and Uniaxial Specimens

6.3.1 General outline of the method for a cubic specimen

The derivation of cell dimensions is best described by reference to a cubic specimen, since here only one dimension is involved—the cell side *a*. Suppose that a photograph has been taken with a small height only of the specimen irradiated, and has been measured as already described. That is, subjective errors have been reduced to a minimum by measuring the positions of the high-angle lines carefully and correcting for film shrinkage. Apart from the refraction correction, which can be made, if necessary, after a has been finally determined, the errors that remain can be treated by extrapolating to $\theta = 90°$. Before extrapolation a value of the cell dimension is derived from the value of θ for each measured line by the relation

$$a = \tfrac{1}{2}\sqrt{N}\,\lambda\,\text{cosec}\,\theta \qquad \ldots 6.2$$

where $N = h^2 + k^2 + l^2$.

The value of a deduced from each measured line is then plotted against the corresponding value of $\tfrac{1}{2}(\cos^2\theta/\theta + \cos^2\theta/\sin\theta)$ read from table 6. If there is no error due to eccentricity of the specimen (6.2.1) the resulting points should be randomly distributed about a straight line. The point where this straight line cuts the vertical line at $\theta = 90°$ is the value of a which, when corrected for refraction (6.2.5), is the value of the cell dimension at the temperature at which the photograph was taken. If the specimen is eccentrically mounted, the resulting error in a will be proportional to $\cos^2\theta$ (6.2.1) and the graph of a against $\tfrac{1}{2}(\cos^2\theta/\theta + \cos^2\theta/\sin\theta)$ will correspond to a curved line. This departure from linearity will make accurate extrapolation more difficult but it does not completely invalidate the procedure.

Analytical methods can be used to fix the best line through the experimental points but normal procedures (Cohen, 1935) fail to allow for the higher accuracy of the points for which θ is near 90°. Hess (1951) worked out a way of giving extra weight to these points but at the expense of additional computational labour to a calculation which is already lengthy. With ready access to computers, this is no longer a drawback and consequently the method of weighted analytical extrapolation is now generally preferred to that of linear graphical extrapolation. There is however much to be said still for graphical extrapolation when the substance is cubic, because of the ease with which the higher orders of reflection can be weighted.

The internal-standard method (6.2.4) can be used to derive accurate cell dimensions for both cubic and non-cubic structures, and need not be discussed further.

6.3.2   Methods of calculation

It is worth while giving some thought to the quickest way of calculating values of $\tfrac{1}{2}\sqrt{N}\,\lambda\,\text{cosec}\,\theta$.

If a large number of nearly equal values has to be determined and the same radiation is used, the same values of N will always occur near $\theta = 90°$. Tables can therefore be drawn up relating a to θ for each line. The value of a for a particular value of θ can then be read off directly.

If, on the other hand, many different values are required, some more general method of calculation must be used. One convenient way is to tabulate $\sqrt{N}\,\lambda/2$ for the different values of λ that are likely to be used, and multiply these values by cosec θ.

6.3.3   Choice of radiation

To get maximum accuracy using this method the extent of extrapolation should be small; in other words, there should be at least one α-doublet with a value of θ near 90°. This condition can usually be satisfied by suitable choice of the radiation. The tables of $\sqrt{N}\,\lambda/2$ can be used to find the value of $\sqrt{N}\,\lambda/2$ which is nearest to, but just less than, the cell dimension. This will correspond to the last line that can appear, since $\sqrt{N}\,\lambda/2 = a\sin\theta$. Repeat this for each of the radiations, and find which gives a possible value of $\sqrt{N}\,\lambda/2$ nearest to a. If the value of $\sqrt{N}\,\lambda/2$ is equal to a within 0·5 per cent, the corresponding line will probably not be recorded on most cameras: if the value is outside 1·5 per cent the accuracy obtainable will be reduced. These limits should, of course, be taken only as a rough guide and each case must be considered by itself.

The use of alloy targets which would give a number of high-angle lines and so make the extrapolation more definite has been suggested. It must be borne in mind, however, that the extrapolation curves for two different radiations may not exactly coincide. Absorption (6.2.2) will be different, and this is one of the most important of the factors that decide the slope of the curve. This objection also applies to the use of α- and β- radiations from the same element.

6.3.4   A practical example

Steps in the calculation of the cell dimensions of a cubic specimen are shown in the next table. The photograph is that of aluminium at 298°C. The third and fourth columns in the table give the readings of the line positions, the positions of the low-angle knife edges (k.e.) being adjusted to readings corresponding to half the distance between them. The sum of the readings for any pair of lines then gives S, the distance between them and the values of S are given in the next column. Each reading is the mean of two or three independent ones, which for well defined lines should agree to within 0·03 mm. The values of θ are then derived from the angle ϕ_k of the camera, the procedure depending on the method used for the particular photograph (see 4.3.1). Cosec θ is then obtained to seven significant figures; this is, of course, more than are physically significant, but it ensures that rounding-off errors in the arithmetic will be small compared with the errors of observation. The product of cosec θ and $\sqrt{N}\,\lambda/2$ then gives a,

Calculation of Lattice Parameter of Aluminium at 298°C from a Powder Photograph taken with CuKα Radiation

| hkl | Radiation | Readings on film | | S | $\theta = S_1\phi_k/S_k$ | cosec θ | $\sqrt{N}\,\lambda/2$ (Å) | a (Å) | $\frac{1}{2}(\cos^2\theta/\theta + \cos^2\theta/\sin\theta)$ |
		Left	Right						
k.e.		(Set to 1·433)							
3 3 1	α_1	18·486	18·554	37·040	55·486°	1·213 610	3·357 46	4·074 64	0·360
3 3 1	α_2	18·558	18·621	37·179	55·695°	1·210 582	3·365 82	4·074 59	0·356
4 2 0	α_1	19·230	19·297	38·527	57·714°	1·182 883	3·444 67	4·074 64	0·311
4 2 0	α_2	19·308	19·371	38·679	57·942°	1·179 926	3·453 26	4·074 59	0·306
4 2 2	α_1	22·586	22·649	45·235	67·763°	1·080 350	3·773 46	4·076 66	0·138
4 2 2	α_2	22·702	22·763	45·465	68·107°	1·077 724	3·782 86	4·076 88	0·134
3 3 3	α_1	26·324	26·388	52·712	78·963°	1·018 846	4·002 35	4·077 78	0·032
5 1 1	α_2	26·575	26·643	53·218	79·721°	1·016 312	4·012 32	4·077 77	0·028
k.e.		28·939	28·933	57·872	86·693				
				$(\phi_k/S_k = 1·498\ 031)$					

Value of a extrapolated to $\theta = 90°$ is 4·078 08 Å.

Refraction correction $= 4·47 \times 10^{-6} \left(\dfrac{1·54}{4·08}\right)^2 \times 4 \times 13·0 = 0·00003$ Å.

$$a = 4·0781_1 \text{ Å.}$$

and this is plotted against $\frac{1}{2}(\cos^2 \theta/\sin \theta + \cos^2 \theta/\theta)$ (6.2.2). The best straight line is then drawn through the points as shown in fig. 6.3, and the value of a at $\frac{1}{2}(\cos^2 \theta/\sin \theta + \cos^2 \theta/\theta) = 0$ is recorded. This, corrected for refraction, is the final result.

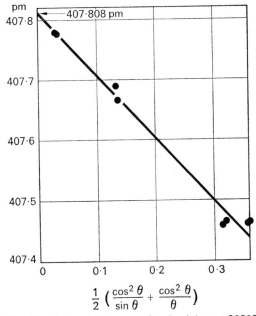

Fig. 6.3   Extrapolation curve for aluminium at 298°C

6.3.5 Method for uniaxial specimens

The graphical method described in the previous section is less easy to apply to systems other than cubic; for such systems Cohen's analytical method is more accurate. Nevertheless, the graphical method is so much quicker than the analytical one that it is still often worth using.

For example, in the tetragonal system the equation relating the cell dimensions to the Bragg angle for the reflection hkl is (2.1.3).

$$\sin^2 \theta_{hkl} = \frac{\lambda^2}{4a^2}(h^2 + k^2) + \frac{\lambda^2}{4c^2}l^2 \qquad \ldots 6.3$$

which may be expressed as

$$\sin^2 \theta_{hkl} = \frac{\lambda^2}{4a^2}\left(h^2 + k^2 + \frac{a^2}{c^2}l^2\right) \qquad \ldots 6.4$$

Thus if the axial ratio c/a is known, the expression $h^2 + k^2 + (a^2/c^2)\, l^2$ can be evaluated for the measured reflections, and a can be derived in the same way as for the cubic system. If reflections for which $h^2 + k^2$ is much greater than $(a^2/c^2)\, l^2$ are used, the error in a due to an error in a/c will be small.

Similarly, to find c the equation may be written as

$$\sin^2 \theta_{hkl} = \frac{\lambda^2}{4c^2} \left\{ \frac{c^2}{a^2} (h^2 + k^2) + l^2 \right\} \qquad \ldots 6.5$$

and c can be evaluated from reflections for which l^2 is much greater than $(c^2/a^2)\,(h^2 + k^2)$.

The axial ratio can then be determined more accurately from the results and, if necessary, the calculations performed again; this usually gives satisfactory results without further refinement.

For the hexagonal system the corresponding equations are

$$\sin^2 \theta_{hkl} = \frac{\lambda^2}{3a^2} \left\{ (h^2 + hk + k^2) + \frac{3a^2}{4c^2} l^2 \right\} \qquad \ldots 6.6$$

and

$$\sin^2 \theta_{hkl} = \frac{\lambda^2}{4c^2} \left\{ \frac{4c^2}{3a^2} (h^2 + hk + k^2) + l^2 \right\} \qquad \ldots 6.7$$

The same equations can be used for rhombohedral crystals if they are referred to a hexagonal unit cell.

Choosing the best radiation for determining the cell dimensions of uniaxial crystals is not so easy as for cubic crystals. Probably the best method is to take a photograph with a short wavelength such as copper $K\alpha$ and to find from it which wavelength will give suitable lines near $\theta = 90°$. It is possible that no single wavelength is suitable for determining both a and c, and it may be necessary to use two different wavelengths.

6.4 Cohen's Analytical Method

6.4.1 Introduction

Personal preference for the graphical method may lead one to choose it rather than analytical extrapolation to determine the accurate cell dimensions of cubic and of uniaxial crystals, but for crystals of symmetry lower than hexagonal there is generally no alternative but to apply Cohen's method. For such crystals—that is, those with orthorhombic, monoclinic or triclinic symmetry—it may not even be possible to apply Cohen's method, because of the presence of too many powder lines for the unambiguous indexing of the high orders of reflection. The same difficulty will be encountered with crystals which have large unit-cell dimensions, even though they may be of high symmetry. The difficulty can be minimized by choosing radiation of long wavelength; in this way the angular separation of the lines will be increased, thus reducing the number of lines and improving the chances of unambiguous indexing.

Cohen's method is a least-squares treatment of the observations to minimize the random errors; at the same time systematic errors are eliminated using appropriate extrapolation functions. As described by Cohen, the method makes no allowance for the increased accuracy of the experimental data as the Bragg angle, θ, tends to 90°; Hess's method (6.4.3) can be used to do this for cubic and tetragonal structures.

6.4.2 Procedure for eliminating systematic errors

Cohen's method will be illustrated by reference to an orthorhombic crystal, beginning with the procedure for eliminating systematic errors. Suppose that the errors, δd, in interplanar spacing which arise from eccentricity, absorption and beam divergence are all proportional to $\cos^2 \theta$. We can write

$$\delta d/d \propto \cos^2 \theta$$

From Bragg's law,

$$\lambda^2 = 4d^2 \sin^2 \theta$$

and if the wavelength of the radiation is considered to be exact, this leads to

$$0 = 8\delta d.d \sin^2 \theta + 4d^2\delta(\sin^2 \theta)$$

Dividing by $d^2 \sin^2 \theta$, we get

$$0 = 2\delta d/d + \delta(\sin^2 \theta)/\sin^2 \theta$$

or

$$2\delta d/d = -\delta(\sin^2 \theta)/\sin^2 \theta$$

But we have assumed that $\delta d/d \propto \cos^2 \theta$. Thus

$$\delta(\sin^2 \theta)/\sin^2 \theta \propto \cos^2 \theta$$

i.e.

$$\delta(\sin^2 \theta) \propto \sin^2 \theta \cos^2 \theta$$

or

$$\delta(\sin^2 \theta) = D \sin^2 2\theta$$

where D is a constant for the film.

If the systematic errors are assumed to be proportional to $\cos^2 \theta/\theta + \cos^2 \theta/\sin \theta$ rather than to $\cos^2 \theta$, then

$$\delta(\sin^2 \theta)/\sin^2 \theta \propto \cos^2 \theta/\theta + \cos^2 \theta/\sin \theta$$

or

$$\delta(\sin^2 \theta) = E(\sin^2 2\theta/\theta + \sin^2 2\theta/\sin \theta)$$

where E is another constant.

For an orthorhombic crystal (2.1.3)

$$\sin^2 \theta_{hkl} = \frac{\lambda^2}{4a^2} h^2 + \frac{\lambda^2}{4b^2} k^2 + \frac{\lambda^2}{4c^2} l^2$$

$$= A_0 h^2 + B_0 k^2 + C_0 l^2$$

where A_0, B_0, C_0 are respectively the correct values of $\lambda^2/4a^2$, $\lambda^2/4b^2$, $\lambda^2/4c^2$. If there is a systematic error proportional to say, $\cos^2 \theta$, then the experimental value, $\sin^2 \theta_{exp}$, for $\sin^2 \theta$ will be in error by $D \sin^2 2\theta$ so that

$$\sin^2 \theta_{exp} = A_0 h^2 + B_0 k^2 + C_0 l^2 + D \sin^2 2\theta$$

If there is also an error proportional to $\cos^2 \theta/\theta + \cos^2 \theta/\sin \theta$ the expression for $\sin^2 \theta_{exp}$ becomes

$$\sin^2 \theta_{exp} = A_0 h^2 + B_0 k^2 + C_0 l^2 + D \sin^2 2\theta + E(\sin^2 2\theta/\theta + \sin^2 2\theta/\sin \theta)$$

or

$$A_0\alpha + B_0\beta + C_0\gamma + D\delta + E\epsilon = \sin^2 \theta_{exp} \qquad \ldots 6.8$$

where

$$\alpha = h^2, \qquad \beta = k^2, \qquad \gamma = l^2, \qquad \delta = \sin^2 2\theta$$

and

$$\epsilon = \sin^2 2\theta/\theta + \sin^2 2\theta/\sin \theta$$

6.4.3   Treatment of random errors

The relation 6.8 eliminates the systematic errors present for a given value of $\sin^2 \theta_{exp}$, that is, for a given hkl reflection. To minimize the effect of random errors the least-squares treatment must be applied. To find the five constants A_0, B_0, C_0, D and E, five normal equations must be set up; namely

$$A_0 \sum \alpha^2 + B_0 \sum \beta\alpha + C_0 \sum \gamma\alpha + D \sum \delta\alpha + E \sum \epsilon\alpha = \sum \alpha \sin^2 \theta_{exp}$$

$$A_0 \sum \alpha\beta + B_0 \sum \beta^2 + C_0 \sum \gamma\beta + D \sum \delta\beta + E \sum \epsilon\beta = \sum \beta \sin^2 \theta_{exp}$$

$$A_0 \sum \alpha\gamma + B_0 \sum \beta\gamma + C_0 \sum \gamma^2 + D \sum \delta\gamma + E \sum \epsilon\gamma = \sum \gamma \sin^2 \theta_{exp}$$

$$A_0 \sum \alpha\delta + B_0 \sum \beta\delta + C_0 \sum \gamma\delta + D \sum \delta^2 + E \sum \epsilon\delta = \sum \delta \sin^2 \theta_{exp}$$

$$A_0 \sum \alpha\epsilon + B_0 \sum \beta\epsilon + C_0 \sum \gamma\epsilon + D \sum \delta\epsilon + E \sum \epsilon^2 = \sum \epsilon \sin^2 \theta_{exp}$$

where the summation is over the high-angle reflections for which values of $\sin^2 \theta_{exp}$ have been obtained. If there are n constants to be determined then n normal equations must be set up; the number of unknowns becomes higher the lower the symmetry of the crystal and the greater the number of extrapolation functions used to eliminate systematic errors. If Hess's method is used to allow for the increased accuracy obtainable with the higher orders of reflection, additional computation is necessary. However, with access to a computer the total computational labour does not present a serious problem or set a limit to the complexity of the arithmetic undertaken. Vogel and Kempter (1961) have, in fact, by computer, increased the accuracy of Hess's calculation and extended his method to cover the hexagonal and orthorhombic systems. To simplify the procedure as far as possible, approximate values of the parameters a, b and c are adopted initially. With these adopted values we have

$$A_{ad}\alpha + B_{ad}\beta + C_{ad}\gamma = \sin^2 \theta_{ad}$$

and subtraction of this equation from (6.8) gives

$$\Delta A_{ad}\alpha + \Delta B_{ad}\beta + \Delta C_{ad}\gamma + D\delta + E\epsilon$$
$$= \sin^2 \theta_{exp} - \sin^2 \theta_{ad}$$
$$= v \text{ (say)},$$

where ΔA_{ad}, ΔB_{ad}, ΔC_{ad} are the errors in the approximate values A_{ad}, B_{ad}, C_{ad}. These errors need to be determined to three significant figures only; the differences v are, in general, required to two significant figures only. The five normal equations now take the form

$$\Delta A_{ad} \sum \alpha^2 + \Delta B_{ad} \sum \beta\alpha + \Delta C_{ad} \sum \gamma\alpha + D \sum \delta\alpha + E \sum \epsilon\alpha = \sum \alpha v$$
$$\text{etc.}$$

from which the corrections ΔA_{ad}, ΔB_{ad}, ΔC_{ad} to the adopted values may be determined.

6.4.4   A practical example

In an actual calculation it is usual to write

$$\delta = 10 \sin^2 2\theta \quad \text{and} \quad \epsilon = 10(\sin^2 2\theta/\theta + \sin^2 2\theta/\sin \theta)$$

so that α, β, γ, δ and ϵ are all of comparable magnitude; further, since A_0, B_0, C_0 (and also A_{ad}, B_{ad}, C_{ad}) have the same values for all reflections whether they arise from $K\alpha_1$ or $K\alpha_2$ radiation, the $K\alpha_2$ wavelength must be expressed in terms of $K\alpha_1$. These modifications are included in the table below which is taken from Peiser, Rooksby and Wilson (1955).

Calculation of Lattice Parameters of CdMg at 18° C by Cohen's Analytical Extrapolation
Copper $K\alpha_1 = 1\cdot54050$ Å, Copper $K\alpha_2 = 1\cdot54434$ Å

hkl	$\sin^2\theta_{exp}$ $K\alpha_2 \to K\alpha_1$	$\sin^2\theta_{ad}$ $K\alpha_2 \to K\alpha_1$	v ($= \sin^2\theta_{exp} - \sin^2\theta_{ad}$)	δ ($= 10\sin^2 2\theta$)	Corrected $\sin^2\theta_{exp}$	Calculated $\sin^2\theta$	Diff.
331	0·74976	0·74920	$+56 \times 10^{-5}$	7·5	0·74913	0·74896	$+17 \times 10^{-5}$
331	0·74933	0·74920	$+13 \times 10^{-5}$	7·5	0·74870	0·74896	-26×10^{-5}
332	0·81360	0·81329	$+31 \times 10^{-5}$	6·1	0·81309	0·81305	$+4 \times 10^{-5}$
134	0·88026	0·88020	$+6 \times 10^{-5}$	4·2	0·87991	0·87992	-1×10^{-5}
430	0·89379	0·89361	$+18 \times 10^{-5}$	3·8	0·89347	0·89338	$+9 \times 10^{-5}$
522	0·90642	0·90629	$+13 \times 10^{-5}$	3·4	0·90614	0·90617	-3×10^{-5}
504	0·93420	0·93392	$+28 \times 10^{-5}$	2·5	0·93399	0·93387	$+12 \times 10^{-5}$
415	0·97026	0·97025	$+1 \times 10^{-5}$	1·2	0·97016	0·97014	$+2 \times 10^{-5}$
514	0·99094	0·99111	-17×10^{-5}	0·4	0·99091	0·99103	-12×10^{-5}
							$\pm10 \times 10^{-5}$

The normal equations are:

$$2631\Delta A_{ad} + 537\Delta B_{ad} + 1370\Delta C_{ad} + 431D = 1835$$
$$537\Delta A_{ad} + 423\Delta B_{ad} + 255\Delta C_{ad} + 277D = 1156$$
$$1370\Delta A_{ad} + 255\Delta B_{ad} + 1427\Delta C_{ad} + 197D = 548$$
$$431\Delta A_{ad} + 277\Delta B_{ad} + 197\Delta C_{ad} + 201D = 913$$

from which

$$\Delta A_{ad} = +0\cdot031 \times 10^{-5}; \Delta B_{ad} = -2\cdot56 \times 10^{-5}$$
$$\Delta C_{ad} = -0\cdot34 \times 10^{-5}; D = +8\cdot34 \times 10^{-5}$$

Hence

$$a = \lambda(K\alpha_1)/2\sqrt{A_0} = 5\cdot0051, \qquad b = \lambda(K\alpha_1)/2\sqrt{B_0} = 3\cdot2217, \qquad c = \lambda(K\alpha_1)/2\sqrt{C_0} = 5\cdot2700 \text{ Å.}$$

The table shows the step-by-step calculation of the lattice parameters of the alloy CdMg. The alloy structure is orthorhombic and four normal equations only are required since, to make the calculations easier, it has been assumed that the systematic errors are proportional to $\cos^2 \theta$. From the differences, $\sin^2 \theta_{exp} - \sin^2 \theta_{ad}$, shown in column four, it is clear that $\sin^2 \theta_{exp}$ is subject to systematic errors. Column six gives the values of $\sin^2 \theta_{exp} - D$, which are $\sin^2 \theta_{exp}$ corrected for systematic errors, and column seven shows the values of $\sin^2 \theta$ calculated from

$$\sin^2 \theta = A_0\alpha + B_0\beta + C_0\gamma$$

where $A_0 = A_{ad} + \Delta A_{ad}, \dots$. The differences between the $\sin^2 \theta$ values listed in columns six and seven are shown in the final column and demonstrate that the effect of systematic errors has been reduced as far as the experimental data will allow.

6.5 Statistical Method

The statistical method described by Beu, Musil and Whitney (1963) does not involve an extrapolation procedure, and applies to both photographic-film and counter measurements. It assumes that each individual reflection, at least three of which have been measured many times, has been corrected for systematic errors, and a test is incorporated which indicates whether the systematic errors have, in fact, been removed in a valid statistical manner. The authors claim that, neglecting the wavelength error, the lattice parameter of a cubic crystal has in this way been determined to about 1 part in 400,000!

6.6 The Use of the Cylindrical Focusing Camera

6.6.1 Introduction

The symmetrical cylindrical focusing camera (fig. 4.24) is essentially a back-reflection camera and as such records high-angle reflections. It is, therefore, particularly suitable for the accurate determination of lattice parameters and indeed, as was stated in (4.8.1), Cohen (1935) has shown that theoretically it is the most accurate powder camera available for this purpose. The incident X-ray beam enters through a hole punched in the film and the mounting is thus similar to the van Arkel mounting for the normal cylindrical powder camera. From fig. 6.4 it can be seen that the separation, S, on the film, of two corresponding reflections is given by $S = (2\pi - 4\theta)R$ for the Debye–Scherrer camera and $S = (4\pi - 8\theta)R$ for the symmetrical focusing back-reflection camera; that is, for the same camera radius, R, the separation obtained with the symmetrical focusing camera is twice that given by the normal cylindrical camera.

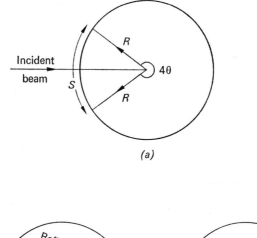

(a)

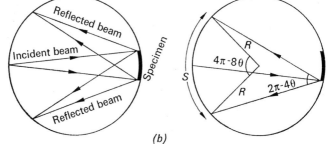

(b)

Fig. 6.4   The reflection geometry for (a) the Debye–Scherrer camera, (b) the symmetrical focusing back-reflection camera

6.6.2   Treatment of errors

The error in the interplanar spacing, d, caused by an error $\delta\theta$ in the measurement of the Bragg angle is given by

$$\delta d = -d \cot \theta \, \delta\theta$$

when it is assumed that there is no error in the wavelength of the radiation. Expressed in terms of ϕ, where $8\phi = 4\pi - 8\theta$, the relation becomes

$$\delta d = d \tan \phi \, \delta\phi$$

Thus correct extrapolation to $\phi = 0$ of the plot of d against an appropriate function of ϕ will eliminate the systematic errors in d except for that introduced by the wavelength error.

The random scatter of the plotted points about the extrapolation line due to the random errors of measurement will be less and less pronounced as ϕ tends to zero and will be exactly zero at $\phi = 0$. The greater the random scatter, the more difficult will it be to fit the correct extrapolation curve.

The systematic error caused by shrinkage of the film is treated in the same way for both the Debye–Scherrer and the symmetrical focusing camera. If the shrinkage is uniform, the angle, ϕ, associated with a particular reflection is given by

$$\phi = S\phi_k/S_k$$

where S is the distance on the film between corresponding reflections and ϕ_k and S_k correspond to the knife-edge shadows printed at the low-angle ends of the film. When S is small, that is at high Bragg angles, an error in ϕ_k will produce only a small error in ϕ so that accurate camera calibration is not essential. Again, the effects of non-uniform shrinkage can be dealt with using either a printed scale or an internal standard (6.2.4).

The effect of an error in ϕ_k on the measured value of the interplanar spacing can be deduced quite simply. Since

$$\phi = S\phi_k/S_k \quad \text{and} \quad S = 8\phi . R$$

for an error $\delta\phi_k$ in ϕ_k we have

$$\delta\phi = S\,\delta\phi_k/S_k$$

which, on substitution for S, leads to

$$\delta\phi = 8\phi R\,\delta\phi_k/S_k$$

But
$$\delta d = d\tan\phi\,\delta\phi$$

so that
$$\delta d = d\tan\phi . 8\phi R\,\delta\phi_k/S_k$$

and the error δd is proportional to $\phi\tan\phi$.

Sources of systematic error other than those also associated with the normal powder camera include the finite size of the entrance slit, E (fig. 4.24) and the displacement of the slit from the camera surface, the displacement of the specimen from the camera surface and penetration of the specimen by the primary beam, the divergence of the beam in a direction parallel to that of the axis of the camera and, finally, the coating with emulsion of both sides of the film. Broadening of the reflection will result from an entrance slit of finite size, and displacement of the slit from the camera surface will effectively change the value of R, thus causing an angular displacement of the reflections; displacement of the specimen from the camera surface will also produce an angular displacement. Penetration of the beam into the specimen will effectively increase R, but will also cause asymmetric broadening of the line. In order to reduce penetration the linear absorption coefficient of the specimen should be as high as possible; this is contrary to the requirements for a sample in the Debye–Scherrer camera. Vertical divergence of the primary beam and double coating of the film will each produce a combination of broadening and displacement of the diffracted beam. Blackening will occur on both the front and the back surfaces of a double-coated film, and since the diffracted radiation strikes the film in a focusing camera obliquely, the two lines will not be suprimposed; use of a single-coated film will eliminate the effect.

These systematic errors can be eliminated by correct extrapolation to $\phi = 0$ of the plot of d against $\phi \tan \phi$ (Cohen, 1935). The function $\phi \tan \phi$ is particularly useful in that the error in interplanar spacing produced by uniform film shrinkage is proportional to $\phi \tan \phi$, and therefore extrapolation with this function automatically corrects for film shrinkage.

By accurate construction of the camera and careful experimental technique it is possible to reduce the systematic errors to negligible proportions so that there is no need to extrapolate and lattice parameters can be determined accurately from a single high-angle reflection. Generally, either with or without extrapolation the accuracy obtainable is of the order of 0·002 to 0·003 per cent.

6.7 The Back-reflection Flat-plate Camera

Because of the nature of the specimen or of the type of investigation undertaken (6.9) it is sometimes necessary to use a back-reflection flat-plate camera for diffraction measurements; with such a camera there may be only one reflection available from which to determine the cell dimensions accurately. The sources of error involved must clearly be very much the same as those associated with the symmetrical focusing camera (6.6.2) and they also can be reduced by attention to design, construction and experimental technique. The Bragg angle, θ, is given by the relation

$$\tan 2\theta = -R/S$$

where R is the radius of the back-reflection powder ring and S is the distance of the specimen from the film.

Alternatively, systematic errors can be corrected by using an internal standard. In this method a coating of fine metal powder is deposited on the surface of the specimen. A composite powder pattern is given by the specimen and the powder and, if the angles of the latter are known, the angles of the former can be found by proportion. This method has the merit of eliminating errors automatically, but it requires longer exposures because of the absorption of the standard powder.

In dealing with lump specimens, it must be remembered that the crystal size may be large and that reflections may be of the Laue type. In order to make sure that the characteristic radiation is being used, the specimen should be oscillated through a few degrees during exposure. In order to produce continuous rings for easy measurement, it is the practice in some laboratories to rotate the film about the axis of the incident beam. This, however, is rather unsafe; if the axis of rotation does not coincide exactly with that of the X-ray beam, the lines will be broadened and so the accuracy will decrease.

6.8 The Counter-diffractometer Method

6.8.1 Measurement of the Bragg angle

The positions of X-ray diffraction lines from powder specimens are obtained from both the photographic-film and the counter-diffractometer methods by examining the intensity distribution in the diffracted radiation. With the photographic film the Bragg angle, θ, is determined from estimates, made either by eye or some form of photometer (7.2.2), of the position of maximum density of blackening of the recorded line; that is, measurements are made of the position of peak intensity. The data recorded by the counter diffractometer give the intensity distribution in the form of a profile which extends over a considerable angular range and it is difficult to decide where to make measurements. Obvious choices are either the peak of the profile or the centre of gravity of the area enclosed by the profile; in general the peak is chosen.

When the profile is sufficiently narrow the peak can be located visually; but if this is not possible, a curve can be fitted to the experimental points in the region of the peak. Alternatively the peak can be defined as the intersection with the profile of the line joining the midpoints of horizontal chords which have been drawn at various heights. Yet another method is to measure to the midpoint of the chord which is drawn at one half the peak height, or at two-thirds the peak height. If the profile is asymmetric then all of these methods will give different values for 2θ.

These difficulties can be avoided if the centroid of the line is used because the centroid calculation is not affected by asymmetry of the line. In order to make a realistic calculation, however, the complete line profile must be obtained, and this takes time. Further, when the reflection is broad, and particularly at high values of 2θ, locating the centroid becomes difficult because of the uncertainty in fixing the points at which the profile merges with the background. Truncation of the profile is one way in which the uncertainty can be removed and several methods have been devised whereby finite limits can be set to the extent of the profile without loss of its characteristics; these methods were compared by Taylor, Mack and Parrish (1964).

Clearly the precision with which both the peak and the centroid can be located will depend upon the precision of the intensity data. Geiger, proportional or scintillation counters (4.5.1) may be used and the intensity recorded by either the continuous-scanning or the step-by-step method. The proportional and the scintillation counters using pulse-height discrimination are to be preferred, and the step-by-step method of recording is the more accurate since, unlike the more popular continuous-scanning method, it does not distort the line profile.

6.8.2   Sources of error

The powder diffractometer is a much more complex instrument than the powder camera and is more difficult to align for optimum performance. Failure to ensure proper alignment will result in errors in the position of the diffracted line, which will be additional to those arising from instrumental factors. The requirements for correct alignment, and the sources of instrumental errors, are now well known and have been discussed by several authors, notably by Parrish (1962). If the diffractometer is to be correctly aligned, then

(1) the distance between the X-ray focus and the specimen must be equal to that between the specimen and the receiving slit of the detector;

(2) The centroid of the primary beam must intersect the diffractometer axis;

(3) the zero position of the instrument must be known, since reflections in powder diffractometry can be measured to one side only of the zero setting;

(4) the angles made by the counter tube and the specimen with the undeviated beam must always be in the ratio $2:1$;

(5) the divergence of the X-ray beam in the plane containing the diffractometer axis must be kept to a minimum. This is usually accomplished by using Soller slits (4.5.2) both before and after diffraction by the specimen. A further requirement is that the planes of the Soller slits must be at right angles to the diffractometer axis.

The major sources of instrumental error are:

(6) the use of a flat specimen rather than of one which conforms to the shape of the focusing circle (4.5.2);
(7) the transparency of the specimen to X-radiation;
(8) the displacement of the specimen from the axis of rotation;
(9) the Lorentz, polarization and dispersion factors.

Except in one important instance, misalignment and instrumental aberrations affect the position of both the peak and the centroid of the profile in much the same way; the exception is the markedly different effect on the positions of the two points produced by the Lorentz, polarization and dispersion factors. This is partly why the systematic errors which arise in the two cases are treated differently.

6.8.3   Treatment of peak-position errors

Two obvious advantages of using the peak of the line profile are that intensity measurements need be taken for only a limited range of angles and that some overlap of the lines can be tolerated. On the other hand, when the profile is asymmetric, it is difficult to decide just where the peak intensity is and further, when the peak is determined by curve fitting, its position varies with the range selected.

Proper adjustment of the instrument will minimize errors arising from inequality of the distances of the focus to the specimen and the specimen to the counter, and from misalignment of the primary beam. The zero error between the straight-through direction of the primary beam and the angular scale reading can be eliminated mechanically. Any error in the angular 2:1 ratio can be eliminated by correctly setting the ratio at $\theta = 0°$ and incorporating a gear train which is capable of maintaining the ratio at all other angles. Alternatively the zero error and 2:1 mis-setting error can be eliminated by calibration either separately with a known standard substance or by the use of an internal standard.

Error in the zero setting will clearly remain constant throughout the entire range of 2θ and from

$$\delta d = -d \cot \theta \, \delta\theta$$

it is evident that the consequent error in d will extrapolate to zero at $\theta = 90°$. Inaccuracy of the 2:1 ratio produces both asymmetry in the profile and shift of the peak.

Axial divergence is limited by the Soller-slit assemblies but produces both asymmetric broadening of the line and a shift in its position. Pike (1957) has described a procedure for correcting the errors caused by these effects. Use of a flat specimen and the transparency of the sample both cause asymmetric broadening and consequently an error in the value of the interplanar spacing in the crystal which is proportional to $\cos^2 \theta$; this error therefore extrapolates to zero at $\theta = 90°$. As with the focusing camera (4.8.1) the linear absorption coefficient of the sample should be as high as practicable so as to reduce the transparency error.

Displacement of the surface of the specimen from the axis of rotation results in an interplanar-spacing error which is proportional to $\cos \theta \cot \theta$ and this function also extrapolates to zero at $\theta = 90°$. Finally the Lorentz, polarization and dispersion factors have very little effect on the position of the peak of the profile.

Thus those systematic errors which cannot be eliminated by alignment of the diffractometer are proportional either to $\cos^2 \theta$ or to $\cos \theta \cot \theta$ and final choice of the extrapolation function depends upon the relative importance of the corresponding effects. Generally the greatest error is introduced by displacement of the sample from the axis of rotation; the extrapolation function is then $\cos \theta \cot \theta$.

After extrapolation the parameters can be corrected for the effects of refraction and of temperature.

Parrish, Taylor and Mack (1964) measured the lattice parameters of silicon and of tungsten with both CuKα and FeKα radiations. They estimated that their cell-dimension value of tungsten was determined to within 1 part in 50,000 for both radiations, and that for silicon the estimated precision was 1 part in 50,000 for CuKα radiation and 1 part in 25,000 for FeKα radiation.

6.8.4   Treatment of centroid-position errors

The centroid of the line profile gives a better representation of the maximum intensity of a reflection than does the peak, which is based on a relatively few measurements in the region of the peak itself. Another advantage in using the centroid is that the centroid can always be located accurately regardless of line asymmetry or breadth; one consequence of this is that broad lines can often give a higher accuracy than one would expect (6.2.6).

Because the centroid method requires that the whole of the profile should be measured, such measurements must be made over a wide range of angles. Again, the centroid is difficult to determine if there is any overlap at all, which means that the method is limited to relatively uncomplicated patterns. A further limitation is imposed by our lack of knowledge of the X-ray wavelengths associated with centroids; at present the centroid wavelengths are known only for CuKα and FeKα radiations. At high Bragg angles the effects of dispersion produce an excessively large range for each reflection; the background slope is no longer linear and truncation is correspondingly more difficult. The Lorentz and polarization factors vary significantly over such a range and this results in profile distortion and a shift in the centroid, which affect d in a way which does not extrapolate to zero at $\theta = 90°$. Extrapolation procedures are therefore useless and no advantage can be gained from the measurement of high-angle reflections.

However, when the centroid is used, extrapolation is unnecessary. It is possible to correct the position of the centroid of each line individually for systematic errors, including that due to refraction. Such corrections are difficult to apply to the peak position. Accurate lattice-parameter determinations are not therefore confined to the high-angle region and of the reflections which are actually measured, each one will give an accurate value; thus, if necessary, only one reflection need be measured.

Delf (1963) has quoted a precision of 1 part in 80,000 for his determinations of the cell dimensions of aluminium, tungsten and diamond. He assumed a value of 1·54176 Å for the centroid wavelength of CuKα radiation and his results do not take into account any error that there may be in this wavelength value. The centroid position of each line was corrected for all known systematic errors and the effectiveness of his corrections may be judged from the fact that the values of the parameters calculated from each line were independent of the Bragg angle, θ.

Taylor, Mack and Parrish (1964) have similarly obtained the cell dimensions of tungsten and of silicon. Because of the difficulty in applying the dispersion correction at high angles they recommend that the most suitable range of 2θ values is from 120° to 150°. They conclude from measurements with both CuKα and FeKα radiations that, arising partly from the limited accuracy to which these wavelengths are known, their precision is not greater than 1 part in 100,000 and may be as low as 1 part in 25,000; under favourable experimental conditions and neglecting the effect of wavelength errors, their precision was as high as 1 part in 500,000.

6.9 Applications

6.9.1 Determination of thermal expansion coefficients

X-ray methods are well suited to the determination of coefficients of thermal expansion because the specimen, being small, can easily be kept at a uniform temperature.

They have the added advantage that they can provide a continuous record of change in lattice parameter with temperature (Endter, 1960). The methods are particularly useful when thermal expansion coefficients of volatile substances are required at high temperatures and when expansion data are required for condensed gases at low temperatures. Even when the absolute values of the lattice parameters are not known, the coefficients can be calculated from changes in interplanar spacing resulting from changes in temperature. X-ray methods also enable the principal coefficients of thermal expansion of uniaxial substances to be determined without the need for single crystals; in this way the lattice parameters of hexagonal ice have been measured over the temperature range from 15K to 200K (Brill and Tippe, 1967).

At one time there was some suspicion that the coefficients determined by X-ray techniques were not identical with those obtained by dilatometric measurements on massive specimens, but it has since been established that, in general, there is reasonably good agreement between X-ray and macroscopic measurements.

6.9.2   Determination of phase boundaries in equilibrium diagrams

Suppose that the line CD in fig. 6.5 represents the boundary between the α phase field and the $\alpha + \beta$ phase field in the binary equilibrium diagram of

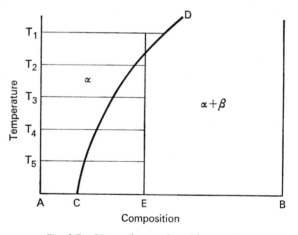

Fig. 6.5   Phase diagram for a binary system

the substances A, B. By quenching specimens from the temperature T_1 it may be possible to maintain them in states of solid solution, and hence to derive the curve of a cell dimension as a function of composition. If now the mixture of gross composition E is quenched from various temperatures T_2, T_3, T_4,...it should always be two-phase, although the composition of the α-phase will not be constant. Measurement of the cell dimensions should enable the compositions of the α-phase to be found from the data obtained.

This method has been much used for studying binary equilibrium diagrams of metals. It is particularly powerful because much information can be obtained from few alloys, but there are some precautions that have to be taken if good results are to be obtained. First, the initial cell dimensions must be those of alloys of which the composition is reliably known; secondly, the curve of cell dimension against composition should have an adequate number of points on it, as there may be discontinuities in slope; thirdly, the curve must not be extrapolated for data that do not lie within the range studied; and fourthly, it must be verified that there is no decomposition of the filings during quenching. The argument that the cell dimensions may be altered by the presence of internal strains is probably not valid for powders, since it is unlikely that such strains would have a preferred direction in the specimen; strains could, however, produce a broadening of the X-ray reflections. An excellent account of the application of lattice-parameter measurements to the determination of equilibrium diagrams is given by Barrett and Massalski (1966).

6.9.3   Measurement of densities and molecular weights

Densities of substances can sometimes be measured most accurately by X-ray methods, the formula (Stockdale, 1940; Foote and Jette, 1940; Bragg, 1947) being

$$\rho = 1 \cdot 66020 \sum Z / V \qquad \qquad \ldots 6.9$$

where ρ is the density, ΣZ is the sum of the atomic weights of the atoms in the unit cell, and V is the volume, in Å^3, of the unit cell.

The X-ray method can be used to determine atomic weights. To do this the density must be measured accurately by a macroscopic method and the value of Avogadro's number must be known together with the lattice spacings of the crystal. Similarly, if the number of molecules in the unit cell of an organic substance is known, the molecular weight of the compound can be derived. It may happen that the space group has been determined, but the number of molecules per unit cell is not known. From the space group we can deduce the possible values of this number and hence a series of possible values of the molecular weight. As we always know the lowest possible value of the number of molecules per unit cell, this gives us the upper limit to the possible values of the molecular weight, which is often very useful information. This procedure will also give the number of atoms in a unit cell of a metal or an alloy and provide information as to whether solid solutions are interstitial, substitutional or subtractive.

6.9.4 Measurement of internal stresses

The detection of small changes in the Bragg angles of X-ray reflections can provide useful information about the presence of internal stresses.

The principle of the method is that the change, δd, in the spacing of a set of planes in a crystal is given by the relation

$$\delta d/d = T/E \qquad \ldots 6.10$$

where E is Young's modulus and T is the applied stress, assumed to be perpendicular to the planes. If T is parallel to the planes, it will produce a Poisson contraction in the spacing, given by

$$\delta d/d = -\sigma T/E \qquad \ldots 6.11$$

where σ is Poisson's ratio. If T is not parallel to the reflecting planes, two components of the stress must be found and this can be done only by taking two or more photographs with the X-ray beam making different angles of incidence with the surface of the specimen.

There are certain difficulties in applying these methods which are sometimes overlooked. For example, the value of E is not necessarily that for a macroscopic polycrystalline specimen; even for a cubic crystal, the value of Young's modulus and of the rigidity modulus, upon which σ depends, are not independent of direction. This does not matter for mechanical testing, but the X-ray method selects special directions in each crystal for measurement. It seems that the most satisfactory way of finding the value of E appropriate for a given reflection is to measure the change in spacing of that reflection produced by a known stress in a similar piece of material; in other words, the method is best used for *comparing* stresses.

Moreover, if the material under test is an alloy, it is always possible that changes in spacing may be due to changes in composition from point to point in the specimen; it is often difficult to produce a large casting, for example, with the same composition throughout. Nevertheless, the change of spacing produced by the solution of certain elements in others (for example, chromium in iron) is so small that differences in composition of one or two per cent are inappreciable; in this case changes in spacing can reasonably be attributed to the presence of internal stress. But if the change in spacing is large, as in the solution of silicon in iron (10.3.2), a test for uniformity of composition must precede any determination of residual stresses from spacing measurements.

Although cell dimensions can be measured with high accuracy, measurement of stress cannot be made with the same degree of accuracy, because only small differences are involved. X-ray reflections given by commercial materials are usually not particularly sharp, and it is probably safe to say that the accuracy in spacing measurement is not better than 0·01 per cent

(6.26). For metals, σ is usually about 0·3, so that the corresponding error in T is $0·0003E$. For steel, $E = 2 \times 10^{11}$ Nm^{-2}, so that the limits of error of measurement of T will be about $\pm 6 \times 10^7$ Nm^{-2}, or about ± 4 ton/in^2. Generally the accuracy will be less than this, but it may, in special cases, be greater.

A comprehensive review of the application of both film and diffracto-meter techniques to the measurement of stress in metals is given by Cullity (1956), Kaelble (1967) and Azároff (1968).

6.9.5 Determination of crystal structures

It may not be possible to grow a suitable single crystal of a substance whose structure is required, and in this case the structure will have to be deter-mined from a powder pattern. The pattern could be complex, with many overlapping reflections, but if the structure is to be found, these reflections must be identified and indexed unequivocally. To do this the lattice para-meters must be evaluated as accurately as the pattern will allow; the higher the precision to which the cell dimensions are known the more likely it is that the correct indices can be assigned to every line. To determine the parameters there must be a sufficient number of resolved reflections at high angles on a photographic plate to enable the extrapolation procedure to be carried out effectively; the same considerations apply if the peak of the profile of a diffractometer trace is to be utilized. The alternative is to measure to the centroids of the lower-angle profiles obtained with a dif-fractometer and to obtain the parameters from single reflections.

One of the earliest and best examples of the application of accurate lattice parameters to the determination of crystal structures from powder photographs is that of Bradley (1935a) for the element gallium; the tech-nique employed also illustrates the value of a suitable choice of radiation. It will be described in more detail in section 8.3.4.

7+

CHAPTER 7

The Measurement and Calculation of the Intensities of X-ray Reflections

7.1 Introduction

The intensities of powder reflections are required for crystal-structure determination (chapter 8), for the investigation of the broadening of the reflections (chapter 9) and for the identification of a substance by comparison with data given in the A.S.T.M. powder diffraction file (chapter 10). For the purposes of identification it is often sufficient to classify by eye the intensities of the reflections as very weak, weak, fairly strong, and so on. In line-broadening measurements and for structure determination, however, more precise intensity data are required; these data are obtained from a microphotometer, or microdensitometer, when the reflections are recorded on a film, and by a counter when diffractometer techniques are employed. Allowance must be made for the general background intensity by measuring this intensity on either side of the line, and unevenness in the intensity of the line itself, arising from unduly large grain size within the specimen, can be averaged out by repeated exposure of different parts of the sample to the incident radiation. Steps must be taken to ensure that accidental preferred orientation does not exist within the specimen.

7.2 Measurement of Intensities by Photographic Methods

7.2.1 Effect of X-radiation on a photographic film

In photographic methods the film is exposed to radiation long enough for the effect of the random emission of quanta to be inappreciable. Because the reflections are recorded simultaneously, there is no need to take special precautions to ensure stability of the output from the X-ray tube. The essential problem is therefore to correlate the degree of blackening of a powder line with the number of incident X-ray quanta. This is not simply a problem in photographic photometry, since the many factors discussed later in this chapter influence the intensity, but we shall first consider the problem of how to find the number of X-ray quanta producing a given line.

The blackening, D, of a photographic film suitable for X-ray crystallography is defined in terms of the intensity, I_0, of the light incident upon it and the intensity, I, transmitted through it, by the relation

$$D = \log_{10}(I_0/I)$$

The quantity, D, has been shown to be proportional to the total number of X-ray quanta falling on unit area of the film in the region examined, provided that D is less than a certain value which depends on the film and the method of processing. Thus to obtain a measure of the number of X-ray quanta which have fallen on a film, it is necessary to find the ratio of the incident beam of light to that of the transmitted beam after traversing a *uniformly* darkened region.

7.2.2   Types of photometer

Three kinds of photometer can be used to find the ratio of the intensity of the light incident upon the darkened photographic film to that transmitted through it. The simplest is that which uses a single light source and only one photocell. The current through the photocell is registered on a meter and the deflection of the meter is observed, first without the darkened film interposed in the light beam and then with it. The ratio of these two deflections gives the value of I_0/I. The success of the measurement depends on the stability of the light source both in position and intensity. In practice, the sagging of the light filament and fluctuations in potential across the filament usually limit the accuracy of this type of measurement.

The light-meter part of the instrument can be made more simple by equalising the intensities of the two beams. This is done by inserting a calibrated neutral wedge in the path of the direct beam when its strength is being determined. Such a wedge is arranged so that the density increases uniformly along its length.

Suppose the distance along the wedge from the thin end to any point is d and the angle of the wedge is α. Then the corresponding thickness of absorbing material is $d\alpha$. If κ is the optical absorption coeficient, we have for the ratio of the intensities of the incident and transmitted light beams at this point of the wedge,

$$I_0/I = \exp \kappa d\alpha$$

or $\qquad\qquad \log_{10} I_0/I = 0{\cdot}434\kappa\alpha d$

The quantity $0{\cdot}434\,\kappa\alpha$ is a constant for the wedge, usually supplied by the maker, and we may replace it by the symbol q. Thus the blackening D of the wedge is given by

$$D = qd$$

Suppose now that the number of X-ray quanta which have fallen on unit area of the line (assuming it to be uniformly blackened) is p_1, and the corresponding number for the background is p_2. Then we have seen (7.2.1) that

$$p_1 = cD_1, \qquad p_2 = cD_2,$$

where c is a constant for the X-ray film and its conditions of processing, and D_1 and D_2 are the degrees of blackening produced.

Thus $\quad p_1 - p_2 = c(D_1 - D_2) = cq(d_1 - d_2),$

where d_1 and d_2 are the distances from the thin end of the optical wedge.

Thus, using a calibrated wedge, the measurement of the X-ray intensity is reduced to a linear measurement of the displacement of the wedge which is required to balance the shifting of the film from line to background.

The curve relating density to exposure varies for most X-ray films with the conditions of development, and from one sample of film to another. It is, therefore, essential to produce on a portion of the film which carries the

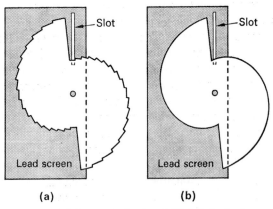

Fig. 7.1   (a) Stepped wheel and (b) spiral wheel

photograph a set of blackened steps the exposures of which are in a known ratio, or a wedge, the exposure at each point of which is known. Provided that the calibration steps cover the range of density found in the line, the blackening at any point can be converted into a quantity proportional to the X-ray exposure. One way of producing such a stepped or continuous wedge is to rotate in front of the film a wheel from which a stepped or spiral sector has been cut out. The steps on the sector are arranged so that the exposures along a radius of the rotating wheel increase in known ratios in passing from one step to the next. A lead screen with a rectangular slot in it is arranged with the length of the slot parallel to the radius of the sectored wheel (fig. 7.1). The film, after exposure to the reflections from the powder, is placed under the lead screen so that a strip, usually near one edge, receives the imprint of the steps or wedge. To obtain a uniform intensity of the X-rays all along the length of the rectangular slot, the rotating wheel is set up several metres from the X-ray tube.

Fluctuations of the light intensity are usually a serious limitation in accurate photometry. If a mains supply is used to work the lamp, smoothing devices have to be introduced to keep the fluctuations down to one per cent.

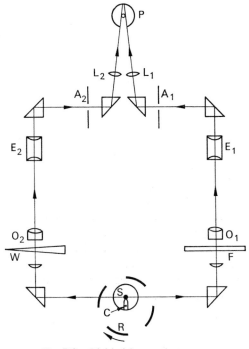

Fig. 7.2   Divided-beam photometer

190   X-Ray Powder Diffraction

If an accumulator is used as the source, it has to be used across a d.c. mains supply so that there is no steady fall in current during use. Finally, lamps blacken gradually and recrystallization of the filament causes sagging and drift of the image across the photometer slits.

For these reasons another type of photometer, called a microdensitometer, is more often used. In this instrument, fig. 7.2, light from a single lamp, S, passes alternately through a rotating shutter, R, to the left-hand or to the right-hand optical train, and finally, to a single photocell, P. The film under test, F, is viewed by the microscope O_1E_1 which casts an image of the film on the narrow aperture A_1. The calibrated wedge, W, is viewed by the microscope O_2E_2 and an image is cast on the aperture A_2. The lenses L_1, L_2 produce overlapping images of the apertures A_1, A_2 on the photocell surface. The alternating part of the signal given out by the photocell is proportional to the difference in intensity of the two light beams and vanishes when the two beams are of equal intensity. Thus the wedge is moved by hand, or by a servo-mechanism operated by the photocell, until the intensity of light passing through the wedge is equal to that passing through the film. This arrangement can be made automatic by arranging that the wedge moves the sliding arm of a potentiometer. The output of the potentiometer is applied to a pen recorder. If the film F is driven under the scanning microscope at a constant speed, an automatic record of density variation is obtained.

7.2.3   Fundamental conditions of measurement

If the darkened area of the film under examination has a small clear space in it, more light may be transmitted through it than through all the rest of the darkened space. This difficulty may arise where there are 'pinhole' blemishes in the photographic emulsion. The genuine photographic lines due to X-ray reflections are small and vary greatly in density from the centre to the outside; areas which are uniformly black may be no more than 0.1 mm $\times$ 0.1 mm. To get accurate results it may then be necessary to examine areas no larger than this at one time, since areas which are not uniformly blackened give inaccurate results. For example, suppose that in an area under examination one half had been irradiated with p X-ray quanta per unit area and the other half with $2p$. Then the number per unit area for the whole area should be $1.5p$. Suppose that p quanta per unit area produce a blackening of unity. Then we have

$$1 = \log_{10} I_0/I_1 = cp, \qquad I_1 = \tfrac{1}{10} I_0$$

$$2 = \log_{10} I_0/I_2 = c(2p), \quad I_2 = \tfrac{1}{100} I_0$$

where I_0 is the intensity of the incident light and I_1, I_2 are the intensities of the beams transmitted through the two darkened areas. If I_3 is the light transmitted through the composite area,

$$\frac{I_3}{I_0} = \frac{1/10 + 1/100}{2} = 0.055$$

Hence the corresponding darkening D_3 is

$$D_3 = \log_{10} 1/0.055 = 1.26$$

Thus we should obtain a number of X-ray quanta equal to 1·26 instead of the true value 1·5.

This type of error can be decreased by using less dense lines. If, for example, within the area of the line examined, there are equal portions of densities 0·1 and 0·2 the observed density would be 0·147 instead of the true value 0·15. This is not a completely satisfactory solution, however, since working with weak lines increases the errors due to the uncertainties in the general background and to the accidental defects due to dust and surface markings of the film. Thus the aperture limiting the light beam reaching the photocell must be small enough to keep the variations of blackening within a prescribed range depending on the accuracy aimed at in the final result. If the aperture is reduced too far, too little light will reach the photocell for the recording mechanism to operate with the required accuracy.

7.2.4 Application to powder photographs

The photographic density of a line on a powder photograph varies rapidly in a direction perpendicular to the length of the line and only slowly along its length. Such lines are examined in the photometer by a rectangular aperture. The length of the aperture is, of course, arranged parallel to the length of the line, and the breadth is made so small that there is no appreciable variation of density within the rectangle. To examine areas of uniform blackening within the line a magnified image of the central section of the line is projected on to the rectangular aperture and the light passing through this aperture is balanced against the light passing through a wedge, such as that described in (7.2.2). By fine adjustments the image of the central section of the line can be made to traverse the fixed aperture at right angles to the length of the aperture; in this way the image is divided into a large number of areas of uniform density. These densities are determined by the photometer and converted (with calibration if necessary) into readings of X-ray intensity at the corresponding points. In addition, the area well outside the line itself must be photometered to get a reliable measure of the general background blackening. A graph may be drawn showing the variation of the X-ray intensity across the line (fig. 7.3) or the variation may be recorded automatically by some commercial instrument.

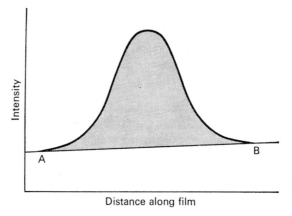

Fig. 7.3   The derivation of the total intensity of a powder line

When the total intensity of the reflection is required, a base line is drawn connecting the background curves on either side of the line as shown at AB in fig. 7.3. The shaded area under the curve and above the base line is proportional to the intensity of reflection corresponding to this powder line.

7.3 Measurement of Intensities by the Counter-Diffractometer Method

7.3.1   Detection of X-ray quanta by counter methods

The counter diffractometer, unlike the photographic film, is not capable of recording simultaneously all the lines of a powder pattern. The lines are measured in succession, each for a limited time only, by either the continuous-scanning or the step-by-step method; this means first that the random emission of X-ray quanta will introduce intensity variations, and secondly, that the output from the X-ray source must be kept stable. Despite these limitations the counter method is now the most accurate and direct way of determining the intensity of diffracted X-radiation.

There are two types of detector for X-ray quanta. Geiger counters and proportional counters detect the radiation by amplifying the initial gas ionization which quanta produce in a discharge tube, and scintillation counters rely on fluorescent photons which result from absorption of the X-ray beam in a suitable material. Semiconductors are potentially the most suitable substances for detecting and counting X-ray quanta but as yet they have not been adapted for use in this field. Electrons and holes are produced by ionization in a solid and since the ionization voltage for a semiconductor is much smaller than that for a gas, the degree of ionization produced in the former by one quantum is correspondingly greater. Further, a solid has a far greater linear absorption coefficient, so that for a given rate of absorption of X-ray quanta a semiconductor is much more compact than is the equivalent gaseous absorber.

7.3.2   The Geiger counter and the proportional counter

When an X-ray quantum is absorbed by an atom, a photoelectron may be ejected from one or other of the inner shells (3.4.2) and then, depending on the nature of the atom, it may either emit its own characteristic (fluorescent) radiation or there may be some reorganization of the remaining electrons inside the atom with a consequent emission of further electrons. These electrons are named after Auger, who was the first to detect them. With gases such as argon and xenon, almost complete conversion occurs and the energy of the incident quanta transforms into photo electrons and Auger electrons. Secondary ionization is then produced by these electrons until insufficient energy is available to continue the process. If characteristic radiation is emitted the only source of further ionization is the single photoelectron.

In the absence of an electric field the ion pairs created by collisions between the energetic electrons and the neutral gas molecules will eventually recombine and the gas will return to its original neutral condition. On the other hand, if a sufficiently large potential difference is applied between two electrodes placed in the gas then the liberated electrons will gain sufficient kinetic energy in their passage towards the anode to produce still further ionization by collision. If the energy of the incident quantum of X-radiation is directly responsible for creating n ion pairs and if Bn ions are ultimately collected at one or other of the electrodes, then the factor B is called the **gas multiplication**. The charge collected at the anode will be $-Bne$ where e is the electronic charge, and if this charge is conducted away the resulting voltage pulse will be $-Bne/C$ where C is the capacitance of the system. It is the magnitude of the applied voltage, and hence the value of B, that determines whether the detector operates as a Geiger (fig. 7.4) or as a proportional counter.

The basic features of the Geiger counter are illustrated in fig. 7.5 where C, the cathode, is a metallic cylinder approximately 100 mm long and 20 mm

7*

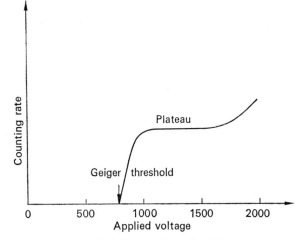

Fig. 7.4   A typical characteristic of a Geiger counter

in diameter; the cylinder, which contains either argon at a pressure of 80 kNm^{-2} (60 cmHg) or krypton at a pressure of 40 kNm^{-2} (30 cmHg), is sealed. Mounted coaxially with the cathode is a tungsten wire, A, which acts as the anode, and at the end of the cylinder X-rays are admitted through the mica window, W, and thence directed along the length of the anode. An applied potential difference of about 1000 volts between the anode and the cathode produces a sufficiently high field round the anode to ensure that, once ionization has been initiated by the absorption of an X-ray quantum, further ionization by collision is so prolific that a continuous discharge is maintained throughout the length of the tube. When the discharge becomes continuous it is not possible to detect further quanta and the discharge must therefore be quenched before the next count can be made. Quenching is achieved either by mixing with the discharge gas a small quantity of a halogen, such as chlorine or bromine, or by incorporating suitable electronic devices in the external circuit. The period during which

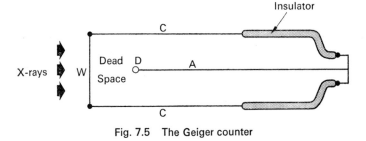

Fig. 7.5   The Geiger counter

incident radiation cannot be detected is known as the dead time of the Geiger counter and this period must be made as short as possible to reduce the counting losses. Counting losses also arise when primary ionization takes place in the dead space between the end D of the anode and the window of the tube. These losses will form a constant fraction of the incident quanta irrespective of the rate of arrival, but the former will be zero at sufficiently low intensities and will rise non-linearly as the counting rate increases, thus making the response of the detector non-linear. Further, at the normal operating voltages of a Geiger tube the gas amplification is so high that all pulses, whatever the energies of the absorbed quanta which produce them, are effectively the same size. It is thus not possible to discriminate between different wavelengths in the incident X-radiation.

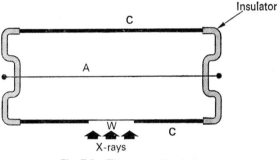

Fig. 7.6   The proportional counter

Once the voltage, called the Geiger threshold, at which pulses are sufficiently large to be counted has been reached, further increase in voltage produces a very rapid increase in counting rate (fig. 7.4). At about 100 volts above the threshold the counting rate becomes almost independent of the applied potential difference and remains so for a range of a few hundred volts. This region is called the plateau and the counter is operated at about the middle of the range. The longer and flatter the plateau the less dependent is the performance of the counter on the stability of the applied-voltage supply.

Although the Geiger counter is robust and simple to use, and capable of producing large pulses, it has now been superseded in popularity by both the proportional counter (Lang, 1956) and the scintillation counter (West, Meyerhof and Hofstadter, 1951). The dimensions of the proportional counter are similar to those of the Geiger type but X-radiation is admitted through a window in the side of the tube and not through the end (fig. 7.6). The window is made of an electrical conductor, such as beryllium, so as to ensure that the electric field applied between the tungsten anode, A, and the cylindrical cathode, C, is uniform throughout the length of the tube at all points at the same distance from the anode. The X-rays are in the electric field as soon as they enter the tube so that the dead space of the Geiger counter is eliminated. Xenon at about a pressure of 80 kN m^{-2} is normally used as the discharge gas because of the requirements for higher linear absorption to compensate for the shorter path length of the X-radiation in the gas. The operating voltage of the proportional counter is below that of the Geiger threshold and is sufficiently low for the gas multiplication, B, to be independent of the number of primary ion pairs, n, produced by the incident quantum; the voltage pulse, $-Bne/C$, is thus directly proportional to n, and hence to the energy of the absorbed quantum. This relationship between the magnitude of the discharge pulse and the energy of the absorbed X-ray quantum means that a pulse-height analyser can be introduced into the electronic circuits to discriminate against the effects of unwanted radiation (Lang, 1951; Arndt and Riley, 1952). The analyser can for example be set so as to reject pulses which are below a predetermined amplitude and also those which are above another predetermined amplitude. Such a device could be set to accept, say, only those pulses produced by the characteristic radiation and would not count quanta of either white radiation or the harmonics of the characteristic radiation.

A further advantage is the shorter dead time of the proportional counter. The counting rate is therefore higher than that of the Geiger counter and the linearity of response of the instrument consequently improved.

7.3.3   The scintillation counter

In the scintillation counter (fig. 7.7) the X-rays are absorbed by a single crystal, SC, of thallium-activated sodium iodide. The energy is converted by the fluorescent material into photons of light which pass through the glass windows, G, on to the cathode, C, of the photomultiplier tube, and are

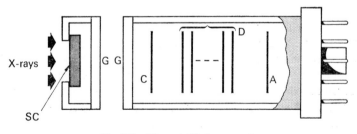

Fig. 7.7   The scintillation counter

there absorbed. This absorbed energy is in turn converted into photo-electrons, which are accelerated towards the anode, A. In their passage from the cathode to the anode, the photoelectrons strike the first of the electrodes, D, which are progressively higher positive potentials with respect to the cathode. At the first electrode further photoelectrons are emitted and the photoelectron multiplication proceeds at each successive electrode until the final amplified pulse is collected at the anode. The magnitude of this pulse is proportional to the energy of the incident quantum so that the device operates as a proportional counter; there is again no dead space and the dead time is about the same as that of the proportional counter.

With all three types of counter the counting efficiency decreases as the wavelength decreases because of the lower absorption of X-ray quanta of shorter wavelength (3.4.2). On the other hand there is less absorption by the window as the wavelength decreases and this will tend to offset the decrease in counting efficiency.

7.3.4 The integrated intensity

The relative intensity of the radiation diffracted in a particular direction is proportional to the rate of emission of X-ray quanta in that direction; provided that there are no counting losses, this intensity is therefore proportional to the corresponding counting rate as measured by the detector. Whether the width of the diffraction line is scanned step by step or continuously, the counting rate will be subject to statistical error due to the random emission of quanta from the X-ray tube. Such errors can be reduced by increasing the number of counts made at each step using the step-by-step method or by reducing the scanning speed when the continuous method is used. The effect of the fluctuations can be further reduced if the separate counting rates determined at each step are plotted against 2θ over the width of the line and a smooth curve drawn through the points. In the continuous-scanning method the output from the rate meter is recorded on a strip chart which is driven in synchronism with the scanner, and the statistical fluctuation will cause the line profile to be jagged. Here, too, a smooth curve will average out the random errors.

The intensity of a line can be represented either by its peak height or by the area under the line profile; the area represents the total diffracted energy in the reflection and is a measure of what is called the **integrated intensity** of the line. Since the integrated intensity is obtained from a considerably greater number of counts than is the peak intensity, it is less susceptible to the effect of fluctuations in the counting rate; this is particularly advantageous when the continuous-scan method is used because

the area under the profile is little affected by the distortion which is inherent in this method. The background on an X-ray diffraction photograph varies with angle. When the integrated intensity of a line is measured
the background count to be subtracted is taken over the same angular
range as that covered by the reflection, and as near to it as possible. If the
peak intensity is required, the background should be taken ideally at the
angle of the peak intensity and if this is not possible, as near the peak as
possible—not over a range of angles. There will be included in the background not only pulses produced by X-radiation, but also contributions
from electronic noise on the measuring circuit, cosmic rays and radio-
activity.

These methods of measuring the intensity assume a linear relationship
between the intensity and the counting rate. This is no real problem with
the proportional and the scintillation counter, where the dead time is short,
but the Geiger-counter dead time does affect the linearity and a correction
must be made. The correction is well known and takes the form

$$n_0/n = 1/(1 + nt)$$

where n_0 is the observed counting rate and n is the actual rate of arrival of
X-ray quanta; t is the resolving time of the detector and is measured from
the instant the counter becomes completely insensitive until full sensi-
tivity is regained.

Large fluctuations in intensity result if a powder specimen which contains
large grains is either stationary or rotated slowly in the X-ray beam. The
variations can be averaged out by sufficiently rapid rotation of a cylindrical
specimen about its axis or of a flat specimen in its own plane. Errors in the
integrated intensity from misalignment of the diffractometer and mis-
setting of the specimen are small and do not require correction. It is sometimes better to transmit the X-rays through a thin plate of a material such
as a plastic or through liquids contained in thin-walled vessels, but if the
sample is cylindrical it may be necessary to correct the integrated intensity
for absorption in the specimen (7.4.4).

7.4 Calculation of Intensities of X-ray Reflections

7.4.1 Introduction

In section 2.2.3 a simple construction was described for finding whether a
particular X-ray reflection would occur with a crystal in a particular
orientation with respect to the X-ray beam. From the considerations discussed in (2.2.2), however, it will be clear that although all observed re-

flections can be accounted for by the construction, not all the reflections permitted by the construction will necessarily be present; further, the intensities of the reflections which are actually observed will vary greatly. Thus Bragg's law gives only the geometrical conditions that must be satisfied if a reflection is to occur; the intensity depends upon many factors, and may be zero.

7.4.2   The expression for relative intensities

The formula for the intensity of reflection from a powder photograph is given in the International Tables, Vol. II (1959), but if we wish to consider only relative values the formula can be greatly simplified to the form

$$I \, \alpha \, \frac{1 + \cos^2 2\theta}{\sin^2 \theta \cos \theta} \, pA \, |F|^2$$

In this expression I is the integrated intensity (7.3.4) and θ is the Bragg angle. Of the other quantities that appear $|F|$ is the **structure amplitude** (7.5.1) which depends upon the types of atoms in the unit cell and their relative positions, A is the **absorbtion factor** (7.4.4) and p is a factor which takes into account the possible number of reflections from a single crystal that are superimposed to form a powder line and is called the **multiplicity factor** (7.4.6). When, as is usual, the specimen in a counter diffractometer is in the form of a flat plate, the focusing introduced by reflection from the plate makes the absorption independent of θ; the absorption decreases the intensities of all diffracted beams by the same factor and therefore does not affect the *relative* intensities. Under these circumstances the absorption factor, A, can be disregarded in the calculations.

7.4.3   The trigonometrical factor

The expression $(1 + \cos^2 2\theta)/\sin^2 \theta \cos \theta$ is called the **trigonometrical factor** and does, in fact, take into account two factors. The term $1 + \cos^2 2\theta$ allows for the partial polarization of the reflected beam and can be used only if the incident beam is unpolarized. This is generally considered to be true for the characteristic emission from an X-ray tube. It is not true if crystal-reflected radiation (3.5.1) is used, and a general expression for the polarization correction for such radiation has been given by Azároff (1955). This expression reduces to $(1 + \cos^2 2\theta_1 \cos^2 2\theta_2)/(1 + \cos^2 2\theta_1)$ for the value of the polarization factor after crystal-reflected radiation has been diffracted by a powder specimen. The Bragg angle of the crystal-reflected radiation is θ_1 and that of the specimen-reflected radiation is θ_2. In this expression it is assumed that intensities are measured in the plane of incidence and that the rotation axis of the specimen is parallel to the reflecting planes of the crystal monochromator.

Three factors contribute to the appearance of the term $\sin^2 \theta \cos \theta$ in the denominator of the trigonometrical factor. The first, which applies to both single-crystal and powder photographs, is that the reciprocal points pass through the surface of the reflecting sphere (2.2.3) at different rates. If the reciprocal lattice is rotating with an angular velocity ω then a point P (fig. 2.14) will move with a velocity ω. OP which is proportional to $2\omega \sin \theta$. The component of this velocity perpendicular to the surface of the reflecting sphere is proportional to $2\omega \sin \theta \cos \theta$ and therefore the integrated intensity, which is proportional to the time spent in the reflecting position, varies as $1/(\sin \theta \cos \theta)$. Secondly, the total intensity of a powder line is spread out over a curve and only a finite length of this curve is measured; this also introduces a factor of $1/(\sin \theta \cos \theta)$ into the expression for the integrated intensity. Finally, the probability that a particular set of planes will pass through the reflecting position is proportional to $\cos \theta$. The product of these three factors is proportional to $1/(\sin^2 \theta \cos \theta)$.

Values of the trigonometrical factor are given in table 7.

7.4.4   The absorption factor

The intensity of reflection depends upon the linear absorption coefficient μ, and for a cylindrical powder specimen the calculations can become quite complicated. The relation between A and μ has been calculated by Bradley (1935a) and put into a form suitable for general use. He evaluated the quantity, A, that appears in the formula in (7.4.2) in terms of μr where r is the radius of the specimen. His results have been verified by graphical methods by Taylor (1945). Values of the absorption factors for cylindrical specimens, together with those for spheres, have been tabulated by Peiser, Rooksby and Wilson (1955). The value of μ to be used is that for the complete specimen and not that for the diffracting material only. If a binding agent is used, the proportion of this in the complete specimen must be known. This proportion can be found, for example, by weighing the specimen, dissolving away the binding agent with a suitable solvent, and weighing the remainder. As explained earlier (3.4.1), the absorption coefficient is the product of the mean density (which can be obtained from the weight and dimensions of the specimen) and the weighted mean of the mass absorption coefficients. From the value of μr so obtained A can be plotted as a function of θ, and the values at any particular angle may be obtained by interpolation.

Taylor (1944a) and Brindley (1945) have shown that this treatment is not perfectly satisfactory unless the specimen is homogeneous. The intensities of the reflections from a particular crystal in a powder specimen depend not only on the mean value of μ but also on the value μ for the crystalline material itself. Taylor calls the latter factor the **microabsorption factor** and gives tables of its values as a function of $(\mu - \bar{\mu})\,a$, where μ is the linear absorption coefficient of the diffracting material, $\bar{\mu}$ is the mean value for the specimen, and a is the effective mean radius of the particles of powder (Taylor, 1944b). If this last quantity cannot easily be determined, it is probably best to use Bradley's figures; the results will then be only approximately correct, but, as Taylor points out, the error is equivalent to a change in the temperature factor (7.5.3) and so will not matter for most work.

7.4.5   Extinction

The accuracy of the calculation of intensities is limited also by an effect known as **extinction**; this occurs because the incident X-ray beam is affected by the process of diffraction as explained below.

X-rays are scattered by an atom because it is set into a state of vibration by the X-rays, and each electron in it then acts as a source of radiation. The wave scattered by a plane of atoms is the resultant of waves scattered with various phase differences, the combined effect being a wave with a phase difference of 90° with respect to the incident beam. Now from fig. 7.8 it can be seen that the X-rays reflected from a given set of lattice planes at an

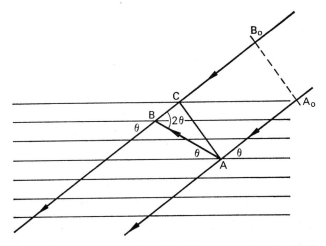

Fig. 7.8   Double reflection of X-rays, which causes primary extinction

angle θ must also be incident at an angle θ on the other side of the same planes, and therefore reflection will again take place. At both reflections there will be a change of phase of 90°, and therefore there will be a total change of phase of 180°. In fig. 7.8, A and B are two points where two such reflections have taken place. Consider an incident ray B_0B which passes through B. If C is the foot of the perpendicular from A on to B_0B, the difference between the paths A_0AB and B_0B is $AB - CB$. This is equal to $AB(1 - \cos 2\theta)$, which equals $4d \sin \theta$ (for $AB = 2d/\sin \theta$). But $2d \sin \theta = \lambda$, and therefore, if there were no change of phase on reflection, the ray reflected at B would be in phase with the incident ray B_0B. This will be true, whatever plane B is in. Since, however, there is change of phase of 180°, the doubly reflected ray will be 180° out of phase with the incident ray and so will reduce its intensity. There will of course be further reflections which may reinforce the incident beam, but since the proportion of radiation each time is less than unity—and usually much less than unity—the total effect is always a decrease in intensity. The effect will be large, however, only if the reflection is an intense one; that is, the incident beam will be reduced in intensity when it passes through the crystal at a reflecting angle. This effect will occur only in regions of a crystal that are perfect; it is known as **primary extinction**.

If the crystal is not perfect, the problem is more difficult. The crystal has to be treated as if it were composed of a number of small perfect blocks slightly different in orientation, as shown in fig. 7.9. Such a crystal is said to

Fig. 7.9   Mosaic structure in a crystal

have a **mosaic structure**. Primary extinction in each block will cause the rays passing through it at a reflecting angle to be weakened, but such rays will not be at the correct angle for reflection by a block below; also, the rays which are reflected by the block below will not have been weakened by primary extinction in the block above. Thus the reflections from a crystal with a mosaic structure should not be greatly affected by primary extinction, if the regions which are effectively perfect are small. The formulae given in 7.4.2 should then be fairly accurately obeyed. Since, however, reflection from a small crystal takes place over a range of angles, rays reflected from the lower blocks of crystal may still be affected somewhat by primary extinction in the upper blocks. This effect, the shielding of the lower blocks by the upper ones, is known as **secondary extinction**. Obviously it will be less important as the crystal becomes less perfect.

One method of allowing for secondary extinction is to add to the absorption coefficient a quantity proportional to the true intensity of reflection. Since the constant of proportionality is unknown and is likely to vary greatly from specimen to specimen, its determination must be carried out on the specimen for which the intensities are to be measured. The experimental work, however, is rather complicated, and it is more usual to try to select a crystal sufficiently small to render extinction negligible. Powder photographs are relatively free from extinction effects, for in general the particles in the specimen can be made so small that no large blocks of ideal crystals are present. If the crystals of a particular substance are plate-like, limited extinction effects may be produced.

7.4.6   The multiplicity factor

The multiplicity factor, p, appears in the formulae for powder photographs because each line is produced by reflections from sets of planes parallel to all the faces of the same crystal form (1.1.4). Each reflection on an oscillation photograph is produced by one set of parallel planes, the crystal being oscillated so that it passes through the reflecting position. Fig. 7.10(a) shows a crystal with one set of planes in the correct orientation to reflect the incident beam. Fig. 7.10(b) shows the same crystal turned through 180°, and it can be seen that the same planes will again reflect the incident beam in the same direction. Thus if the crystal is being rotated through 360° about an axis perpendicular to the plane of the paper, the reflection in this direction will occur twice. In other words, for a single set of parallel planes the multiplicity factor for a rotation photograph is 2. This is the simplest case. If the symmetry is high and the crystal is rotated about an axis of symmetry, several reflections of the same form will be superimposed.

The value of p depends upon the indices of the reflecting planes and on the symmetry of the crystal. For example, the reflections of the form {100} for a cubic crystal are 100, 010, 001, $\bar{1}$00, 0$\bar{1}$0 and 00$\bar{1}$; the multiplicity factor here is 6. The number of sets of reflecting planes is of course only 3, since the faces such as (100) and ($\bar{1}$00) are parallel. Nevertheless the multiplicity factor is 6 because, as shown in the previous paragraph, each set of planes gives two reflections. For {111} the possible arrangements are 111, $\bar{1}$11, 1$\bar{1}$1, 11$\bar{1}$, 1$\bar{1}\bar{1}$, $\bar{1}$1$\bar{1}$, $\bar{1}\bar{1}$1 and $\bar{1}\bar{1}\bar{1}$; the multiplicity factor is 8. The factor is greater when indices are non zero or unequal and reaches a maximum of 48. For crystal systems other than cubic the value of p will generally be lower. Values for all the 11 Laue symmetry groups are given in table 8. It should be noted that the multiplicity factor is always equal to the number of faces in the crystal form.

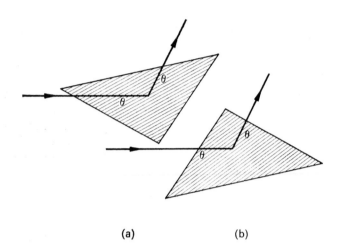

(a)   (b)

Fig. 7.10(a) and (b)   Crystal in two positions differing by 180°; X-rays reflected from the same set of lattice planes

There is a complication in the case of cubic, tetragonal, hexagonal and trigonal systems. Within some Laue groups of these systems, planes of more than one form (1.2.5) have the same spacing (1.3.3) and on a powder photograph the reflections from these forms will be superimposed. The contributions to the intensity of the resulting powder line will thus arise from sets of planes not all of which produce the same intensity of reflection. Under these circumstances it is necessary to know how many of the co-operating planes belong to each of the contributing forms.

In table 8 the number of sets of cooperating planes, that is, the multiplicity factor, in each form for every Laue symmetry group is given. Enclosing two numbers in a rectangle means that, although the numbers correspond to different forms, the spacing of the two forms is the same. Thus there are two different general forms in the crystal class $\bar{3}$, each with six sets of cooperating planes all with the same spacing, but there is only one general form of twelve sets of planes in the class $\bar{3}$m.

As an illustration consider the two sets of general forms {01$\bar{2}$}, {10$\bar{2}$} and the two sets {25$\bar{4}$}, {52$\bar{4}$} of the crystal class $\bar{3}$ in conjunction with the general forms {10$\bar{2}$} and {25$\bar{4}$} of the class $\bar{3}$m. Rhombohedral indices *pqr* are assumed and the six forms are shown in the stereograms of fig. 7.11. As shown in fig. 7.12(a) the plane (01$\bar{2}$) will cut the *x, y* and *z* axes at points which are distant ∞, *b*, $-c/2$ from the origin respectively and similarly the points of intersection of the plane (10$\bar{2}$) with the axes are *a*, ∞, $-c/2$ from the origin (fig. 7.12(b)); for the rhombohedral lattice *a = b = c*. The $\bar{3}$

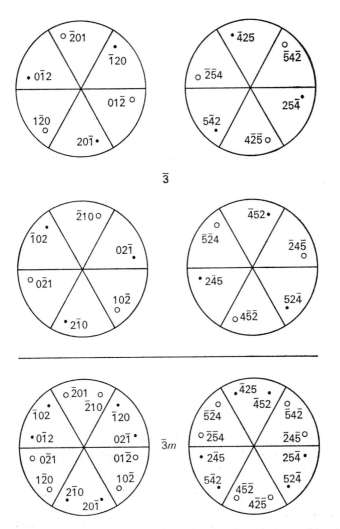

Fig. 7.11   Stereograms of general forms in two trigonal classes (rhombohedral indexing)

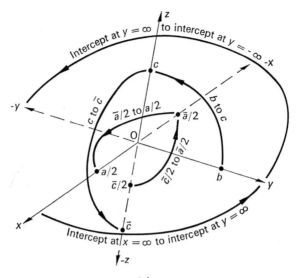

(a)

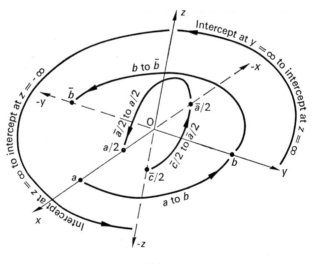

(b)

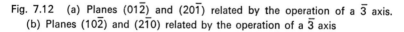

Fig. 7.12   (a) Planes (01$\bar{2}$) and (20$\bar{1}$) related by the operation of a $\bar{3}$ axis.
(b) Planes (10$\bar{2}$) and (2$\bar{1}$0) related by the operation of a $\bar{3}$ axis

axis is the body diagonal [111] through 0 making equal angles with Ox, Oy, Oz; operation of the three-fold rotation axis will bring b to c, $-c/2$ to $-a/2$ and the infinite intercept on the x axis to an infinite intercept on the y-axis (fig. 7.12 (a)). From the same figure it can be seen how inversion then brings c to $-c$, $-a/2$ to $a/2$ and $\infty(y)$ to $-\infty(y)$; hence the plane which cuts the axes in these points has indices $(20\bar{1})$ and therefore the planes $(01\bar{2})$ and $(20\bar{1})$ belong to the same form. Continued repetition of the $\bar{3}$ operation will give the four other planes (or faces) of this form. Similar considerations (fig. 7.12(b)) show that $(10\bar{2})$, $(2\bar{1}0)$ and the other four related faces belong to the second form (fig. 7.11), and that it is impossible by the operation of a $\bar{3}$ axis to bring the faces of the first form into coincidence with those of the second. However, if as in class $\bar{3}m$, a mirror plane

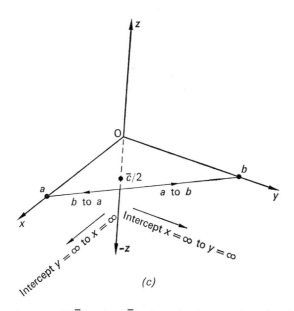

Fig. 7.12   (c) Planes $(01\bar{2})$ and $(10\bar{2})$ related by the operation of a mirror plane in $\bar{3}m$. The mirror plane contains [111] and Oz; $\bar{c}/2$ is not affected

contains both the $\bar{3}$ axis and the z axis then $(01\bar{2})$ will reflect into $(10\bar{2})$, and so on, giving in all twelve faces of the same form (figs. 7.11 and 7.12(c)).

   Again from fig. 7.11 and figs. 7.13(a), (b) and (c) it can be seen that in class $\bar{3}$ the sets of faces $(25\bar{4})$, $(4\bar{2}5)$ . . . and $(52\bar{4})$, $(4\bar{5}2)$ . . . cannot be related to each other by the $\bar{3}$ axis, but that the insertion of the mirror plane in $\bar{3}m$ relates $(25\bar{4})$ and $(52\bar{4})$, and so on, and therefore brings all twelve faces into the same form.

   Finally, from table 8 it can be seen that in the Laue group, m3, of the cubic system, $\{hkl\}$ and $\{lkh\}$ constitute two different forms of twenty-four faces (that is, twenty-four sets of planes); all forty-eight sets of planes will have the same spacing. However, in the Laue group m3m there is just one

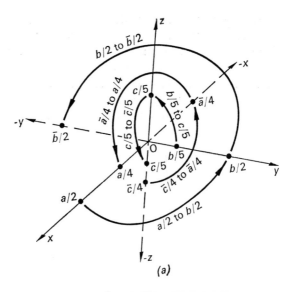

(a)

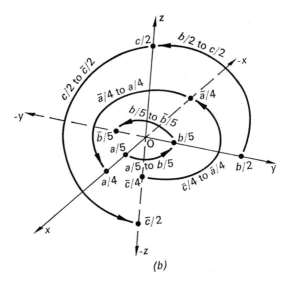

(b)

Fig. 7.13   (a) Planes (25$\bar{4}$) and (4$\bar{2}\bar{5}$) related by the operation of a $\bar{3}$ axis.
(b) Planes (5$\bar{2}$4) and ($\bar{4}\bar{5}$2) related by the operation of a $\bar{3}$ axis

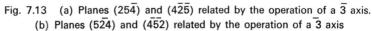

form, but this form has forty-eight related sets of planes. The diagram
(fig. 7.14) shows, for example, that the faces (321) and ($\bar{1}$23), which are

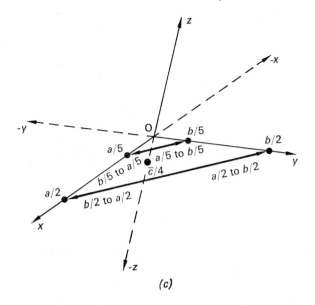

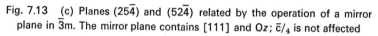

Fig. 7.13   (c) Planes (25$\bar{4}$) and (52$\bar{4}$) related by the operation of a mirror plane in $\bar{3}$m. The mirror plane contains [111] and Oz; $\bar{c}$/4 is not affected

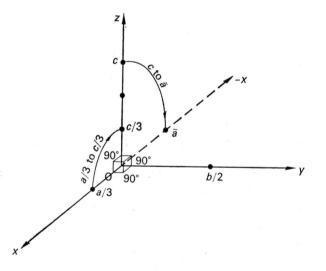

Fig. 7.14   {321} and {123} are not related in the Laue group m3, but are related in m3m by the operation of the tetrad symmetry element. The face (321) becomes ($\bar{1}$23)

parallel to planes of the same spacing, belong to different forms of the Laue group m3 but to the same form of the group m3m. In the Laue group m3, Oy is not a tetrad axis and therefore {321} and {123} are unrelated to each other; in the Laue group m3m, Oy is a tetrad axis and the faces of {321} and {123} are therefore related by the operation of this particular symmetry element.

It should also be noted that in the cubic system all reflections with the same value of $h^2 + k^2 + l^2$ have the same Bragg angle. For example, a line with $h^2 + k^2 + l^2 = 9$ may be composed of the reflections 300 (with a multiplicity factor of 6) and 221 (with a multiplicity factor of 24).

7.5 The Structure Factor

7.5.1 Definition of structure factor and structure amplitude

In the general equation for the relative intensity of powder reflections given in (7.4.2), a quantity $|F|$ appeared which we called the structure amplitude. This is the quantity that is measured experimentally and is, in fact, the modulus of the **structure factor**, F. The structure factor itself is generally complex with real and imaginary parts A and B such that

$$F = A + iB$$

The modulus $|F|$ is therefore given by

$$|F|^2 = A^2 + B^2$$

and the **phase angle,** α is given by the relation

$$\tan \alpha = B/A$$

The physical significance of the structure factor is that its modulus (the structure amplitude) is the numerical value of the amplitude of a scattered wave and that its phase angle is the phase of the wave relative to that of a wave scattered by a point at the origin. Of these two quantities the modulus is observed but the phase angle is not.

The structure amplitude is defined as the ratio of the amplitude of the radiation scattered in the direction of a reflection hkl by all the atoms in one unit cell to that scattered by one electron under the same conditions. In order, therefore, to calculate the structure amplitude for a given reflection we need to know (i) to how many electrons each atom is equivalent and (ii) how the waves from the different atoms combine. That is, we need to know the structure factor F. The quantity (i) is the **atomic scattering factor** and (ii) leads to the **geometrical structure factor.**

The geometrical structure factor is also sometimes called the structure factor and we have, therefore, added the word geometrical to make the term more precise. The geometrical structure factor is an expression which is characteristic of a given space group and is, in general, a complex quantity which can be represented by real and imaginary components

$$A' = \sum \cos 2\pi(hx + ky + lz)$$
$$B' = \sum \sin 2\pi(hx + ky + lz)$$

where x, y and z are the co-ordinates of a set of points related by the space-group symmetry elements. The product of the geometrical structure factor and the atomic scattering factor of atoms lying on a set of equivalent positions in the space group is also a complex quantity and gives the contribution of that set of atoms to the structure factor. The modulus of this product will give the contribution of the set of atoms to the structure amplitude and hence to the intensity. The quantities introduced here are discussed in section 7.5.4.

7.5.2   The atomic scattering factor

If atoms were merely points their scattering factors would be equal to the number of electrons they contain, that is to their atomic numbers; since they are not points, their scattering factors are less than their atomic numbers, the value depending on the spacing of the planes giving the reflection concerned; this can be seen from fig. 7.15. The circle represents a

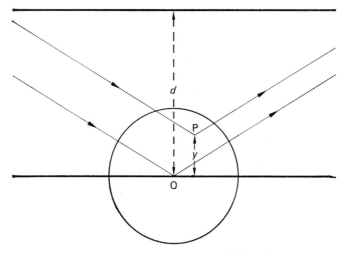

Fig. 7.15   Scattering by an atom of finite size

section of an atom lying on one of a set of planes of spacing d. A ray scattered from a point such as P will have a phase difference of $2\pi y/d$ with respect to a ray scattered from a point at the centre. The total scattering from an atom of known electron-density distribution can be obtained by integrating the scattering for all values of y. The calculations are difficult but it can be seen that, since the distribution of values of y is fixed for any one atom, the atomic scattering factor must be a function of d and hence of $\sin\theta/\lambda$ since $1/2d = \sin\theta/\lambda$. Atomic scattering factors are usually given in tabular form, and graphical interpolation may be used to obtain the values for particular reflections (fig. 7.16).

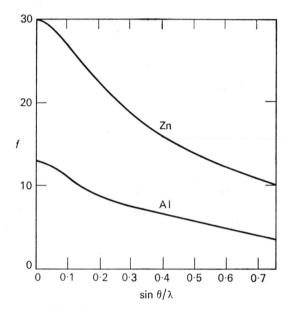

Fig. 7.16   Typical curves of atomic scattering factor

When the wavelength of the incident radiation is far removed from an absorption edge the atomic scattering factor, f_0, of an atom at rest depends, as we have just seen, upon $\sin\theta/\lambda$; that is the scattering factor decreases as θ increases. However, when the wavelength of the incident X-ray beam is close to an absorption edge (K, L, M...) of a scattering atom, the atomic scattering factor becomes a complex quantity and can be written in the form

$$f_0' = f_0 + \Delta f' + i\,\Delta f''$$

In this expression, f_0 is the normal scattering with a wavelength far removed from any absorption edge and the quantities $\Delta f'$ and $\Delta f''$ are correction terms which arise from **anomalous dispersion** effects in the neighbourhood of an absorption edge. The values of the terms $\Delta f'$ and $\Delta f''$ depend upon the wavelength of the incident radiation but may be considered to be independent of θ, because the tightly bound electrons which are responsible for their presence are concentrated in a small volume near the atomic nucleus. Thus, since $\Delta f'$ and $\Delta f''$ are independent of θ and since f_0 falls off rapidly as θ increases, the effect of the correction terms becomes relatively greater at the higher scattering angles.

The correction term $\Delta f'$ applies to the real part of the complex scattering factor f_0' so that the value of the real component is $f_0 + \Delta f'$, the phase of the imaginary component $\Delta f''$ leading that of the real component by $\pi/2$. Hence if $\Delta f''$ is not zero the anomalously scattered radiation undergoes a phase change. In this case it can be shown that Friedel's law breaks down for non-centrosymmetric structures; that is

$$|F(hkl)| \neq |F(\bar{h}\bar{k}\bar{l})|$$

for non-centrosymmetric crystals. The change of phase does not affect the validity of Friedel's law as far as centrosymmetric structures are concerned. When $\Delta f''$ is zero the anomalous scattering takes place without a phase change and this occurs when the wavelength of the anomalously scattered radiation is longer than that of an absorption edge. Under these conditions Friedel's law is obeyed for both centrosymmetric and non-centrosymmetric structures. It should be noted that, even if $\Delta f''$ is zero, $\Delta f'$ may still be significant.

Values of $\Delta f'$ and $\Delta f''$ for the elements calcium to curium have been compiled for the $K\alpha$ radiations of chromium, copper and molybdenum by Dauben and Templeton (1955), and an extended list is given in International Tables, Vol. III (1962). Cromer (1965) later tabulated the values for the elements from neon to californium for the $K\alpha$ radiations of chromium, iron, copper, molybdenum and silver.

One of the first applications of anomalous dispersion was by Jones and Sykes (1937) in distinguishing between atoms of nearly equal atomic number in the alloy CuZn. The copper atoms scatter $ZnK\alpha$ radiation anomalously without phase change, and with this radiation Jones and Sykes detected the superlattice (7.6.2) lines of CuZn on a powder photograph of the alloy; with other radiations the difference between the scattering factors of copper and zinc is so small that the superlattice lines cannot be detected.

With the accuracy attainable by counter techniques it is now routine to measure the effects of anomalous dispersion in X-ray diffraction studies. In such studies, usually with single crystals (Ramachandran, 1964; Kaelble, 1967), anomalous scattering without phase change can be used to solve both centrosymmetric and non-centrosymmetric structures. If the phase does change, the phase angles of the reflections from non-centrosymmetric structures can be determined.

7.5.3   The temperature factor

Atomic scattering factors are derived from the electron distributions of atoms at rest. At ordinary temperatures, however, the thermal vibrations of the atoms will cause them to occupy a larger volume than they would at rest and will make the scattering still smaller. It can be shown theoretically (Debye, 1914) that the relation between the scattering factor f of a vibrating atom to that of an atom at rest, f_0, is

$$f = f_0 \exp\left(-B_t \sin^2 \theta/\lambda^2\right) \qquad \ldots 7.1$$

The quantity B_t is not strictly a constant; it will be different for different atoms in the same crystal and will vary with direction in the crystal. Attempts to calculate B_t for particular cases have not led to any great degree of success, and for most work it is sufficient to regard it as a constant which can be used to make the calculated intensities of reflection agree with those observed over the complete range of θ. The thermal parameter is usually denoted by B but to avoid confusion with the imaginary part B of the structure factor, we have referred to it as B_t.

7.5.4   The geometrical structure factor

We have already seen that the structure factor is the quantity that determines how the intensities of the reflections from a crystal depend on the atomic arrangement within the unit cell. This dependence can be calculated precisely as a function of the atomic positions and the indices of the particular reflection concerned. The general formula for the structure amplitude $|F|$ is

$$|F|^2 = \{\sum_n f_n \cos 2\pi(hx_n + ky_n + lz_n)\}^2$$
$$+ \{\sum_n f_n \sin 2\pi(hx_n + ky_n + lz_n)\}^2 \qquad \ldots 7.2$$

where x_n, y_n, z_n are the co-ordinates of an atom expressed as fractional values of the cell edges, which are taken as axes of reference, and f_n is the atomic scattering factor for that atom for the value of $\sin \theta/\lambda$ associated with the indices hkl. The summation is taken over all the atoms in the unit cell. The quantities x_n, y_n, z_n, are called the **structural parameters** of the atom n.

Suppose there are two equal atoms in the unit cell, at $(x_1 y_1 z_1)$ and

$(x_2 y_2 z_2)$. Then, since the atomic scattering factor is the same for both atoms, equation (7.2) reduces to

$$|F|^2 = f^2[\{\cos 2\pi(hx_1 + ky_1 + lz_1) \\ + \cos 2\pi(hx_2 + ky_2 + lz_2)\}^2 \\ + \{\sin 2\pi (hx_1 + ky_1 + lz_1) \\ + \sin 2\pi(hx_2 + ky_2 + lz_2)\}]^2 \qquad \ldots 7.3$$

As an example of such a structure, we may take α-iron; the co-ordinates of the atoms in this case are (000) and $(\tfrac{1}{2}\tfrac{1}{2}\tfrac{1}{2})$. Inserting these values in equation (7.3), we get:

$$|F|^2 = f^2\left[\left\{1 + \cos 2\pi\, \frac{(h + k + l)}{2}\right\}^2 \\ + \left\{0 + \sin 2\pi\, \frac{(h + k + l)}{2}\right\}^2\right]$$

or, since $\sin \pi (h + k + l) = 0$,

$$F = f\left\{1 + \cos 2\pi\, \frac{(h + k + l)}{2}\right\} \qquad \ldots 7.4$$

It can be seen that if $h + k + l$ is odd, the expression is zero so that such reflections will be absent. When $h + k + l$ is even, $|F| = 2f$, which means that the two atoms are scattering in phase for these reflections and the intensity is the maximum possible.

Equation (7.2) is often written as

$$|F|^2 = A^2 + B^2 \qquad \ldots 7.5$$

where $\qquad A = \sum_n f_n \cos 2\pi(hx_n + ky_n + lz_n) \qquad \ldots 7.6a$

and $\qquad B = \sum_n f_n \sin 2\pi(hx_n + ky_n + lz_n) \qquad \ldots 7.6b$

Suppose that we consider only one set of equivalent points, x, y, z, in a space group, and that we take the atomic scattering factor f as unity. Then

$$A' = \sum \cos 2\pi(hx + ky + lz)$$

and $\qquad B' = \sum \sin 2\pi(hx + ky + lz) \qquad \ldots 7.7$

the summations being over all the atoms in the set. The expressions (7.7) are the geometrical structure factors for the space group. Lonsdale (1936) has found it convenient to denote them by A and B, although A and B usually represent the components of the structure factor F, as shown in equation (7.5), and are therefore physical quantities, not trigonometrical expressions.

The importance of this factor is that, although it is always possible to calculate structure amplitudes from equation (7.2), it is usually much simpler to collect together all the atoms related by symmetry. For example, suppose there is a centre of symmetry at the origin; that is, for an atom with co-ordinates (x, y, z), there is another with co-ordinates $(\bar{x}, \bar{y}, \bar{z})$. Then $A = 2f \cos 2\pi(hx + ky + lz)$ and $B = 0$. If there are several pairs of such atoms,

$$A = 2 \sum_n f_n \cos 2\pi(hx_n + ky_n + lz_n)$$

and $B = 0$. This expression is much simpler than (7.2) and it will be seen that the presence of a centre of symmetry at the origin is a great help in simplifying the calculations. Before the advent of X-ray analysis, there was no advantage to be gained by choosing a centre of symmetry as the origin of the unit cell, and the conventional choice of origin for some of the centrosymmetric space groups did not coincide with a centre of symmetry. While X-ray workers have attempted as far as possible to maintain the older crystallographic conventions, for these space groups it is convenient to transfer the origin to a centre of symmetry in order to make use of the fact that B is then equal to zero.

If there are also other elements of symmetry, the formulae may be further simplified. For example, if there is a plane of symmetry through the origin, parallel to (010), there must be equivalent atoms at (x, y, z) and $(x, \bar{y}, z)$. The centre of symmetry will add atoms at $(\bar{x}, \bar{y}, \bar{z})$ and $(\bar{x}, y, \bar{z})$, and if we evaluate A and B for unit atoms at these points, we find that $B = 0$, as before, and

$$A = 2f\{\cos 2\pi(hx + ky + lz) + \cos 2\pi(hx - ky + lz)\}$$
$$= 4f \cos 2\pi(hx + lz) \cos 2\pi \, ky \qquad \qquad \ldots 7.8$$

We can thus use one expression for four atoms at a time.

As the number of equivalent points becomes larger in space groups of higher symmetry, the geometrical structure factors also become more complicated. Expressions for the geometrical structure factors for the 230 space groups are given by Lonsdale (1936) and in the International Tables, Vol. I (1952). These tables should not be used uncritically; Lonsdale (1950) has pointed out several mistakes that occurred in her own tables. Users should use the tables as a check on their own derivations, and therefore should understand the general principles as set out in this chapter and in Lipson and Cochran (1966).

A slight difficulty arises in applying these formulae when atoms lie in special positions. For example, if an atom lies on the plane of symmetry in the arrangement considered above, its position can be expressed as $(x, 0, z)$ and it might be expected that the value of A will be obtained by putting

$y = 0$ in expression (7.7). This is not so; this expression is valid only for four atoms in general positions, and the atom on the plane of symmetry is in a special position and has only one other equivalent to it. The correct expression is therefore

$$A = 2 \cos 2\pi(hx + lz).$$

Care must be taken over this if a crystal contains atoms in both special and general equivalent positions; a useful check is that the maximum possible value of the geometrical structure factor for any set of atoms must not be greater than the number of atoms in the unit cell, and is usually equal to this number.

7.5.5 Absent reflections due to lattice type

In general, one cannot easily deduce from the atomic arrangement in a crystal structure whether a particular reflection will be strong or weak. It will be strong if the contributions of the various sets of atoms, as calculated from equations (7.5) and (7.6), are large and of the same sign; and it will be small if these contributions tend to cancel each other. There are, however, some general rules, examples of which have been given in (2.2.2), connected with the lattice and with the symmetry operations of the space group, which can be used to predict the complete absence of certain reflections; and, conversely, it is often possible to determine the space group by knowing which reflections are absent. These absences can be recognized because they are systematic; that is, they are related in a definite way to the indices of the reflections: absences due to the small structure amplitudes will not be related in any simple way to the indices.

We shall, here, confine ourselves to two examples of systematically absent reflections, the first of which will be those produced by a body-centred lattice as in (5.3.1). These particular absences have already been derived in (2.2.2), but now the same result will be obtained using an analytical method.

For a body-centred lattice there will be equivalent atoms at points x_n, y_n, z_n and $\frac{1}{2} + x_n, \frac{1}{2} + y_n, \frac{1}{2} + z_n$. Substituting these values in equation (7.6a), we arrive at the expression

$$A = \sum_n f_n[\cos 2\pi\{hx_n + ky_n + lz_n\}$$
$$+ \cos 2\pi\{h(\tfrac{1}{2} + x_n) + k(\tfrac{1}{2} + y_n) + l(\tfrac{1}{2} + z_n)\}]$$
$$= \sum_n f_n[\cos 2\pi\{hx_n + ky_n + lz_n\}$$
$$+ \cos 2\pi\{hx_n + ky_n + lz_n + \tfrac{1}{2}(h + k + l)\}] \quad \ldots 7.9$$

If $h + k + l$ is odd, the second term will be equal and opposite to the first, and A will be zero. It can be shown in a similar way that B is also zero

8+

when $h + k + l$ is odd, so that the intensities of the reflections with $h + k + l$ odd are zero. Similar rules governing the absence of reflection from other lattices can be worked out from equations (7.6). Note that such absences are always in reflections of the general form hkl, and if examples of all the various types of such reflections are present, the lattice must be primitive.

The second example refers to systematically absent reflections which also occur in reflections of special types—that is, in those which have one or two indices zero. These absences indicate the presence of certain symmetry elements in the space group (2.22). The subject is discussed in International Tables (Vol. I, 1952) and by Buerger (1942, 1956) but as our example we shall consider only the systematically absent reflections arising from the presence of an a-glide plane perpendicular to the c axis. As a result of the operation of this symmetry element an atom at x_n, y_n, z_n will have an equivalent atom at $\frac{1}{2} + x_n, y_n, \bar{z}_n$ (fig. 7.17); it is assumed that the origin of coordinates lies in the glide plane. Substitution of these values in (7.6a) gives

$$A = \sum_n f_n [\cos 2\pi(hx_n + ky_n + lz_n) + \cos 2\pi(h/2 + hx_n + ky_n - lz_n)]$$

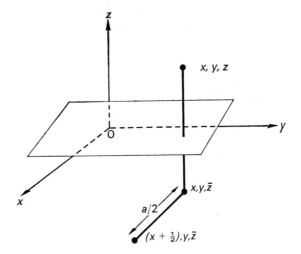

Fig. 7.17   Atoms related by an a-glide plane through the origin and perpendicular to c

When $l = 0$ this reduces to

$$A = \sum_n f_n [\cos 2\pi(hx_n + ky_n) + \cos 2\pi(h/2 + hx_n + ky_n)]$$

If h is odd the cosine terms are equal in magnitude but opposite in sign and

A is therefore zero. Similarly B is zero when $l = 0$ and h is odd, and so the $hk0$ reflections are absent when h is odd.

7.6 Applications

7.6.1 Structure determination

The main reason for measuring the intensity of X-ray reflections is to determine crystal structures. For complicated structures, intensity data from single crystals are essential, but it is possible to use the powder method to analyse comparatively simple structures such as those of many metals and alloys (chapter 8). Normally, it is not possible to obtain much evidence about a crystal structure without a complete analysis, but occasionally it is possible to obtain some information about the structure from a knowledge of the space group and the contents of the unit cell. For example, if a monoclinic organic compound is found to give no reflections of the type $h0l$ with h odd and no reflections of the type $0k0$ with k odd, the space group must be $P2_1/a$. This has four general equivalent positions; hence if the unit cell contains four molecules, these molecules may have no symmetry whatsoever. But if the unit cell contains only two molecules, then in this space group these must lie on centres of symmetry. The molecule must therefore have a centre of symmetry. Such information can be extremely useful to the chemist investigating an unknown compound. It is interesting that although, in general, the presence or absence of a centre of symmetry cannot be inferred from X-ray evidence alone, in special cases such as that just described it can be. Glide planes and screw axes can both be found by systematic absences, and then it follows that there must also be centres of symmetry.

Sometimes it is possible to obtain further information from the shape of the unit cell. Bernal and Crowfoot (1934) and Hargreaves and Taylor (1941) have given some examples of information obtained in this way, without complete structure determination.

7.6.2 Study of superlattice formation in alloys

Many alloys are solid solutions of one metal in another, the different atoms being distributed at random on the available equivalent positions, and the X-ray photograph is similar to that of a pure metal of the same structure. The atomic scattering factor used in calculating the theoretical intensity of reflection would be a weighted mean of the atomic scattering factors of the atoms present. It is possible, however, that the atoms may not be distributed at random; they can still occupy the same sites, but with one type of atom in a particular position within each unit cell. For example, iron is cubic with atoms at 000 and $\frac{1}{2}\frac{1}{2}\frac{1}{2}$; that is, it has a body-centred cubic lattice. The alloy FeAl also has atoms at 000 and $\frac{1}{2}\frac{1}{2}\frac{1}{2}$, but the iron atoms tend to occupy the positions 000 and the aluminium atoms the positions $\frac{1}{2}\frac{1}{2}\frac{1}{2}$. (The structure could equally well be described as having iron atoms at $\frac{1}{2}\frac{1}{2}\frac{1}{2}$ and aluminium atoms at 000; the difference lies only in the choice of origin. The structure has a primitive lattice because it has unlike atoms at 000 and $\frac{1}{2}\frac{1}{2}\frac{1}{2}$; nevertheless it is usually spoken of as body-centred cubic with a **superlattice**. If it is described in this way, it must be remembered that the presence of the superlattice means that the lattice is primitive. Another example occurs in the alloy $AuCu_3$, which is cubic with gold atoms at 000 and copper atoms at $0\frac{1}{2}\frac{1}{2}$, $\frac{1}{2}0\frac{1}{2}$, $\frac{1}{2}\frac{1}{2}0$. (Again, alternative descriptions are possible, depending upon the choice of origin.) This alloy is often described as face-centred cubic with a superlattice; the lattice is, of course, primitive.

The effect of superlattice formation on powder photographs is shown on plate III the first photograph being that of an ordinary body-centred cubic structure; only lines for which $h^2 + k^2 + l^2$ is even (5.3.1) appear, since if $h^2 + k^2 + l^2$ is even, $h + k + l$ is also even. The second photograph, however, is that of an alloy with a superlattice; the structure is not truly body-centred, and lines with other values of $h^2 + k^2 + l^2$ can appear. These lines, which are fainter than the main lines, are called **superlattice lines**.

Superlattice formation often occurs in alloy systems. In order to interpret any photographs showing the effect, it is necessary to be able to calculate both the positions and intensities of the superlattice lines, and to do this, the formulae given in this chapter must be applied.

7.6.3   Determination of proportions of constituents

It is sometimes possible to estimate the relative amounts of two or more constituents in a mixture by comparing the intensities of the lines of each pattern on a powder photograph. These intensities will, however, often depend upon factors which cannot be accurately allowed for (7.4.4), and the safest way of using the method is to measure the relative intensities of lines given by a mixture of the constituents in known proportions. Occasionally, however, it may be possible to obtain absolute values of the proportions directly from the intensities. Edwards and Lipson (1943) for example, found the relative amounts of cubic and hexagonal cobalt in a solid specimen by the direct comparison of the intensities of two chosen lines.

CHAPTER 8

The Determination of Crystal Structures

8.1 Comparison of Single-crystal and Powder Methods

It is usual to determine crystal structures with the aid of single crystals but when these are not available solution by powder methods can be attempted. For this reason it is important to be aware of the limitations of powder methods and of the positive advantages that may accrue from their use.

In the first place, let us consider unit-cell dimensions; they can be determined purely objectively from single crystals, but from a powder photograph the determination is largely an art (chapter 5). Nevertheless, if the unit cell can be found, high accuracy is far easier to obtain from powder photographs (chapter 6) and this may sometimes be of invaluable significance.

Space-group information is also obtained much more directly by single-crystal methods. Lattice absences and glide-plane absences (8.2.1) may be reasonably clear from powder photographs but screw-axis absences are difficult to observe because there are so few of them and they may be concealed by other reflections.

It is possible to obtain a complete set of intensities from single-crystal photographs that do not contain any overlapping reflections. This will not be possible from a powder photograph which contains seriously overlapped reflections. However, those reflections that are clearly resolved, whether they are lines on a powder photograph or records from a diffractometer, can be assessed much more accurately than can single-crystal reflections; a one-dimensional analysis will give all the information required (8.2.2, 8.3.3). Moreover the measurements may have higher physical significance since, with a fine powder, extinction (7.4.5) can be eliminated. Absorption corrections (7.4.4) can also be more easily made.

Finally, the powder method has the advantage that, although individual reflections may be lost, it gives the overall diffraction pattern; in principle, all the interatomic distances in the crystal can be obtained from it whatever the resolution. Considerations of this sort led to the idea of the Patterson synthesis for crystal structures. In his synthesis Patterson used as Fourier coefficients the squares of the structure amplitudes of the reflections; these quantities are directly related to the observed intensities and can always be measured (7.5.1). Patterson showed that this synthesis could give direct evidence about atomic positions with no preliminary assumptions (Lipson and Cochran, 1966).

8.2 Preliminary Procedures

8.2.1 Determination of unit cell and space group

The first step in solving a crystal structure is to find the unit-cell dimensions; this will also define the crystal system to which the substance belongs. If, on the powder pattern, the individual reflections are either resolved or do not seriously overlap, the cell dimensions can be obtained by applying

Conditions for General Reflections hkl to be Present

Lattice type	Condition
A face-centred	$(k + l) = 2n$
B face-centred	$(h + l) = 2n$
C face-centred	$(h + k) = 2n$
F (all face-centred)	h, k and l homogeneous (that is, all odd or all even)
I (body-centred)	$(h + k + l) = 2n$
R (rhombohedral, indexed on hexagonal axes)	$(-h + k + l) = 3n$ for obverse orientation of rhombohedron $(h - k + l) = 3n$ for reverse orientation of rhombohedron
C or P (hexagonal indexed on rhombohedral axes)	$(p + q + r) = 3n$
P (primitive)	No restriction. But the lattice can be called primitive only after the absence of all of the possible restrictions has been demonstrated

one or other of the methods described in chapter 5. However, when there is considerable overlapping it may be necessary to obtain accurate values of the lattice parameters in order to index such reflections unambiguously; since every line must be accounted for, indices must be assigned to all the reflections whether or not there is overlap.

When indexing has been completed determination of the space group (1.3.1, 1.3.2) can be attempted. On a powder photograph both general and special reflections are recorded, so that information on lattice type and translational elements of symmetry (2.2.2) is included. If there are no systematic absences from the hkl reflections then the lattice is primitive, if there are no hkl reflections with $h + k + l$ odd then the lattice is body-centred, and so on. The analytical method of determining these conditions is illustrated in (7.5.5) and the complete list for all lattices, of the conditions for the general reflections to be *present*, is given in the table on page 222. Note that if the C (hexagonal) and R lattices are indexed on hexagonal and rhombohedral axes respectively, there will be no systematically absent hkl reflections since in both cases the unit cell is primitive.

Similarly, translational elements of symmetry can be obtained by noting systematic absences from special reflections. For example, if there are no $hk0$ reflections with h odd then a glide plane perpendicular to the c-axis giving rise to a translation of $a/2$ parallel to the a axis (7.5.5) is present. Again, a two-fold screw axis parallel to the c axis will be indicated by the systematic absence from the $00l$ reflections of all lines with l odd. A summary of the commonly occurring translational elements of symmetry and the conditions for the associated special reflections to be present is given in the following table. For a complete list, a standard work such as International Tables, Vol. I (1952) should be consulted.

Conditions for Special Reflections to be Present

Element of translation	Symbol	Condition
Glide plane $\perp$ to c, translation $a/2$	a	$hk0$ with h even
Glide plane $\perp$ to c, translation $a/2 + b/2$	n	$hk0$ with $h + k$ even
Glide plane $\perp$ to c, translation $a/4 + b/4$	d	$hk0$ with $h + k$ divisible by 4
Two-fold screw axis $\parallel$ to a, translation $a/2$	2_1 (may be 4_2)	$h00$ with h even
Four-fold screw axis $\parallel$ to c, translation $c/4$	$4_1, 4_3$	$00l$ with l divisible by 4
Three-fold screw axis $\parallel$ to c, translation $c/3$	$3_1, 3_2$ (may be $6_2, 6_4$)	$000l$ (hexagonal) with l divisible by 3

Systematic absences will reveal the presence of glide planes and of screw axes, but they will give no information about mirror planes, rotation axes or centres of symmetry; these symmetry elements do not involve translation. Thus, although the crystal system will be known from the indexing procedure and the lattice type will be known from a study of the general reflections, the space group as determined from the systematic absences only may still be incomplete. In this case it may still be possible to find the crystal structure but longer calculations will be necessary and it is an advantage to know the space group without ambiguity. Methods, such as those described by Lipson and Cochran (1966), for resolving space-group ambiguities are available, but not all are applicable to powder data.

The 230 space groups, together with the conditions that general and special reflections should be present, are listed in the International Tables, Vol. I (1952).

8.2.2   The scaling and temperature factors

It has already been stated (7.4.2) that the relative intensities, I, of the lines on a powder photograph obey the relation

$$I \propto \frac{1 + \cos^2 2\theta}{\sin^2 \theta \cos \theta} pA \, |F|^2$$

In this expression the structure amplitude, $|F|$, is calculated in terms of atomic scattering factors f (7.53) where $f = f_0 \exp\left(-B_t \sin^2 \theta / \lambda^2\right)$. The values of f refer to atoms which are vibrating at the temperature at which the diffraction data are obtained, whereas the values of f_0, which are derived theoretically, refer to atoms at rest. In order that f_0 values, which are known and tabulated, may be used we can write $|F| = |F_0| \exp\left(-B_t \sin^2 \theta / \lambda^2\right)$ and the expression for the relative intensities becomes

$$I \propto \frac{1 + \cos^2 2\theta}{\sin^2 \theta \cos \theta} pA \, |F_0|^2 \exp\left(-2B_t \sin^2 \theta / \lambda^2\right)$$

If we call $|F_0|^2$ the ideal or calculated intensity, I_c, then the observed relative intensities on a powder photograph are given by

$$I \propto \frac{1 + \cos^2 2\theta}{\sin^2 \theta \cos \theta} pA I_c \exp\left(-2B_t \sin^2 \theta / \lambda^2\right)$$

or by

$$\frac{I}{pA} \cdot \frac{\sin^2 \theta \cos \theta}{1 + \cos^2 2\theta} \propto I_c \exp\left(-2B_t \sin^2 \theta / \lambda^2\right)$$

The relation can be rewritten in the form $I' \propto I_c \exp\left(-2B_t \sin^2 \theta / \lambda^2\right)$, that is, in the form I' . constant $= I_c \exp\left(-2B_t \sin^2 \theta / \lambda^2\right)$ where I' are the observed relative intensities of the reflections after corrections have been applied for multiplicity and absorption, and account has been taken of the trigonometrical factor. The constant term is the scaling factor by which I' must be multiplied in order to put the relative intensities on an absolute scale. Expressed in logarithmic form the relation between I' and I_c becomes

$$\log_e (I'/I_c) = \frac{-2B_t \sin^2 \theta}{\lambda^2} - \log_e S$$

where the symbol S now represents the scaling factor. Thus if I_c is known— that is if the structure is known—we can plot $\log_e (I'/I_c)$ against $\sin^2 \theta/\lambda^2$ and the slope of the resulting straight-line graph will give the value of $2B_t$; the intercept on the $\log_e (I'/I_c)$ axis will give the scaling factor, S.

Often, however, at the start of a structure determination not even an approximate solution is known and it would be useful if the thermal para- meter, B_t, could be determined at the outset. Wilson (1942a) showed how this can be done. According to Wilson the average value, $\bar{I}_c$, of the calcu- lated intensities of a large number of reflections is, to a close approximation, given by

$$\bar{I}_c = \sum f_0^2$$

where the summation is taken over all the atoms in the unit cell and the values of f_0 are the scattering factors of the atoms at rest. Consequently if the reflections are divided into five or six groups, containing roughly equal numbers of reflections and covering successive ranges of $\sin^2 \theta$, then for each group $\bar{I}_c = \sum f_0^2$ approximately, where the values of f_0 are those appropriate to the middle of the corresponding range of $\sin^2 \theta$. Further, if $\bar{I}'$ is the mean value of the corrected relative intensities in a group, we can now write down for that group the approximate relation

$$\log_e (\bar{I}'/\sum f_0^2) = -2B_t \sin \theta/\lambda^2 - \log_e S$$

The points obtained by plotting $\log_e (\bar{I}'/\sum f_0^2)$ against the mean value of $\sin^2 \theta/\lambda^2$ for each group will lie roughly on a straight line the slope of which will give the value of $2B_t$; from the intercept on the axis of $\log_e$ $(\bar{I}'/\sum f_0^2)$ the value of the scaling factor, S, can be determined and from this the intensities can be placed on an absolute scale. In calculating $\bar{I}'$ all possible reflections must be included, whether they are present or acci- dentally absent, but not if they are systematically absent.

Strictly, the theory is valid only for three-dimensional data, and it can therefore be justifiably applied to reflections on a powder photograph. Primary and secondary extinction effects introduce errors, but they do not seriously affect the intensities of powder lines. Finally, the method is not entirely satisfactory when the structure contains atoms in special positions

8*

(Foster and Hargreaves, 1963a,b); this may be a serious problem in powder work since many of the structures which can be solved from powder data do have atoms in special positions.

With simple structures in which some or all of the atoms are in special positions it is sometimes possible to find the temperature factor by an alternative method. Consider, for example, the structure of the alloy CdMg (Steeple, 1952); the symmetry is orthorhombic with space group Pmma, and there are two cadmium and two magnesium atoms in the unit cell. The cadmium atoms are at $\frac{1}{4}, 0, z$ and $\frac{3}{4}, 0, z$, and the magnesium atoms are at $\frac{1}{4}, \frac{1}{2}, z$ and $\frac{3}{4}, \frac{1}{2}, z$. For atoms in the eight general positions produced by the space group Pmma the geometrical structure factor of the $hk0$ reflections is, for h even,

$$A' = 8 \cos 2\pi hx \cos 2\pi ky \cos 2\pi lz, \qquad B' = 0$$

and for h odd,

$$A' = -8 \sin 2\pi hx \sin 2\pi ky \sin 2\pi lz, \qquad B' = 0$$

But the cadmium and the magnesium atoms in the CdMg structure occupy special positions at junctions of the two mirror planes so that the multiplicity of each atom is two, and not eight, and therefore for these atoms the geometrical structure factor reduces to a quarter of these expressions. Substitution of the atomic coordinates of the two different types of atoms, presumed to be at rest, gives

$$F_0(hk0) = 2(f_0{}^{Cd} + f_0{}^{Mg}) \quad \text{when } h \text{ and } k \text{ are both even}$$

and

$$F_0(hk0) = 2(f_0{}^{Cd} - f_0{}^{Mg}) \quad \text{when } h \text{ is even and } k \text{ is odd}$$

Thus $I_c(=|F_0|^2)$ can be determined for each of these reflections and the corresponding values of $\log_e(I'/I_c)$ can be plotted against $\sin^2 \theta/\lambda^2$; as we have already shown, values of both B_t and the scaling factor can be obtained from the graph drawn through these points. As an example, a set of data is shown in the table below and is plotted in fig. 8.1.

$hk0$	$\sin^2 \theta/\lambda^2$ Å^{-2}	$|F_0|$	I_c	I'	$\log_e (I'/I_c)$
010	0·024	62·4	3894	3600	1·9200
200	0·040	92·6	8575	10609	0·2126
210	0·065	54·2	2938	2401	1·7979
020	0·096	77·6	6022	3136	1·3485
220	0·137	70·6	4984	2500	1·3110
400	0.160	67·0	4489	2704	1·4922

From the graph $B_t = 3 \cdot 2$ Å^{-2} and the scaling factor $= 1 \cdot 27$.

As the refinement of a structure proceeds it may be necessary to re-determine the values of both B_t and the scaling factor, S. If the values of $S.I' - I_c \exp(-2B_t \sin^2 \theta/\lambda^2)$ drift steadily as θ increases from $0°$ to $90°$ the value of B_t requires modification. When refinement has been completed the corrected observed intensities I' can be placed on an absolute scale by calculating the scaling factor, S, from the relation

$$S \sum I' = \sum I_c \exp(-2B_t \sin^2 \theta/\lambda^2)$$

where I_c is derived from the atomic scattering factors of the atoms at rest, and the thermal parameter, B_t, is now known. The summation includes all reflections except those affected by extinction.

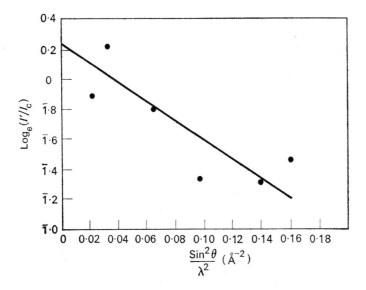

Fig. 8.1   Determination of the thermal parameter and the scaling factor for CdMg

Other methods for determining absolute intensities have been described elsewhere and in particular in the reference work by Peiser, Rooksby and Wilson (1955).

8.2.3 The determination of approximate structures

Although it is possible to solve a crystal structure with the aid of relative intensities only, it is more convenient to use the methods just described to place the observed and calculated values on the same scale. At the outset we therefore have intensity data in the form

$$\text{scaling factor. } I' = I_o \exp\left(-2B_t \sin^2 \theta/\lambda^2\right) \qquad \ldots 8.1$$

for each of the available reflections, and the problem is to determine I_o for each reflection so that there is satisfactory agreement between the left-hand and right-hand sides of these relations. The expressions are in suitable form for manipulation when the reflections are resolved, but if overlapping reflections are included the relations must be rewritten. Clearly if there is any more than slight overlap, individual observed relative intensities cannot be assigned to each reflection in the group affected, and therefore the corrections for absorption and multiplicity and for the Lorentz and polarization factors cannot be applied to each observed intensity. Instead the factor must be applied to the separate calculated intensities in the form

$$\text{scaling factor. } I = \frac{1 + \cos^2 2\theta}{\sin^2 \theta \cos \theta} pAI_o \exp\left(-2B_t \sin^2 \theta/\lambda^2\right) \qquad \ldots 8.2$$

which is given, without the scaling factor, in 8.2.2. For a group of overlapping reflections I is the relative integrated intensity of the whole group, and the corresponding calculated intensity is the sum of the right-hand side expressions taken over all reflections in the group.

Now

$$I_c = |F_0|^2 = A^2 + B^2$$

where

$$\left. \begin{array}{l} A = \sum f_0 \cos 2\pi(hx_n + ky_n + lz_n) \\ B = \sum f_0 \sin 2\pi(hx_n + ky_n + lz_n) \end{array} \right\} \qquad \ldots 8.3$$

and the summation includes all atoms in the unit cell; as usual f_0 is the atomic scattering factor of an atom considered to be at rest at (x_n, y_n, z_n). The problem of determining I_c is therefore the problem of determining the fractional coordinates of every atom in the unit cell. Before this can be undertaken, however, the number of atoms in the unit cell must be known; this number is found from density calculations (6.9.3). The density is measured experimentally and is compared with that calculated from a knowledge of the unit-cell dimensions and the masses and relative numbers of the different kinds of atom present in the structure; the comparison gives directly either the number of each kind of atom or the number of formula units in the unit cell.

In general these atoms will not all be independent. Some will be related by operation of the symmetry elements of the space group and atoms related in this way are said to be **equivalent**; the coordinates of all such atoms can be derived from those of any one of the set. As discussed in 7.5.4 the intensity calculations can often be greatly simplified by collecting together all the atoms related by symmetry.

Once the minimum number of structural parameters, x_n, y_n, z_n, required for the solution of the problem, has been established, the next step is to postulate a feasible approximate structure, and with powder data this is possible, in general, only by trial and error. From recorded intensities, whether from single crystals or from polycrystalline materials, only the strengths of the respective diffracted beams are obtained; all knowledge of the phases of the beams relative to the incident beam, and hence relative to each other, is lost. When one adds to this the inaccuracy which is inherent in any measurement of X-ray intensity, it becomes clear that in order to postulate even an approximate structure all possible information about the crystal must be collected. The various procedures are described in detail in books such as those by Bunn (1961) and by Lipson and Cochran (1966); some of the techniques described apply only to single crystals and others have only limited application to polycrystalline materials because of the limited number of reflections available.

The techniques which do apply to polycrystalline materials include those which give information about atoms in special positions. If the space group can be determined from the systematic absences (8.2.1) the number, n, of equivalent general positions is then known, and it follows that if there are fewer than n atoms of a particular kind in the unit cell these atoms must occupy special positions. The special positions are on centres of symmetry, planes of symmetry and rotation axes (1.2.1); an atom on any of these symmetry elements cannot be multiplied by the operation of that element. Thus an atom on a four-fold rotation axis is not affected by the four-fold rotation whereas one not on that axis would be multiplied four times. Similarly an atom on the line of intersection of two mirror planes is multiplied by neither symmetry element and in the absence of further elements of symmetry the position is one-fold.

The presence of atoms in special positions automatically reduces the number of structural parameters to be determined. An atom on a centre of symmetry cannot be moved arbitrarily and its coordinates are fixed by those of the centre; that is, no variable parameters are involved. An atom situated on a mirror plane has one parameter fixed, and one on the intersection of two mirror planes has two parameters fixed; the only variable is that associated with position on the line of intersection. An atom located on a rotation axis also has only one variable parameter.

In many simple structures all of the atoms occupy special positions with the result that only one, or even none, of the atomic parameters needs to be found; under these circumstances determination of the structure is easy. In more complicated structures it may be that a special position is occupied by a heavy atom; if so, the contribution of that atom to the intensities of the line of the powder pattern can usually be determined. An approximate structure might then be found by applying either the heavy atom or the isomorphous replacement technique as described by Lipson and Cochran (1966).

8.3 Some Examples of Structure Determination

8.3.1 General survey of intensities

Useful information may be obtained by examining the intensities of the lines on a powder photograph. A very strong reflection at low angle indicates that the atoms of the crystal lie close to this plane. If the second-, third-, ..., nth-order reflections are also strong but decrease regularly in intensity (caused by the decrease in the atomic scattering factors as θ increases) then the atoms almost certainly lie close to the plane giving rise to the first-order strong reflection. At the other extreme it may be that certain reflections which are accidentally absent have calculated values that are too large; adjustment of the structural parameters may rectify this (Steeple, 1952).

Additional hints to the approximate atomic configuration may be given by the packing of the atoms into the unit cell. The volume of the unit cell and the number of atoms in the cell will be known; average interatomic and intermolecular distances will also be known, and from these considerations various atomic arrangements may be postulated. Another possible source of information is structures which have already been solved. An unknown structure may be a modification of a known phase, and the known may be used as a trial structure for the unknown.

Without doubt, the finest example of the use of the property that powder photographs have, of giving an overall view of the diffraction pattern, is the determination of the structure of γ-brass by Bradley and Thewlis (1926). The unit cell had been found from single-crystal methods, but it could have been found by powder methods (5.3.1) since it is body-centred cubic with $a = 8.85$ Å. The problem was, however, reckoned to be too difficult to solve at the time, since the unit cell contains 20 copper atoms and 32 zinc atoms.

The powder photographs, however, showed clearly that the structure must be simply related to the structure of CuZn with two atoms in the unit cell, for the photograph of γ-brass looks like that of CuZn with a large number of weaker lines. The cell dimension of Cu_5Zn_8 is roughly three times that of CuZn, and so an approach to the structure can be obtained by stacking 3^3 small unit cells together. This arrangement should contain 54 atoms; to make a body-centred structure from this, we merely have to remove the two atoms $(0, 0, 0)$ and $(\frac{1}{2}, \frac{1}{2}, \frac{1}{2})$ from the unit cell. Two holes are left, and the remaining atoms can then be moved to fill up these holes as well as possible.

By these means, Bradley and Thewlis found a structure, depending on five variable parameters, that gave very good agreement between calcu-lated and observed intensities.

It is interesting that the original 110 reflection of the two-atom structure, which becomes 330 of the 52-atom structure, can be accompanied by another line, 411 (7.4.6). It turned out that this reflection was extremely strong—about half that of 330. The influence of this fact on the develop-ment by Jones (1934) of the Brillouin-zone theory of metals, and the sub-sequent development of solid-state physics, may well mean that the deter-mination of the structure of γ-brass by Bradley and Thewlis was the most significant of all the determinations of crystal structures that have ever been made!

8.3.2 Determination of space groups

If the unit cell of a substance has been determined by powder methods, it is often worth while searching for systematic absences (8.2.1) that may indi-cate the space group. For example, Bradley and Taylor (1937) found the structure of $NiAl_3$ without using single crystals at all.

The structure is orthorhombic, and the unit cell has $a = 6\cdot6$, $b = 7\cdot4$, $c = 4\cdot8$ Å. The first seven lines were unambiguously indexed as 011, 101, 020, 111, 210, 201 and 211. The lattice is thus primitive, for none of the absences—such as reflections with $h + k + l$ odd, indicating an I lattice, or those with $h + k$ odd indicating a C lattice (8.2.1)—is observed. To test for the presence of glide planes, we extract those clearly resolved reflections with one index zero, and these are shown in the following table.

0kl	h0l	hk0
011	101	020
020	201	210

Looking for even single indices or sums of indices we notice that in the first column $k + l$ is even, and in the third column h is even. The absences indicate an n-glide plane perpendicular to the a axis and an a-glide plane perpendicular to the c axis.

Of course, we could not generalize from so few reflections. But a complete survey supported these deductions and a structure based upon the space group was easily found. It contained 4 Ni atoms and 4 Al atoms in special positions on mirror planes, and 8 Al atoms in general positions.

8.3.3 Accurate intensity determination

If accurate intensities are required, they are much more easily obtained from powder photographs than from single-crystal photographs since the photometry of powder photographs (7.2.2) is only one-dimensional. (For this reason, if sufficient reflections are resolved, the powder method could help to relate the different layers given by Weissenberg or precession photographs.)

An example is the determination of the positions of the carbon atoms in the important compound Fe_3C which can occur in iron–carbon alloys (Lipson and Petch, 1940). Up to this time, no very good photographs had been obtained, but the structure had been found to be orthorhombic,

Fig. 8.2   Section of a three-dimensional Fourier synthesis showing the positions (c) of the carbon atoms in Fe_3C. Positions suggested by previous workers are shown at W

with $a = 4·5$, $b = 5·1$, and $c = 6·7$ Å, the space group being Pnma. The positions of the iron atoms has been found but the positions of the carbon atoms—comprising only 7 percent of the total electron content—could only be conjectured.

A powder photograph was taken with $MnK\alpha$ radiation, in a camera with a diameter of 35 cm, and of the 23 lines observed only 4 were not uniquely indexed. The unresolved lines were assumed to be composed of reflections whose intensities were proportional to the values given by the iron atoms alone. In this way a complete set of three-dimensional data was obtained.

It was clear, however, that the contributions of the carbon atoms would not be much greater than the experimental error, even with accurate photometry. Since the carbon atoms were known to lie in planes of symmetry (8.2.3), a section of a three-dimensional synthesis was carried out—one of the first three-dimensional computations—and showed the position of the carbon atom quite clearly (fig. 8.2). The residual (8.4.1) was about 0·08—much better than the usual values for single-crystal work, but quite normal for powder work.

A more recent example is the determination by Grigorivici, Mănăilă and Vaipolin (1968) of $CdGeP_2$. The powder photograph was taken with monochromatic $CuK\alpha$ radiation and a flat specimen was used so that an exact absorption correction could be made. The final residual, obtained from observed and calculated intensities, was 0·06; this would have been about 0·03 if structure amplitudes had been used.

8.3.4   Detection of pseudo-symmetry

The overlapping that is so great a disadvantage for most purposes can be turned to advantage to examine slight departures from apparent symmetry; for example reflections hkl and khl overlap if the specimen is tetragonal but not if it is orthorhombic. Slight changes are much more easily detected, and accurately measured, on a powder photograph than on a single-crystal photograph.

Such a problem arose in the determination of the structure of gallium (Bradley, 1935a). It was thought to be tetragonal and a structure based on a tetragonal unit cell gave quite good agreement for intensities determined from a powder photograph, but Laves (1933), from a single-crystal photograph, deduced that the structure was really orthorhombic. His measurements indicated that the cell had exactly equal a and b axes (4·506 Å) with $c = 7·642$ Å.

Bradley (1935b) took photographs of gallium with Fe, Cu and $NiK\alpha$ radiation. The first produced no reflections with high angles and therefore the resolution was not sufficient to show any fine structure in the lines; the second showed some blurring of the lines with indices 1, 5 and 3, which could be a mixture of reflections 153 and 513; and the third, $NiK\alpha$, moved these reflections to the end of the film, two pairs of α doublets being clearly shown. The unit-cell dimensions deduced from the photograph were $a = 4\cdot5167$, $b = 4\cdot5107$, $c = 7\cdot6448$ Å.

It is also worth noting that the intensity residual for the original tetragonal structure was only about 0·20, corresponding to about 0·10 for the amplitude residual; the intensity residual for the final structure was about 0·11—a considerable reduction. (Incidentally, the tetragonal structure gave atomic environments that agreed well with a rule known as the '$8N$-rule'—that in the structures of the elements each atom has $8N$ closest neighbours, where N is the number of the group of the element. The

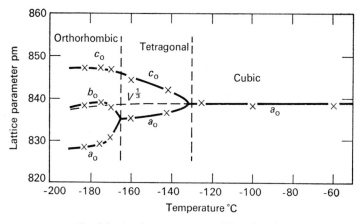

Fig. 8.3   Lattice parameters of $Fe_{1\cdot4}Cr_{1\cdot6}O_4$

discovery that the structure was orthorhombic unfortunately made gallium an exception!)

There are now many more examples of this phenomenon, some of considerable physical interest. The class of compounds known as spinels, with a formula of the type AB_2O_4 where A is bivalent and B is trivalent, are usually cubic. But some compounds—for example the iron chromites, $FeCr_2O_4$—become tetragonal at low temperatures as shown by Francombe (1957). The compound $Fe_{1\cdot4}Cr_{1\cdot6}O_4$ is even more surprising; as shown in fig. 8.3, it becomes tetragonal on cooling to $-130°C$ and orthorhombic on cooling to $-166°C$. Explanations of these effects in terms of antiferromagnetic ordering have been put forward by Rooksby and Willis (1953).

It can thus be seen that these changes, which would hardly be detected by single-crystal methods, become quite clear by powder methods. They represent small variations crystallographically but it is possible that their physical implications, in terms of ordering of magnetic moments of the individual atoms, are very great.

8.4 The Refinement of Crystal Structures

8.4.1 Introduction

When an acceptable approximate atomic arrangement has been obtained for a crystal structure, the parameters x, y, z, are adjusted until their values are as accurate as the experimental data will allow. There will then be a set of observed intensities and a set of calculated intensities, each depending upon x, y, z, and we are required to determine the values which give the best collective agreement between the observed and calculated intensities. In practice, when calculations are performed with resolved reflections only, the square root of expression (8.1) is used and the best general agreement for all reflections is sought between sides of the equation

$$|\sqrt{(\text{Scaling factor}.I')}| = |\sqrt{I_c}| \exp(-B_t \sin^2 \theta / \lambda^2)$$

The left-hand and right-hand terms are, respectively, the observed and calculated structure amplitudes of the reflections and may be expressed as $|F_{obs}|$ and $|F_{calc}|$. Taken over all reflections the measure of agreement between $|F_{obs}|$ and $|F_{calc}|$ is called the **agreement residual** and is usually taken as

$$\frac{\sum ||F_{obs}| - |F_{calc}||}{\sum |F_{obs}|}$$

where the summation covers all reflections considered.

If the data include overlapping reflections then equation (8.2) is used. To calculate the residual, the square roots of the left-hand and right-hand sides of (8.2) may be substituted for $|F_{obs}|$ and $|F_{calc}|$ respectively.

The structure amplitudes are usually calculated from the simplified geometrical structure factors published in International Tables, Vol. I (1952). As discussed in (7.5.4) and (8.2.2), these structure factors apply to atoms in general positions which, of course, have the maximum multiplicity allowed by the space group. Atoms in special positions do not possess the maximum multiplicity, and this must be taken into account when the contribution by these atoms to the structure amplitude is calculated from the simplified geometrical structure factor.

We have so far discussed the solution of crystal structures in terms of the modulus of the structure factor, and in fact the modulus is all that is required for powder methods. If however, as with single-crystal data, sufficient reflections are available, refinement may proceed by successive Fourier synthesis for which the phases of the structure factors are required. The phases are necessary because a Fourier synthesis essentially produces an image of the electron-density distribution inside the crystal from the diffraction data and with any image-forming process the relative phases of the terms included in the syntheses are required. If the temperature factor is isotropic and the same for all atoms in the unit cell then the phase angle, α, for a particular reflection must satisfy

$$\alpha = \tan^{-1} B/A$$

where A and B are defined in equation (8.3).

If the temperature factor varies with the type or with the position of an atom then the components A and B must be expressed in the form

$$A = \sum f_0 \exp\left(-B_t \sin^2 \theta/\lambda^2\right) \cos 2\pi(hx_n + ky_n + lz_n)$$

$$B = \sum f_0 \exp\left(-B_t \sin^2 \theta/\lambda^2\right) \sin 2\pi(hx_n + ky_n + lz_n)$$

where B_t is now not the same for all atoms in the unit cell.

For a centrosymmetric structure it is convenient to choose a centre of symmetry as the origin of coordinates; B is then zero and the phase angle is either 0 or π, so that all that is required is the sign of the structure factor—either plus or minus. When the origin of coordinates is not placed on a centre of symmetry, or when the structure is non-centrosymmetric, then B is not zero, and finding both structure amplitudes and phase angles involve longer calculations.

8.4.2   Refinement of parameters by the method of least squares

Suppose that we have a set of s linear equations in the m unknown variables $x, y, z \ldots$ and suppose that these equations are of the form

$$a_1 x + b_1 y + c_1 z + \ldots = u_1$$
$$a_2 x + b_2 y + c_2 z + \ldots = u_2$$
$$\vdots$$
$$a_s x + b_s y + c_s z + \ldots = u_s$$

where $a_1, b_1, \ldots, a_2, b_2, \ldots, a_s, b_s, \ldots$ are constants and s is greater than m. Values of $x, y, z, \ldots$ can clearly be found which will satisfy any m of the equations but which will not satisfy all s equations unless these are strictly compatible with each other. Under such circumstances the best we can do is to find values of $x, y, z, \ldots$ which satisfy all of the equations as nearly as possible. That is, we must find values of $x, y, z \ldots$ which make as small as possible the errors $\epsilon_1, \epsilon_2, \epsilon_3, \ldots \epsilon_s$ expressed by

$$\epsilon_1 = a_1 x + b_1 y + c_1 z + \ldots - u_1$$
$$\epsilon_2 = a_2 x + b_2 y + c_2 z + \ldots - u_2$$
$$\vdots$$
$$\epsilon_s = a_s x + b_s y + c_s z + \ldots - u_s$$

Provided the errors follow the normal or Gaussian law, one way in which this may be done is to use the principle of **least squares** and minimize

$$R = \sum_{i=1}^{s} w_i \epsilon_i^2$$

The weight w_i of a particular term should be inversely proportional to the square of the probable error in u_i; in the expanded form the expression for R is

$$R = w_1(a_1 x + b_1 y + c_1 z + \ldots - u_1)^2$$
$$+ w_2(a_2 x + b_2 y + c_2 z + \ldots - u_2)^2 + \ldots$$
$$+ w_s(a_s x + b_s y + c_s z + \ldots - u_s)^2$$

R will be a minimum when

$$\frac{\partial R}{\partial x} = 0; \qquad \frac{\partial R}{\partial y} = 0; \qquad \frac{\partial R}{\partial z} = 0; \ldots$$

there being in all m partial derivatives—one for each unknown parameter. Partial differentiation of R with respect to x gives for the first condition

$$2a_1 w_1(a_1 x + b_1 y + c_1 z + \ldots - u_1) + \ldots$$
$$+ 2a_s w_s(a_s x + b_s y + c_s z + \ldots - u_s) = 0$$

Partial differentiation in turn with respect to $y, z, \ldots$ gives the remaining conditions

$$2b_1 w_1(a_1 x + b_1 y + c_1 z + \ldots - u_1) + \ldots$$
$$+ 2b_s w_s(a_s x + b_s y + c_s z + \ldots - u_s) = 0$$
$$2c_1 w_1(a_1 x + b_1 y + c_1 z + \ldots - u_1) + \ldots$$
$$+ 2c_s w_s(a_s x + b_s y + c_s z + \ldots - u_s) = 0$$
$$\text{etc.}$$

of which there are m in all.

After division by 2 the conditions reduce to

$$\sum_{i=1}^{s} w_i a_i^2 x + \sum_{i=1}^{s} w_i a_i b_i y + \sum_{i=1}^{s} w_i a_i c_i z + \ldots = \sum_{i=1}^{s} w_i a_i u_i$$
$$\sum_{i=1}^{s} w_i b_i a_i x + \sum_{i=1}^{s} w_i b_i^2 y + \sum_{i=1}^{s} w_i b_i c_i z + \ldots = \sum_{i=1}^{s} w_i b_i u_i$$
$$\sum_{i=1}^{s} w_i c_i a_i x + \sum_{i=1}^{s} w_i c_i b_i y + \sum_{i=1}^{s} w_i c_i^2 z + \ldots = \sum_{i=1}^{s} w_i c_i u_i$$
$$\text{etc.}$$

These m relations are called the **normal equations** and they determine the values of the m unknowns x, y, z, ... such that $R = \sum_{i=1}^{s} w_i \epsilon_i^2$ is a minimum.

The first normal equation is obtained simply by multiplying the first of the linear equations by $w_1 a_1$, the second by $w_2 a_2$, the third by $w_3 a_3$ and so on, and then summing the products. The second normal equation is obtained by multiplying the first linear equation by $w_1 b_1$, the second by $w_2 b_2$, the third by $w_3 b_3$ and so on, and then summing the products. The other normal equations are derived similarly.

In many problems the original data cannot immediately be expressed as a set of linear equations. One such problem is that of relating observed and calculated structure amplitudes where the relationship between the structure amplitudes, the atomic parameters and the atomic scattering factors is trigonometric. Suppose that, in general, the data are in the form of the s equations

$$f_1(p_1, p_2, p_3, \ldots) = u_1$$
$$f_2(p_1, p_2, p_3, \ldots) = u_2$$
$$\vdots$$
$$f_s(p_1, p_2, p_3, \ldots) = u_s$$

where $p_1, p_2, p_3, \ldots$ are m unknown parameters with $m < s$. The functions $f_1, f_2, \ldots, f_s$ are known and $u_1, u_2, \ldots, u_s$, which are the values of the corresponding observed quantities, are subject to accidental error. We may postulate, initially, approximate values of $p_1, p_2, p_3, \ldots$ of the unknown coordinates, which will be in error by amounts $\Delta p_1, \Delta p_2, \Delta p_3, \ldots$. We then have the approximate relation for the function f_1

$$f_1(p_1 + \Delta p_1, p_2 + \Delta p_2, p_3 + \Delta p_3, \ldots) = f_1(p_1, p_2, p_3, \ldots)$$
$$+ \frac{\partial}{\partial p_1} f_1(p_1, p_2, p_3, \ldots) \Delta p_1$$
$$+ \frac{\partial}{\partial p_2} f_1(p_1, p_2, p_3, \ldots) \Delta p_2$$
$$+ \ldots$$

Now $f_1(p_1 + \Delta p_1, p_2 + \Delta p_2, p_3 + \Delta p_3, \ldots) - f_1(p_1, p_2, p_3, \ldots)$ is the difference between the observed value, u_1, of the quantity under consideration and the corresponding approximate calculated value. This difference we may express as Δu_1 and hence

$$\Delta u_1 = \frac{\partial}{\partial p_1} f_1(p_1, p_2, p_3, \ldots) \Delta p_1 + \frac{\partial}{\partial p_2} f_1(p_1, p_2, p_3, \ldots) \Delta p_2 + \ldots$$

Similarly

$$\Delta u_2 = \frac{\partial}{\partial p_1} f_2(p_1, p_2, p_3, \ldots) \Delta p_1 + \frac{\partial}{\partial p_2} f_2(p_1, p_2, p_3, \ldots) \Delta p_2 + \ldots$$
$$\vdots$$
$$\Delta u_s = \frac{\partial}{\partial p_1} f_s(p_1, p_2, p_3, \ldots) \Delta p_1 + \frac{\partial}{\partial p_2} f_s(p_1, p_2, p_3, \ldots) \Delta p_2 + \ldots$$

These are s linear equations from which we can set up the normal equations to find the corrections $\Delta p_1, \Delta p_2, \Delta p_3, \ldots$ which will make the differences $\Delta u_1, \Delta u_2, \Delta u_3, \ldots, \Delta u_s$ as small as possible.

Applying this to structure refinement, the functions $f_1(p_1, p_2, p_3, \ldots)$, $f_2(p_1, p_2, p_3, \ldots)$ and so on are the calculated values of the structure amplitudes, $|F_{calc}|$, of the reflections; $|F_{calc}|$ for each reflection will depend upon the atomic scattering factor, the temperature factor, the atomic coordinates of the N atoms whose positions are required, and perhaps some other factors. Corresponding to the quantities $u_1, u_2, \ldots$ are the observed values, $|F_{obs}|$, of the structure amplitudes, and these values will be subject to accidental error. If we write $|F_c|$ for $|F_{calc}|$ and $|F_o|$ for $|F_{obs}|$ the observational relations are

$$[F_c]_1 = [F_o]_1$$
$$[F_c]_2 = [F_o]_2$$
$$\vdots$$
$$[F_c]_s = [F_o]_s$$

There is one equation for each reflection and the number of independent equations, s, must be considerably greater than the number of independent parameters, m, to be determined. The unknowns, m, may include coordinates x, y, z, for each of several atoms.

If $|F_c|$ is expressed in terms of approximate parameters $p_1, p_2, p_3, \ldots, p_m$ and if $|F_o|$ is the observed structure amplitude expressed in terms of the correct coordinates $p_1 + \Delta p_1, p_2 + \Delta p_2, p_3 + \Delta p_3, \ldots p_m + \Delta p_m$ then for the first reflection

$$|F_o|_1 - |F_c|_1 = \Delta F_1 = \frac{\partial F_1}{\partial p_1} \Delta p_1 + \frac{\partial F_1}{\partial p_2} \Delta p_2 + \ldots + \frac{\partial F_1}{\partial p_m} \Delta p_m$$

for the second reflection

$$|F_o|_2 - |F_c|_2 = \Delta F_2 = \frac{\partial F_2}{\partial p_1} \Delta p_1 + \frac{\partial F_2}{\partial p_2} \Delta p_2 + \ldots + \frac{\partial F_2}{\partial p_m} \Delta p_m$$

and finally

$$|F_o|_s - |F_c|_s = \Delta F_s = \frac{\partial F_s}{\partial p_1} \Delta p_1 + \frac{\partial F_s}{\partial p_2} \Delta p_2 + \ldots + \frac{\partial F_s}{\partial p_m} \Delta p_m$$

From these s linear relations we are required to find values of Δp_1, Δp_2 ... Δp_m such that $R = \sum_{i=1}^{s} w_i |\Delta F_i|^2$ is as small as possible; the summation is over all the observed reflections and w_i is the weight assigned to the ith reflection.

It was shown earlier in this section that in order to minimize R we set up normal equations, and to form the first of these equations we multiply the first of the linear equations by $w_1 \, \partial F_1/\partial p_1$, the second by $w_2 \, \partial F_2/\partial p_1$, and so on to the sth which is multiplied by $w_s \, \partial F_s/\partial p_1$; the products must then be summed. Ultimately m normal equations will have been formed, the first of which will be

$$\sum_{i=1}^{s} w_i \frac{\partial F_i}{\partial p_1} \Delta F_i = \sum_{i=1}^{s} w_i \frac{\partial F_i}{\partial p_1} \frac{\partial F_i}{\partial p_1} \Delta p_1 + \sum_{i=1}^{s} w_i \frac{\partial F_i}{\partial p_1} \frac{\partial F_i}{\partial p_2} \Delta p_2$$
$$+ \ldots + \sum_{i=1}^{s} w_i \frac{\partial F_i}{\partial p_1} \frac{\partial F_i}{\partial p_m} \Delta p_m$$

The jth normal equation will be

$$\sum_{i=1}^{s} w_i \frac{\partial F_i}{\partial p_j} \Delta F_i = \sum_{i=1}^{s} w_i \frac{\partial F_i}{\partial p_j} \frac{\partial F_i}{\partial p_1} \Delta p_1 + \sum_{i=1}^{s} w_i \frac{\partial F_i}{\partial p_j} \frac{\partial F_i}{\partial p_2} \Delta p_2$$
$$+ \ldots + \sum_{i=1}^{s} w_i \frac{\partial F_i}{\partial p_j} \frac{\partial F_i}{\partial p_m} \Delta p_m$$

Solution of the m normal equations will provide the corrections Δp_1, Δp_2, ..., Δp_m to the approximate coordinates $p_1, p_2, \ldots, p_m$; the new values can then be taken as approximate coordinates and the procedure repeated until no further significant reduction in the value of R can be obtained. It is clear that a problem of this magnitude can be reasonably undertaken only by automatic computation.

The trial structure must be sufficiently accurate to fix the signs of both the structure factors and their derivatives for a large proportion of the reflections, otherwise the least-squares refinement may converge to an incorrect structure. It is not usually necessary to make a precise estimate of the weighting factors, w_i. In general those spectra which suffer from extinction and those which are too weak to be recorded are given zero weight, and all other reflections are considered to have a weight of unity.

8.4.3   Refinement by the method of steepest descents

The application of the principle of least squares to the refinement of structural parameters was first proposed by Hughes (1941). A second method of refinement applicable with a limited number of reflections was used by Booth (1947, 1949) and is called the method of **steepest descents**. The procedure consists in finding the minimum value of the residual

$$R = \sum w^2 (\phi_o - \phi_c)^2$$

where ϕ represents any single-valued differentiable function of the atomic coordinates $p_1, p_2, p_3 \ldots, p_m$. ϕ_o and ϕ_c are, respectively, the observed and calculated values of ϕ, and w is the weight assigned to any particular value of $(\phi_o - \phi_c)$.

Since R is a function of the m variables $p_1, p_2, p_3, \ldots, p_m$, it will be represented by a surface in m-dimensional space; the value of R will, of course, depend upon the values of the m co-ordinates. In terms of say two variables, p_1 and p_2, the curves of constant R will be as shown in fig. 8.4. As the value of R becomes smaller with the increasing accuracy of p_1 and p_2 so the area enclosed will become smaller and the contour will become more like an ellipse. Essentially successive approximations are obtained by moving along the normal to an R contour towards a lower value of R and then along the normal to this contour to a new lower value until the minimum value of R is reached. The same argument applies in m-dimensional space for the refinement of the m co-ordinates $p_1, p_2, \ldots, p_m$. By motion along successive normals, the shortest path is taken to the minimum value of R.

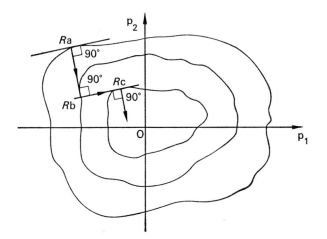

Fig. 8.4 Curves of constant R for the parameters p_1 and p_2. The shortest path to minimum R is of the form R_a to R_b to $R_c \ldots$

The steepest-descents formula given by Booth (1947, 1949) for the correction Δp_i to the co-ordinate p_i is

$$-\Delta p_i \simeq \frac{2R\, \partial R/\partial p_i}{\sum (\partial R/\partial p_i)^2}$$

where R is either

$$\sum (|F|_o - |F_c|)^2$$

or

$$\sum (|F_o|^2 - |F_c|^2)^2$$

summed over all reflections. This formula gives values of the corrections which depend on the scale of representation of the different parameters, and as it stands it does not give maximum convergence. Thus in two dimensions—that is with only two variable parameters—the corrections would depend upon the axial ratio of the elliptical contours. Qurashi (1949) has modified Booth's treatment so that in effect the surfaces of constant R are as nearly equiaxial as possible; under these conditions the parameters are all equivalent and the method provides maximum convergence to the minimum value of R. In two dimensions the contours would then be circles, and in theory one descent along the normal to any one of them would reduce the value of R to zero.

The formula derived by Qurashi for the correction Δp_i to the parameter p_i is

$$\Delta p_i \simeq \frac{\sum w^2(\phi_o - \phi_c)\, \partial\phi_c/\partial p_i}{\sum w^2(\partial\phi_c/\partial p_i)^2}$$

where w is the weight allotted to $(\phi_o - \phi_c)$, ϕ is some suitable function of the atomic coordinates, usually $|F|$, and the summations are over all reflections. This relation involves derivatives with respect to p_i only and thus each parameter can be refined individually. A knowledge of a majority of the phases is assumed; if the phases are not known, the structure at this stage is only rough and convergence to the correct minimum may not occur.

8.4.4 Refinement by the minimum-residual method

A third method is the **minimum-residual** method. The technique was used before computers were in general use—for example, by Thewlis and Steeple (1954) in refining the structure of β-uranium—but its application has become much more widespread since the method was first programmed by Bhuiya and Stanley (1963).

The structural and thermal parameters are varied one at a time by specified increments which are limited to a predetermined range, and the best value of each parameter is taken to be that which gives the lowest value for the residual R as defined by

$$R = \frac{\sum ||F_{obs}| - |F_{calc}||}{\sum |F_{obs}|}$$

It is assumed that all the parameters can be varied independently. When the structure is only approximate, the steps Δp_i by which the coordinate p_i is to be changed can be made large and the steps can be decreased as the structure becomes more refined.

In practice the contribution to each structure amplitude by all the atoms except that being adjusted are first computed and summed. The first coordinate, p_1, to be refined is then set equal to $p_1 - n\,\Delta p_1$, where $2n$ is the total number of steps by which p_1 is to be varied, and the contribution from this atom to each structure amplitude is then calculated and added to that already determined for all other atoms; the value of R is then determined. The procedure is repeated with $p_1 - (n - 1)\,\Delta p_1, \ldots, p_1 + n\,\Delta p_1$ and each time the value of R is determined. From the $2n$ values of R so obtained, the minimum is noted and the value of p_1 which gives that minimum is retained as the correct one. The other coordinates are refined in the same way, after which the thermal parameters can be refined. If necessary further cycles of refinement may be undertaken.

With the minimum-residual method it is again assumed that most of the signs of the structure factors are known. From general experience, however, it would appear that this third method is capable of refining rough structures with a greater degree of certainty than either the least-squares method or the method of steepest descents.

CHAPTER 9

Analysis of the Broadening of Powder Lines

9.1 Introduction

Measurement of crystal size in polycrystalline specimens by means of X-rays is based on two quite distinct effects. First, there is the general appearance of the X-ray photographs, from which it is possible to tell immediately whether the specimen consists of large or of small crystals; secondly, there is the broadening of the powder lines that is produced when the crystals become very fine indeed. The first method is difficult to apply if the crystals are less than 10^{-3} mm in linear dimension, and the second effect becomes measurable only if the crystals are less than about 10^{-4} mm. There is thus a sort of no-man's land between 10^{-3} and 10^{-4} mm, but on the whole X-ray methods cover the range of possible crystal sizes very well.

9.2 The Spottiness of Powder Photographs

9.2.1 Cause of spottiness

Suppose that a transmission or back-reflection photograph of a stationary polycrystalline specimen is taken with the usual mixture of characteristic and white radiations. The size of the crystals in the specimen will have a profound influence on the appearance of the photograph (4.1). For example, the crystals might be so large that only one lies in the path of the beam. Since the specimen is stationary, a Laue photograph would be produced, which may or may not contain spots due to the characteristic radiations. Suppose now that several crystals are irradiated; the pattern will consist of several superimposed Laue photographs, a mixture easy enough to recognize but impossible to index. Spots due to the characteristic radiations are more likely to be present because there is now a greater chance that some crystals will have planes in the correct orientation to reflect these radiations. Such spots will be outstandingly strong.

As the number of crystals in the beam increases, the number of these characteristic spots increases, and they can easily be recognized because they lie on the positions of the powder lines of the substance for the particular characteristic radiation. In other words, the photograph will begin to look like a spotty powder photograph, but with a spotty background of Laue reflections. As the crystals become smaller, the spots become smaller and closer together. The characteristic powder lines become more even and the background more uniform; finally a perfect powder photograph results. Beyond this stage the method is not capable of distinguishing any changes in crystal size and the limit is usually taken to be about 10^{-3} mm.

9.2.2   Deducing grain size from spottiness

It is not easy to make quantitative deductions from the photographs. The best method is to take standard photographs of specimens whose grain size has been found using a microscope. Photographs of the specimen under test are then taken with the same apparatus, and from the appearance of the photograph a fairly good idea of its grain size can be obtained. This is particularly easy when the specimen gives a spotty powder photograph; the number of spots in a given ring is then a quantitative index of the grain size.

Since the same apparatus must be used for standard and test photographs, direct comparison of photographs taken in one laboratory should not be made with those taken in another. Even from photographs taken with the same apparatus, misleading results may be obtained. Clark (1955) has shown two photographs of quartz which were known to have the same average grain size, yet one gave smoother lines than the other. The explanation was that the *distribution* of grain sizes was different. The one that had a mixture of large and small grains gave the spottier photograph, because the reflections from the large grains were more prominent than those from the very small grains, which contributed mainly to the background.

The specimens compared must also be in the same physical condition. Internal strains, for example, will spread the reflections (9.3.2) so that the powder lines appear continuous and may seem to come from a specimen of smaller grain size.

With these difficulties the method is probably only of value for specimens which cannot be examined by the microscope, and even then great care must be taken in interpreting the results.

With the introduction of microbeam techniques, it has been possible to extend the method to include average grain sizes of the order of 0·5 mm (Peiser, Rooksby and Wilson, 1955).

9.3 The Broadening of Powder Lines

9.3.1 Definitions of breadths

Though broadening of X-ray lines can be obvious enough to the eye, it is less simple to describe quantitatively, mainly because the extent of the base of the line (fig. 9.1) is usually not clearly defined. A more useful con-

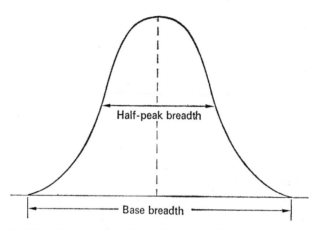

Half-peak breadth

Base breadth

Fig. 9.1   Half-peak breadth and base breadth for a powder line

cept is that of **half-peak breadth,** which is the distance between the two points at which the intensity is half the maximum; this is illustrated in fig. 9.1. This definition does not take into account the lower part of the line profile, and Laue (1926) proposed another concept, namely, the **integral breadth**. This is the breadth of a line with a square-topped profile of the same total area and peak height as the line profile being measured (fig. 9.2). It is obtained by dividing the total area by the peak intensity. This defini-tion is used in this book, but in practice it is much the same as the half-peak breadth.

In measuring the area of a line profile it is necessary to decide on some convention for converting distances along the film (or along the diffracto-meter trace) into quantities independent of the radius of the camera (or of the diffractometer circle). A position on the film could be specified by the

Bragg angle, θ, of a reflection that would occur at that place, but it has become conventional to specify each point by the angle of deviation, 2θ (Laue, 1926).

9.3.2 Causes of broadening

Even with a perfect crystal some broadening of the reflections occurs; this instrumental broadening, as it is sometimes called, is independent of pure diffraction effects and arises from a variety of causes such as divergence of the incident beam, dimensions of the specimen, and the natural width (3.3.2) of the X-ray lines themselves. In addition, the finite width of the slit of the microphotometer (or the detector of a counter diffractometer)

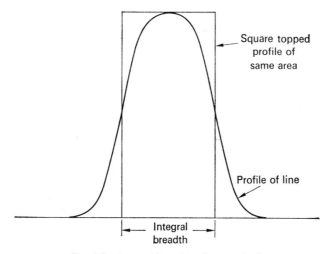

Square topped
profile of
same area

Profile of line

Integral
breadth

Fig. 9.2   Integral breadth of a powder line

used will cause the measured line profiles to be slightly broader than the lines themselves.

Over and above the instrumental broadening, the width of powder lines may be increased by diffraction effects caused by small crystallite size, by lattice distortion and by structural faults. In order to understand how X-ray reflections from small crystals are broadened, we have to extend Bragg's law to cover the *incomplete* reinforcement of the waves scattered by successive lattice planes. Bragg's law was deduced by finding the conditions under which the waves reflected from all planes in a crystal are in phase with each other. There will, however, be an appreciable amount of radiation scattered even when the law is not precisely obeyed. In the following paragraph it is shown that the possible deviation from the law is greater the smaller the crystal. For small crystals the deviations may be quite large; the reflections appear over a range of angle and are thus broadened.

The order of magnitude of the broadening can be found by a simple method due to A. R. Stokes. We consider a beam of X-rays falling on a set of $2m$ lattice planes at an angle $\theta + \delta\theta$ and scattered at the same angle; $\delta\theta$ is the deviation from the correct Bragg angle θ for a particular reflection from the lattice planes. As shown in fig. 9.3, the path difference PBQ for

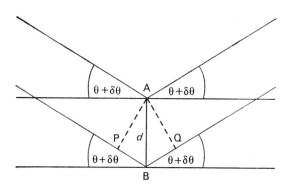

Fig. 9.3   X-rays incident upon lattice planes with slight deviation from the true Bragg angle

waves scattered from successive planes is $2d \sin (\theta + \delta\theta)$. The condition for total reinforcement of the waves is of course $\lambda = 2d \sin \theta$; but suppose that $\delta\theta$ is such that the plane $m + 1$ scatters 180° out of phase with the first plane; that is, that

$$2md \sin (\theta + \delta\theta) = (m + \tfrac{1}{2})\lambda \qquad \ldots 9.1$$

instead of the Bragg relation

$$2md \sin \theta = m\lambda \qquad \ldots 9.2$$

If equation (9.1) is true for the first plane and the plane $m + 1$, it will also be true for any two planes with the corresponding separation, up to the planes m and $2m$. Thus the crystal can be divided into two parts, the scattering from each of which will be 180° out of phase and so will cancel exactly. The value of $\delta\theta$ given by equations (9.1) and (9.2) should therefore correspond to zero intensity of scattering.

The value of $\delta\theta$ can be found by subtracting equations (9.2) from equation (9.1). This gives

$$2md \cos \theta \; \delta\theta = \lambda/2$$

or
$$\delta\theta = \lambda/2t \cos \theta \qquad \ldots 9.3$$

where t ($= 2md$) is the thickness of the crystal. The scattering from the crystal will also be zero when $\delta\theta = -\lambda/2t \cos \theta$, and so the curve of reflected intensity against angle of scattering must be roughly like that shown in fig. 9.1. The angular separation of the two directions in which the scattering is zero is $\lambda/t \cos \theta$. This expression is not precise in view of the assumptions made in deriving it, but it gives the same order of magnitude as expressions obtained with more rigorous assumptions (9.5.1).

Von Laue (1926) has expressed equation (9.3) in terms of the integral breadth, B, of a line broadened only by small crystallite size. The relation is

$$B = K\lambda/t \cos \theta \qquad \ldots 9.4$$

where K is a numerical constant of the order of unity.

The broadening is particularly simple to express in terms of the reciprocal lattice. In earlier chapters the reciprocal lattice was regarded as an array of points. Since we are not now regarding the reflection condition as perfectly precise, the reciprocal points must be considered as extended; for a spherical crystal with diameter t, each reciprocal point will be a small sphere of diameter $\delta\theta = \lambda/t \cos \theta$. Now distances from the origin in reciprocal space are equal to $\lambda/d = 2 \sin \theta$ (2.2.1) and therefore the diameter of a reciprocal point, $\delta(2 \sin \theta)$, is equal to $2 \cos \theta \, \delta\theta$. From the value of $\delta\theta$ given above, the diameter of each reciprocal point is

$$(\lambda/t \cos \theta) \, 2 \cos \theta = 2\lambda/t$$

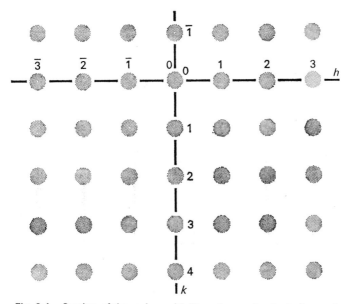

Fig. 9.4   Section of the reciprocal lattice of a small spherical crystal

9+

This is independent of θ and so all the reciprocal points will be broadened to the same extent. A small spherical crystal would thus be represented by a reciprocal lattice of which a section is shown in fig. 9.4; it will be noted that all the points, even the zero order 000, are broadened.

To determine the diffraction broadening caused by lattice distortion we differentiate Bragg's equation to obtain

$$\delta\theta = -(\delta d/d)\tan\theta$$

or
$$-\delta d/d = \delta\theta\cot\theta$$

For a line, $\cot\theta$ may be regarded as constant and thus any measure of the spread of lattice spacings is equal to the product of $\cot\theta$ and the corresponding spread in the diffraction angle. Stokes and Wilson (1944) have shown that in this case the integral line breadth, B, produced by lattice distortion is given by

$$B\cot\theta = K \qquad\qquad ...9.5$$

where K is a quantity that is independent of λ but which may be a function of h, k and l.

The third source of diffraction broadening of X-ray reflections, structural faults, is considered in section 9.5.3. The faults lead to broadening of some of the lines but not of others, and from measurement of the breadths of the various reflections information about the type of fault and its frequency of occurrence may be obtained.

9.4 The Measurement of Broadening

9.4.1 Introduction

Inevitably, any broadening of an X-ray reflection by one or more of the diffraction effects discussed in (9.3.2) will be accompanied by instrumental broadening, so that allowance must be made for this before the pure diffraction effects can be investigated. The instrumental broadening can be calculated, but this is tedious and an experimental correction is preferable. To do this it is necessary to measure lines which are subject to both instrumental and diffraction broadening and also those which suffer from instrumental broadening only. The information can be obtained in one of two ways: either a substance which shows only instrumental broadening can be mixed with the substance under investigation and the sharp and the diffuse patterns obtained at the same time, or the broadened lines can first be obtained and then the sharp lines recorded separately with the same substance after the sources of diffraction broadening have been removed. The first method is sounder, since it does not involve maintaining constant conditions for two photographs, but it has the disadvantages that it involves plotting the breadths of the sharp lines against θ and interpolating to find the corresponding breadths at the positions occupied by the broad lines, that the introduction of a second substance may involve some overlapping of the lines of the two patterns and that the lines from either pattern will be weaker relative to the background than they would be if they were recorded separately.

9.4.2   Distribution of intensity across a broadened powder line

Peiser, Rooksby and Wilson (1955) derived a relationship for the intensity distribution across a broadened line in terms of the separate distributions produced by instrumental and diffraction broadening.

Let the intensity of the line broadened only by diffraction effects be $f(x)$ at a distance x from the position of peak intensity; the intensity associated with the element of the line lying between x and $x + dx$ will then be $f(x)\,dx$. Let the intensity produced by instrumental broadening alone at the same distance x be $g(x)$. If the instrumental and the diffraction broadening occur together, the intensity $f(x)\,dx$ at a point A distant x from the peak and caused by pure diffraction broadening will be further spread out by the instrumental broadening to give an intensity

$$f(x)\,dx\,g(x')$$

at a point Q distant x' from A; the distance of the point Q from the position of peak intensity will therefore be $x + x'$. If this distance is represented by q then $x' = (q - x)$ and the intensity at the point Q distant q from the position of peak intensity is given by

$$f(x)\,dx\,g(q - x)$$

The total intensity, $h(q)$ at Q will be obtained by summation over all values of x and will therefore be

$$h(q) = \sum_{x=-\infty}^{\infty} f(x)\,g(q - x)\,dx \qquad \qquad \ldots 9.6$$

The derivation of $f(x)$ from this relation gives a complete account of the line broadening produced by the pure diffraction effects; it will not, as it stands, distinguish between the different types of diffraction broadening that may be present.

9.4.3   Stokes's method

Stokes's method (1948) for eliminating the effect of instrumental broadening is of general application and does not rely on simplifying assumptions.

Stokes's treatment involves the use of Fourier transforms and so $h(q)$, $f(x)$, and $g(q - x)$ are all expressed as Fourier series. Let $h(q)$ be a function of period 2π so that it can be expressed as a Fourier series in the form

$$h(q) = \tfrac{1}{2}a_0 + \sum_{t=1}^{\infty} (a_t \cos qt + b_t \sin qt)$$

In terms of complex quantities

$$\cos qt = \frac{e^{\imath qt} + e^{-\imath qt}}{2} \quad \text{and} \quad \sin qt = \frac{e^{\imath qt} - e^{-\imath qt}}{2i}$$

whence, by substitution, the series for $h(q)$ becomes

$$h(q) = \tfrac{1}{2}a_0 + \sum_{t=1}^{\infty} \left[\frac{a_t}{2}(e^{\imath qt} + e^{-\imath qt}) + \frac{b_t}{2i}(e^{\imath qt} - e^{-\imath qt}) \right]$$

$$= \tfrac{1}{2}a_0 + \sum_{t=1}^{\infty} \left[\left(\frac{a_t - ib_t}{2}\right) e^{\imath qt} + \left(\frac{a_t + ib_t}{2}\right) e^{-\imath qt} \right]$$

$$= \tfrac{1}{2}a_0 + \sum_{t=1}^{\infty} (c_t e^{\imath qt} + d_t e^{-\imath qt})$$

where

$$c_t = \frac{a_t - ib_t}{2} \quad \text{and} \quad d_t = \frac{a_t + ib_t}{2}$$

$h(q)$ can be written in the alternative form

$$h(q) = \sum_{t=-\infty}^{\infty} H(t) e^{\imath qt} \qquad \ldots 9.7$$

where $c_t = H(t)$ and $d_t = H(-t)$.

We can express $a_t \cos qt + b_t \sin qt$ in the form $A_t \cos (qt + \theta_t)$, so that

$$a_t \cos qt + b_t \sin qt \equiv A_t (\cos qt \cos \theta_t - \sin qt \sin \theta_t)$$

from which, by equating coefficients,

$$a_t = A_t \cos \theta_t \quad \text{and} \quad b_t = -A_t \sin \theta_t$$

But since $c_t = (a_t - ib_t)/2$,

$$c_t = A_t (\cos \theta_t + i \sin \theta_t)/2 = \tfrac{1}{2}A_t e^{\imath \theta_t}$$

The coefficients $H(t)$ can be expressed in integral form by multiplying both sides of (9.7) by $e^{-\imath qn}$ and then integrating between the limits $-\pi$ and $+\pi$. This procedure gives

$$\int_{-\pi}^{\pi} h(q) e^{-\imath qn} dq = \sum_{t=-\infty}^{\infty} H(t) \int_{-\pi}^{\pi} e^{\imath q(t-n)} dq \qquad \ldots 9.8$$

But

$$\int_{-\pi}^{\pi} e^{i(t-n)q} dq = \int_{-\pi}^{\pi} [\cos(t-n)q + i \sin(t-n)q] dq$$

$$= 0 \text{ if } t \neq n \quad \text{and} \quad 2\pi \text{ if } t = n$$

Thus all the terms on the right-hand side of (9.8) vanish except when $t = n$. When $t = n$ we have simply

$$\int_{-\pi}^{\pi} h(q) e^{-iqt} dq = H(t) 2\pi$$

from which

$$H(t) = \frac{1}{2\pi} \int_{-\pi}^{\pi} h(q) e^{-iqt} dq$$

If the function has a period m instead of 2π then the quantity $2\pi q/m$ increases by 2π as q increases by m and the expressions for $h(q)$ and $H(t)$ may be written

$$h(q) = \sum_{t=-\infty}^{\infty} H(t) \exp(it2\pi q/m)$$

where

$$H(t) = \frac{1}{m} \int_{-m/2}^{m/2} h(q) \exp(-it2\pi q/m) dq$$

Even if a function is not periodic it can be expanded in a Fourier series simply by constructing a periodic function, of period m, which is identical over the range m with the non-periodic function. Thus, in considering the distribution of intensity across a diffuse line on a powder photograph, let the distance x measured from the position O of maximum intensity range from $-m/2$ to $m/2$. Further, let $h(q)$, the measured intensity at a point in the broadened line distant q from O, be zero for values of q which lie outside this range, and let the same conditions apply also to $g(q)$, the intensity at a distance q from O arising from instrumental factors only. Then $h(q) g(q)$, and the intensity $f(q)$ produced by diffraction only, can each be expressed as a Fourier series such as

$$h(q) = \sum_{t=-\infty}^{\infty} H(t) \exp(2\pi i q t/m)$$

where the coefficients are given by

$$H(t) = \frac{1}{m} \int_{-m/2}^{m/2} h(q) \exp(-2\pi i q t/m) dq$$

We have already seen that $h(q)$ can also be represented by

$$h(q) = \int_{-\infty}^{\infty} f(x)\, g(q - x)\, dx$$

but now, to express $f(x)$ and $g(q - x)$ as Fourier series, the limits of integration must range from $-m/2$ to $m/2$ because outside these limits the functions are assumed to be zero. When $f(x)$ and $g(q - x)$ are expressed as Fourier series the relation for $h(q)$ becomes

$$h(q) = \int_{x=-m/2}^{m/2} \left[\sum_{t=-\infty}^{\infty} F(t) \exp \frac{2\pi i x t}{m} \right]$$

$$\times \left[\sum_{t'=-\infty}^{\infty} G(t') \exp \left\{ \frac{2\pi i q t'}{m} - \frac{2\pi i x t'}{m} \right\} \right] dx$$

$$= \sum_t \sum_{t'} F(t)\,(Gt') \int_{x=-m/2}^{m/2} \exp \left\{ \frac{2\pi i q t'}{m} + \frac{2\pi i x (t - t')}{m} \right\} dx$$

But $\int_{x=-m/2}^{m/2} \exp 2\pi i x(t - t')dx$ is zero when $t - t' \neq 0$ and is equal to m when $t - t' = 0$, i.e. when $t = t'$.
Therefore

$$h(q) = m \sum_t \sum_t F(t)\, G(t) \exp (2\pi i q t/m)$$

We have, however, already shown that $h(q) = \sum_t H(t) \exp (2\pi i q t/m)$

and hence, by comparing coefficients, $F(t) = \dfrac{1}{m} \dfrac{H(t)}{G(t)}$

This result can also be derived by Fourier-transform and convolution theorems (Lipson and Lipson, 1969).

In practice the factor $1/m$ is omitted since it is a constant and does not affect the shape of the curve.

The procedure for determining $f(q)$ is first to determine the Fourier coefficients of the intensity distributions $h(q)$ and $g(q)$ of a line subject to both diffraction and instrumental broadening and of one subject to instrumental broadening. Each coefficient of $h(q)$ is then divided by the corresponding coefficient of $g(q)$ and the quotients are inserted in a Fourier synthesis to find $f(q)$. Now the Fourier coefficients are complex quantities and hence

$$F(t) = F_r(t) + iF_i(t) = \frac{H_r(t) + iH_i(t)}{G_r(t) + iG_i(t)}$$

where the suffixes r and i refer to the real and imaginary parts of these quantities. Multiplying the numerator and denominator by $G_r(t) - iG_i(t)$ gives

$$F_r(t) + iF_i(t) = \frac{\{H_r(t) + iH_i(t)\}\{G_r(t) - iG_i(t)\}}{G_r^2(t) + G_i^2(t)}$$

from which

$$F_r(t) = \frac{H_r(t)\,G_r(t) + H_i(t)\,G_i(t)}{G_r^2(t) + G_i^2(t)}$$

and

$$F_i(t) = \frac{H_i(t)\,G_r(t) - H_r(t)\,G_i(t)}{G_r^2(t) + G_i^2(t)}$$

Both $F_r(t)$ and $F_i(t)$ can thus be calculated.

Finally,

$$f(q) = \sum_{t=-\infty}^{\infty} F(t)\exp(2\pi iqt/m)$$

$$= \sum_{t=-\infty}^{\infty} [F_r(t) + iF_i(t)](\cos(2\pi qt/m) + i\sin(2\pi qt/m))$$

$$= 2\sum_1^{\infty} F_r(t)\cos(2\pi qt/m) - 2\sum_1^{\infty} F_i(t)\sin(2\pi qt/m)\ tF_r(0)$$

All other terms vanish because $F(-t)$ is the complex conjugate of $F(t)$ and therefore $F_r(-t) = F_r(t)$ and $F_i(-t) = -F_i(t)$.

Stokes (1948) showed that the mean square error in $F(t)$ is given by

$$\overline{|\Delta F(t)|}^2 = \frac{G^2(t)\{\Delta H(t)\}^2 + H^2(t)\{\Delta G(t)\}^2}{G^4(t)}$$

Since $G^4(t)$ appears in the denominator the mean square error in $F(t)$ increases rapidly as $G(t)$ becomes smaller. This is not serious provided that $H(t)$ has fallen to zero before $G(t)$ has decreased very much from its maximum value so that the terms in which $G(t)$ is small can be neglected in the final synthesis. A broad line gives Fourier coefficients which fall off more rapidly with increasing t than do those for a narrow line and it is therefore necessary for $g(q)$ to be narrow compared with $h(q)$. This means that the reflection which gives the instrumental broadening only must be as sharp as possible compared with that produced by instrumental and diffraction broadening together.

Often small particle size and strain in the lattice contribute simultaneously to the broadening of reflections. Warren and Averbach (1950) and Warren (1959) have applied the Fourier method of Stokes to the analysis of line profiles in order to separate the contribution to the broadening of small particle size from that resulting from strain. No assumptions as to line shapes are required, but the methods do require accurate line profiles which are generally obtainable only by diffractometer techniques.

9.4.4   The measurement of line breadths

The elaborate methods of Stokes (1948), Warren and Averbach (1950) and Warren (1959) give more information about the distribution of intensity across the diffraction line than do some other methods which have been designed for speed and convenience rather than for high accuracy and great detail. These quicker methods make use of the breadth of the diffraction line both for the correction for instrumental broadening and for distinguishing between contributions from small particle size and lattice distortion. Usually it is the integral breadth rather than the half-peak breadth that is considered. Generally, but not always, line shapes are assumed to be either all Gaussian or all Cauchy. If the shapes are Gaussian then the square of the total breadth is equal to the sum of the squares of the contributing breadths; if, however, the shapes are Cauchy profiles then the individual breadths are added to give the total breadth.

In making measurements on the broadening produced by small crystals Warren (1941) assumed that the distribution of intensity across a line was, under all conditions, Gaussian and of the form $I = I_{max} \exp(-\alpha\phi^2)$, where I is the measured intensity at an angular deviation ϕ from the true value of θ, and α is a constant. This leads to the relation

$$B_T^2 = B_p^2 + B_e^2$$

where B_T is the total breadth of the line, B_e is the breadth arising from instrumental broadening, and B_p is the required breadth produced by the small crystallite size. It is obvious that the natural breadth of the emission line does not fit in with the assumption of a Gaussian shape, since it contains both α_1 and α_2 peaks, and so its effects have to be allowed for separately. Amongst others, Smith and Stickley (1943) and Rachinger (1948) have described methods for making this allowance, and the quantity B_T mentioned above must be taken as the observed breadth corrected for the presence of the $\alpha_1\alpha_2$ doublet.

Jones (1938) devised a more general method based on the direct comparison of broadened and sharp lines produced under the same experimental conditions. The theory is fully described by Jones and we do not propose to give more than an outline of it here.

From his experimental work Jones could determine the form of the functions $h(x)$ and $g(x)$ in equation (9.6). The form of $h(x)$ was, of course, determined from the broad lines called by Jones the m-lines, and the form of $g(x)$ was obtained from the sharp or s-lines. By numerical integration of equation (9.6) Jones then found the most suitable function, $f(x)$, for broadening due to small crystallite size, which would fit this relation when $h(x)$ and $g(x)$ had the forms that had been determined experimentally. Jones found that the best form for $f(x)$ was the Cauchy curve $1/(1 + k^2x^2)$.

The measured integral breadths, B_0, of the s-lines and the m-lines have first to be corrected for the fact that they are produced by the α-doublet. Jones gave a curve (fig. 9.5) whereby this correction can be made in terms of

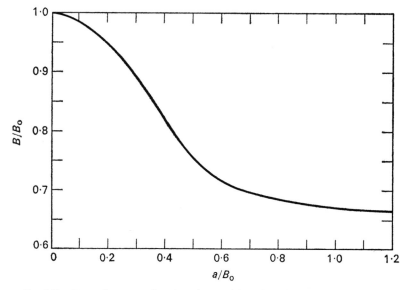

Fig. 9.5   Curve for correcting the observed breadth of a line formed by an α-doublet

a/B_0, where a is the doublet separation in the same units as those in which B_0 is measured; a can be obtained by calculation. From this graph B/B_0 can be read off and the required value of B obtained.

The corrected breadths of the s- and m-lines are called b and B respectively. B is the result of applying a broadening β to b; that is, each element of the sharp line is converted into an element with a breadth β, as shown in fig. 9.6, and the ordinates for all the elements are added. Jones took a typical line and found graphically the effect of operating upon it with different values of β. He expressed his results in the form of a relation between b/B and β/B as shown in fig. 9.7. In this way the integral broadening β of $f(x)$ can be derived from the integral broadening b of $g(x)$ and B of $h(x)$.

9*

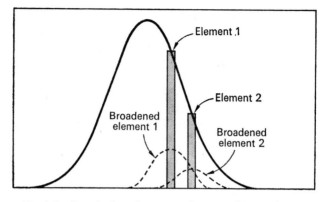

Fig. 9.6   Broadening of separate elements of a powder line

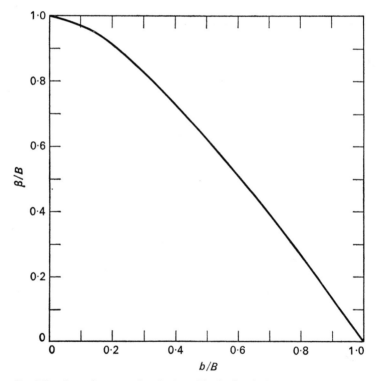

Fig. 9.7   Curve for correcting the breadth of a line for instrumental broadening

The method has certain limitations. The exact form of the curve in fig. 9.6 will depend upon the shape of the s-lines. Since Jones's work is based on a cylindrical camera of the type described in chapter 4, his results are not necessarily applicable to the lines given by other sorts of cameras such as the focusing camera (6.6) or the back-reflection plate camera (6.7). Moreover, since the line shape will change with angle, it is not correct to use the same curve for all reflections. As Jones points out, the errors introduced in this way are likely to be less for specimens with high absorption coefficients than for those with low absorption coefficients; for the latter, as shown in (4.3.3), the line profile changes greatly with θ.

For these reasons the line broadening cannot be measured to any great accuracy. The measurements of breadths themselves are not likely to be more accurate than 5 per cent, even in the most favourable cases. If the broadening to be measured is large, the total error should not greatly exceed this; if the broadening is small, so that the lower part of the curve in fig. 9.7 has to be used, the error may be much bigger. It should be assessed separately for each line, and if possible the mean of several results used.

In making use of line breadths for estimating strain and small-particle-size effects, Schoening (1965) has devised a method which does not rely on the assumption that the profiles resulting from these two effects must have the same form. He points out that the broadening produced by small particle size tends to produce a Cauchy profile and that by strain broadening a Gaussian; his method makes use of both profiles.

Schoening uses equation (9.6) in the form

$$I_t(q) = \int_{-\infty}^{\infty} I_p(x)\, I_s(q - x)\mathrm{d}x$$

where $I_t(q)$ represents the total broadening produced by small particle size ($I_p(x)$) and lattice distortion ($I_s(q - x)$). Two examples are considered. In the first, it is assumed that

$$I_p = C_p/(1 + K_p^2 x^2) \quad \text{and} \quad I_s = C_s/(1 + K_s^2 x^2)^2$$

and in the second example the assumed profiles are

$$I_p = C_p/(1 + k_p^2 x^2) \quad \text{and} \quad I_s = C_s \exp(-k_s^2 x^2)$$

From a considerably simplified Fourier-transform treatment, relations between the integral breadths B_t, B_p and B_s are derived where B_t, B_p and B_s refer to the total broadening, the small-particle broadening and the strain broadening. The first example gives

$$B_t = (2B_s + B_p)^2/(4B_s + B_p)$$

and the second

$$B_t = \frac{\frac{1}{2} B_s \exp\{-(B_p/B_s)^2/\pi\}}{\frac{1}{2} - \mathrm{erf}\,(\sqrt{(2/\pi)}B_p/B_s)}$$

An easy separation of strain broadening from particle broadening is possible if two orders of reflection from the same set of planes are considered.

Yet another method of using line breadths to analyse line profiles has been developed by Willets (1965); standard deviations of the profile functions, and not integral line breadths, are used. The procedure is based on preliminary work by Pitts and Willets (1961) in which it was pointed out that if the breadth of a diffraction line was defined as 2σ, where σ is the standard deviation of the profile function, then it may be shown from equation 9.6 that

$$\sigma_h^2 = \sigma_f^2 + \sigma_g^2$$

σ_h, σ_f and σ_g are the standard deviations of $h(x)$, $f(x)$ and $g(x)$ respectively. The relationship is independent of the form of the functions and is exact.

When considering diffraction broadening by small crystallite size Willets expressed equation 9.4 in the form

$$2\sigma_p \cos \theta = K'\lambda/L \qquad \qquad \ldots 9.9$$

where σ_p refers to the particle broadening and replaces B; λ is the wavelength of the radiation and L is the cube root of the mean grain volume. For broadening arising from strain caused by the distortion of the lattice, Willets rewrote equation 9.5 in the form

$$2\sigma_s \cot \theta = K'' \qquad \qquad \ldots 9.10$$

where σ_s is the standard deviation of the strain-broadening profile and K'' is constant; Willets referred to K'' as the apparent strain.

Willets measured ten diffraction lines in the range of Bragg angles from $15°$ to $74°$. Microdensitometer traces of line profiles were made and σ_h for each observed curve was determined. The standard deviation, σ_g, for each of the line profiles produced by instrumental broadening was derived from patterns produced by large crystallites. The deviation σ_f, which included contributions from both small crystallite size and lattice distortion, was then calculated from $\sigma_h^2 = \sigma_f^2 + \sigma_g^2$. Thus $\sigma_p^2 + \sigma_s^2$ was known since $\sigma_f^2 = \sigma_p^2 + \sigma_s^2$.

By substitution from equations 9.9 and 9.10

$$4\sigma_f^2 \cos^2 \theta = (K'\lambda/L)^2 + (K'' \sin \theta)^2$$

Thus, if a plot of $4\sigma_f^2 \cos^2 \theta$ against $\sin^2 \theta$ for a particular diffraction pattern is a straight line, the intercept and the slope will lead to the values of L and K'' respectively. That is, the mean crystallite size and the apparent lattice strain will be known.

9.5 Applications

9.5.1 Apparent crystal size

The broadening of X-ray lines has been much used to determine the size of crystals of colloidal dimensions. It must be remembered, however, that the quantity determined is the size of *crystal* and not the size of *particle*, since a particle may contain a number of crystals.

The formula for the linear dimensions of a crystal that gives a broadening B is given in equation 9.4 as

$$t = K\lambda/(B \cos \theta)$$

The reasoning in (9.3.1) and (9.3.2) would make $K = 1.0$ but more detailed theory gives 0·94 (Scherrer, 1920), 0·89 (Bragg, 1933), 0·92 (Seljakow, 1925) and 1·42 (Laue, 1926; Jones, 1938). Since these values are all of the order of unity, Jones (1938) has suggested that the quantity t obtained from the formula

$$t = \lambda/(B \cos \theta)$$

should be called the apparent crystal size.

Since 1938 the problem of the value of K has been solved (Waller, 1939; Stokes and Wilson, 1942). The value depends, however, on the quantity that is taken as the linear dimension of the crystal. Taking this as the cube root of the volume, Stokes and Wilson give $K = 1.0747$ for spherical particles; it will have other values for other definitions of size, for other particle shapes, and for different reflections if the crystals are not spherical. For these reasons it seems desirable to retain the definition of apparent crystal size.

9.5.2 Small-angle scattering

It was pointed out in 9.3.2 that all the reciprocal points representing the diffraction of X-rays by a small crystal are broadened to the same extent, and that this applies even to the zero-order point 000. This, however, is true only for an isolated single crystal; it does not apply to the crystal grains in a polycrystalline mass. This can be seen by considering the reflections from a particle of matter consisting of several crystals (fig. 9.8). The grain boundaries are those surfaces on either side of which there are atomic arrangements with different orientations, although there may be no actual space between grains. There will be no phase relationships between the rays scattered from different grains in a general reflection *hkl*, and so each grain will give independent reflections and the broadenings will be characteristic of the shape and size of the crystals grains. But for the zero-order reflection 000, the atoms in *different* grains will scatter in phase, because the condition is then that the phase difference between the waves scattered by different atoms is zero; and this is so if the atoms are rigidly connected together in the same particle, whether they are in the same crystal grain or not. The whole particle will thus behave as a single crystal for this reflection, and so the broadening is a measure of the particle size.

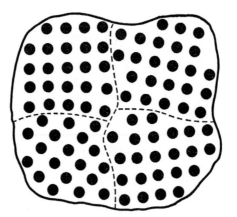

Fig. 9.8   Atomic position in a particle consisting of four crystals

The study of the zero-order reflection—called **small-angle scattering** or **low-angle scattering**—calls for careful work, because of the proximity of the undeviated transmitted beam. Crystal-reflected radiation must be used, and scattering from the air in the camera must be eliminated. Moreover, a fine beam must be used since the broadening may be small; focusing mono-chromators (3.5) are, therefore, useful.

9.5.3   Structural faults

The presence of faults on the atomic scale within a crystal can lead to broadening of some of the X-ray reflections. An example is the alloy $AuCu_3$ (7.6.2), which is cubic with the atoms on the points of a face-centred lattice (000), $(0\frac{1}{2}\frac{1}{2})$, $(\frac{1}{2}0\frac{1}{2})$ and $(\frac{1}{2}\frac{1}{2}0)$. At high temperatures the atoms are distributed at random and so the alloy is truly face-centred, but at low temperatures a superlattice develops; the gold atoms occupy one of the four sets of positions and the copper atoms the three others. There are thus four different ways in which the structure can form on the same lattice and, in any one crystal, regions may exist with these different positions for the gold and copper atoms. These regions are called **anti-phase nuclei**. If the nuclei are of the same order of size as the crystal in which they are formed, the superlattice lines will be as sharp as the main lines on the photograph; but if the nuclei are small, the superlattice lines will appear as if they were produced by small crystals and so will be broadened (Jones and Sykes, 1938; Wilson, 1943). The main lines, which are not affected by the re-arrangement of atoms on the lattice points, remain sharp because the lattice is perfect over the whole crystal. Powder photographs of $AuCu_3$ are shown in plate II(a) to (f); (a) is the pattern of the disordered face-centred cubic phase; (b) to (e) include the broadened superlattice lines produced by small anti-phase nuclei (the smaller the nuclei, the broader the lines) and in (f) the superlattice lines are sharp and have been produced by large anti-phase nuclei.

Another effect of the same sort occurs in cobalt, which has a hexagonal close-packed structure. That the structure is not perfect is shown by the broadening of some of the X-ray reflections, and this broadening can be explained on the theory that there are faults that simulate the cubic close-packed structure (Edwards and Lipson, 1942; Wilson, 1942b). Certain reflections are common to the two structures, and these are sharp; but those that are due only to the hexagonal structure are broadened.

9.5.4 Distorted crystals

The measurement of the broadening of reflections is also important in the study of cold-worked metals. Although it has been suggested that the broadening is mainly due to the smallness of the crystals into which the cold-worked metal is broken down (Wood, 1941, 1943), experimental evidence suggests that it is almost entirely due to distortion of the lattice (Brindley, 1940; Stokes, Pascoe and Lipson, 1943; Smith and Stickley, 1943). Equations 9.4 and 9.5 distinguish between the effects of small crystal size and strain broadening, and since $\cos \theta$ and $\cot \theta$ differ greatly only at small angles the distinction between the two equations requires careful work. The distortion hypothesis has been amply confirmed by the work of Smith and Stickley (1943), Warren and Averbach (1950; 1952), Warren (1959), Schoening (1965) and Willets (1965); it has also received independent support from the results of the detailed examination of metal surfaces carried out by Hirsch (1952) and Gay, Hirsch and Kelly (1954) by means of fine-focus tubes (3.6.5).

CHAPTER 10

Identification of Crystalline Materials

10.1 Introduction

X-ray powder methods are widely used for identifying crystalline substances and may even be used sometimes when the chemical composition of the substance is not known.

From measurements on a powder photograph, the spacings of the reflecting planes and the relative intensities of the reflections are obtained, and in general, **the powder pattern is uniquely characteristic of a particular crystalline substance.** (The term powder pattern is used here to denote both the relative positions and the relative densities of the powder lines.) With simple substances, of course, there may be accidental similarities of powder pattern, and this emphasizes the importance of independent checking in identification work. In alloys and in minerals, one type of atom is often partially replaced by another type, and this complicates the task of identification by means of powder patterns. If the variation of unit-cell dimensions with the amount of atomic replacement is known, then, by making accurate measurements of the spacings of the powder lines, we can sometimes obtain an approximate value of the relative proportions of the end members. But, if more than one type of atom replaces a given type, as is common in minerals, these proportions cannot be obtained in this way.

The powder method is invaluable when the substance occurs only in the form of a powder or of a disorientated crystalline aggregate. Since a powder specimen can with care be prepared from less than 1 mg of material, this method is very useful when only a very small quantity of material is available. Nevertheless, the powder method should be regarded as one method among several, although an important one. A combination of the most suitable methods is best in identification work.

In the following sections general methods of using powder patterns are discussed, along with certain difficulties which arise in practice. When the substance belongs to a known group the use of the powder pattern for identification is most effective. When the number of possibilities is small, powder photographs should be taken of all likely substances, the same radiation and radius of camera being used. The most generally suitable radiation is that from a copper target, but for a particular group of substances it may be preferable to use another radiation. When such photographs have been taken, the unknown is then compared with each of them. Even if it does not fit any of them, certain possibilities can thus be eliminated, and it is usually clear what substances should next be tried. If the unknown is a mixture of two of the standard substances, some idea of the proportions in the mixture can be obtained (7.6.3).

In comparing powder photographs, either directly or by means of derived data, the difficulties and pitfalls discussed in (10.2.3) should always be kept in mind. In the last section of this chapter the application of the powder method of identification to refractories, alloys and minerals is discussed in outline. These groups of substances were chosen because they cover a large part of the field in which powder patterns are useful in identification work and they also provide examples of most of the different types of problem that occur in this work.

10.2 General Methods of Identification

10.2.1 Descriptions of powder photographs

When little is known of a substance, it is obviously difficult to identify it by direct comparison of its powder photograph with those of other substances, and some more general treatment is required. The identification must be based on the comparison of one powder photograph with a large number of others; therefore some simple way of describing a powder photograph must be devised which must lend itself to systematic classification. The description of any one photograph should be independent of the radiation and the size of camera with which it was taken.

The description which has been adopted in practice is that each line is defined by the spacing ($d = \lambda/2 \sin \theta$) of the planes producing it, and by its intensity expressed as a fraction of that of the strongest line of the pattern. It is perhaps unfortunate that d and not some function such as $1/d$ has been chosen, since d varies rapidly with θ at low angles and so does not lend itself to graphical methods or to interpolation in tables.

Although the interplanar spacings are independent of the radiation used to determine them, the intensities of the lines are not; the value of θ for a particular reflection depends upon λ and, as shown in (7.4.3), the formula for the intensity of a powder line contains a quantity which is a function of θ. It is possible to correct each individual intensity for this factor, but this is not often necessary; the trigonometrical factor varies rapidly only near $\theta = 0°$ and $\theta = 90°$, and in practice discrepancies are likely to arise between the intensity data obtained with different radiations only if a value of θ near 90° occurs.

The intensities of the lines should, in theory, be those determined either photometrically or with a counter diffractometer, but in practice it is found that visual estimation is accurate enough for most purposes. One difficulty that arises is that different observers use different scales, but so long as the intensities of the lines are recorded in the correct order, it does not greatly matter whether the numbers used are proportional to the intensities, or merely represent a class of intensity such as medium or strong.

10.2.2   The powder diffraction file

Originally indexes based on the principles outlined were produced by Hanawalt, Rinn and Frevel (1938), Boldyrev, Mikheiev, Kovalev and Dubinina (1939) and Harcourt (1942). Until 1963 the system developed by Hanawalt, Rinn and Frevel formed the basis of all the indexes to the powder diffraction file, a file which has proved to be very useful and which is continually being revised and supplemented.

When the file was instituted by the American Society for Testing Materials each substance was catalogued according to the value of d for the first, second and third strongest lines. Each substance had three cards in order of d, so that each substance appeared in three places in the index. The cards of the file initially served as the index with regard to the d values although a separate index book listing the names and chemical formulae of the compounds in the file was also provided.

A typical set of three cards used in earlier days of the file is illustrated in fig. 10.1. Only on the card which gave the strongest line first were all the available data given. The values of d and the relative intensities of the first, second and third strongest lines were given in the top left-hand corner, and below these data were quantities I which represented roughly the absolute intensities of the lines, information which may be useful after completion of identification. Then followed the formula and sometimes the name of the substance, and below this was space for certain crystallographic data which was, however, rarely filled in. The right-hand side was devoted to a complete list of lines in order of spacing, together with their relative intensities, but this was included only on the cards that corresponded to the strongest line. The heading of the spacing column implies that the values were obtained with molybdenum $K\alpha$ radiation, but this seems unnecessary since other radiations must also have been used for some substances and the spacings should be independent of the wavelength with which they have been determined. As we have seen in chapter 7, the *intensities* vary with change in wavelength, though the relative values are not greatly affected.

The identification of a simple compound, if contained in the file, was quite easy. The values of d for all the lines were measured and that of the strongest noted. The cards in the file for which the strongest line had this value of d could then be consulted to determine whether the pattern on one

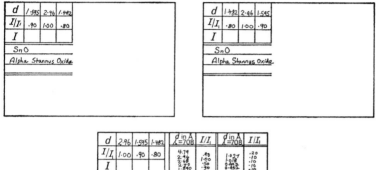

Fig. 10.1   Set of three cards from the powder diffraction file

of them agreed with that of the unknown. If it was not obvious which was the strongest line, the fact that there were three cards for each substance often proved helpful.

The difficulties in identifying all the constituents in a mixture, were greater but the three-card system was again particularly helpful. One method which was found useful was to make a list of the strongest six lines and to extract from the file any card that contained two of these six as the first, second or third strongest lines; the cards so extracted could then be examined more closely to see if all the lines on any one of them were in the observed pattern. If all the strongest lines could be accounted for in this way, it would then be necessary to see if the weaker ones belonged to a single pattern in the file; to be sure of the identification, it was necessary to account for all the lines.

By 1950 the third set of data cards had been published and it became time consuming to search through the data cards directly. The index book was consequently expanded to contain a numerical index in which the d values and the relative intensities of the three possible arrangements of the three strongest lines were listed in groups in the same order as they appeared in the file. The index also gave the chemical formula of the substance and the number of the card in the file. The groups, called Hanawalt groups, were chosen by dividing the spacing range from 10 Å upwards to 1 Å downwards into eighty-seven intervals so that the number of the three strongest lines falling into each interval was about the same. Each card was assigned to its group according to the d value of its strongest line and the cards were arranged in order within each group according to the d values of the second lines. Subsequently the practice of preparing three cards for each substance was abandoned and for many years the Hanawalt book index formed the main method by which a search through the file could be conducted.

Eventually the expansion of the file and its coverage of new fields meant that even the Hanawalt book became less useful. For example, difficulties were encountered when the index was used for identifying powder patterns obtained by electron-diffraction techniques. These arose mainly because X-ray intensities, which are those given on the cards, often differ in order of strength from the intensities produced by electron diffraction by the same substance. Similar problems arise with mixtures when lines from the different patterns overlap.

With these difficulties in mind a new system was devised which was not directly dependent on the order of the strongest relative intensities. The new index was called the Fink index to the powder diffraction file and is the one in current use. It is in book form and uses the d values, less than 9·99 Å, of the eight strongest lines on each powder pattern contained in the file. Relative-intensity values are *not* listed. Lines with spacings greater than 10 Å are given only if the intensities are greater than a certain maximum value. If fewer than eight lines are given on the data card, the rest of the eight are labelled as 0·00 Å.

The organization of the data in the Fink index is in some respects similar to the arrangement in the Hanawalt index. All eight d values are listed in eight different places in the index. In the first entry the d values are given from the left to right in decreasing numerical order and by cyclic permutation the second entry has the second longest d value on the extreme left, the third entry begins with the third longest d value, and so on. Again the d values are arranged in groups and entry in a group is determined by the first d value on the left; order within each group is also on a strictly numerical basis and is determined by the magnitude of the second d value from the left. Further information against each entry includes the chemical and, if appropriate, the mineral name of the substance, and the number of the card in the powder diffraction file from which the book data were taken.

To identify an unknown substance all possible information about the substance should be collected and from a powder pattern of the material the spacings of the eight strongest lines noted and arranged in descending numerical order. From the largest spacing the Fink group can be chosen and a systematic search through this group undertaken using in turn each of the other d values as the second line. If unsuccessful, a new choice of the largest d value can be made and the operation repeated. This sequence can be repeated until identification is achieved.

Indexes, other than those in book form, that have been devised include punched-card systems. For mechanical searching, in which a knowledge of chemical composition can be coordinated with numerical diffraction data, both Keysort punched cards (Matthews, 1949) and IBM cards (Kuentzel, 1951), are available. With the increasing number of patterns, alternative methods of punched-card indexing have become necessary, just as a new kind of book indexing ultimately became desirable, and one such system has been described by Matthews (1962).

The new index is based on the coordinate principle. In a punched-card coordinate index each card represents a subject, and on the card holes are punched the coordinates of which refer to places in the file where information relating to the subject of the card can be found. Thus with reference to the powder diffraction file as described by Matthews (1962) the coordinate index card for a given element will contain holes indicating where in the file the powder patterns of substances containing that element can be found. Another card of the index will have holes the coordinates of which give the cards in the file which have patterns with a strong line in a given range of d. Another index card will have holes giving the location of patterns with a strong line in another range of d, and so on. Another card might carry holes showing where in the file are patterns of substances which have a given chemical property. When all the relevant index cards have been extracted and superimposed, the hole, or holes, common to all the index cards will give the data-file card, or cards, which carries all the details which are the subjects of the individual index cards.

10.2.3 Some practical difficulties

So far we have assumed perfect accuracy both in the measurement of the powder photographs and in the spacings recorded in the file. In practice considerable errors can occur, chiefly on account of absorption, which displaces the low-order lines (4.3.3). For some of the data in the file this error has been eliminated, either by calculating the spacings from accurate cell dimensions, or by mixing the specimen with a substance that gives lines of accurately known spacings, which can then be used as reference lines. This, however, is not true for all the cards. The spacings of the unknown substance may also be corrected by means of a standard; but, if they are not, it is advisable to allow a liberal margin of error, and to consult an extended range of cards. The range depends greatly on the spacing concerned. At about 3 Å, errors of 2 per cent are possible, so that cards from 2·94 to 3·06 should be consulted. For smaller spacings the error decreases rapidly and should not be greater than 0·2 per cent at 1 Å.

It may happen that certain lines recorded in the file do not appear on a powder photograph, although it is known that the particular substance is present in the specimen. If these lines are the weakest of the pattern, the details given in the file are probably obtained from a better photograph than the one under consideration. This may easily be so if the substance forms only a small part of the specimen. On the other hand, certain observed lines may not be recorded in the file. It is most likely that they are due to another constituent; but it is not impossible that they are a true part of the pattern. Many of the data in the file were obtained with molybdenum Kα radiation; the low values of θ which result give inaccurate values for the higher spacings, and the crowding together of the lines makes it difficult to detect weak ones. If it is suspected that some lines are omitted from the pattern recorded in the file, the only satisfactory way of making sure of one's conclusions is to take a powder photograph of a pure sample of the material. The file is likely to contain errors but it is subject to continual revision, in which users are asked to take part.

In general, if a material is thought to consist of a mixture of certain constituents in certain proportions, it is advisable to take a photograph of a similar artificial mixture. This will check whether the deduction is completely sound, and it may also show whether there are any complications such as solid solution.

A further difficulty is that the powder pattern of a substance is not necessarily unique; many alloys, for example, could be made that would give exactly the same powder photograph. This is, of course, due to the large ranges of solid solution which occur in metals, and the existence of such solid solutions is a complication that cannot be satisfactorily dealt with in the file. Solid solutions can also occur to a great extent in minerals.

Apart from this trouble, there is the possibility that quite different substances may have similar powder patterns. Frevel (1944) has given a list of some that have been discovered; lithium fluoride and aluminium are a striking example. This is another reason why the file should not be used independently of other methods, such as chemical analysis and optical investigation.

In its present form, the file is an extremely useful tool and its usefulness increases from year to year as the number of compounds it includes is increased. The file now contains data for about 15 000 materials.

10.3 Identification within a Limited Range of Possibilities

10.3.1 Refractories

The application of X-ray methods to the study and classification of refractory (heat-resisting) materials has proved extremely useful. The number of compounds involved is not large, and it is a simple matter to make a collection of powder photographs of them. Moreover the compositions of the substances are usually fairly definite, and it is therefore not to be expected that there will be variations due to solid solution in the intensities and positions of the X-ray reflections. On the other hand, many of the substances have fairly complicated patterns, and to detect them in small amounts, when other substances are present, good photographs are required.

One of the most important of the refractory materials is silica, SiO_2. At ordinary temperatures this can exist in four forms, depending on the thermal history of the specimen–amorphous silica, and the low-temperature forms of quartz, tridymite and cristobalite. X-ray methods cannot be used to identify amorphous silica, although this can be easily done by a combination of optical and simple chemical methods. The powder patterns of the other three forms are, however, very distinctive; this is extremely useful because the properties of refractory materials of this kind depend greatly on the form or forms of silica present.

10.3.2 Alloy systems

The identification of the phases present in alloys is not always as straightforward as the identification of refractories. For example, it is not always possible to know what structures can occur in an alloy of given composition. If the compositions of possible phases are known from previous work on the equilibrium diagram, alloys of these compositions can be made and their powder patterns should then be representative of the phases that can exist in the alloys; the methods outlined in the previous sections can then be used. There are, however, some difficulties; the positions of the lines may change owing to the variation, with solid solution, of the cell dimensions, and extra lines may appear due to the formation of a superlattice (7.6.2). For these reasons, one should begin by finding whether any regions of solid solution exist. Next, cell dimensions are determined as functions of composition within the ranges of solid solution, and the occurrence of superlattice formation is noted. These difficulties in the study of alloys are often offset by the comparatively simple powder patterns of the structures that occur.

The general principles of the powder method applied to the study of a binary alloy system are illustrated by the set of photographs shown in plate III which are those of alloys in the iron-silicon system (Farquhar, Lipson and Weill, 1945). The photographs of the α- η-, ϵ-, ζ- and θ-phases were obtained from alloys of the compositions given by the known equilibrium diagram. The pure η-phase was not obtained easily, because its composition was not accurately known, and because it does not appear to have an appreciable range of composition. In dealing with a phase of this sort, it must always be borne in mind that the photograph may show traces of neighbouring phases.

From the photographs of the single-phase alloys, the phases present in any alloy can be identified. For example, the photograph of the 30 per cent alloy obviously contains lines belonging both to the pattern of the ϵ-phase (immediately below) and to the pattern of the η-phase (immediately above). Changes due to heat treatment can also be followed. For example, the photograph of the η-phase was obtained from an alloy heated to about 900°C, and quenched in cold water; if such an alloy is annealed at about 700°C it gradually dissociates into $\alpha + \epsilon$, and the change can be followed quantitatively by measuring the relative intensities of chosen lines.

The photographs also illustrate other points. There is an extensive solution of silicon in iron, as is shown by the shifts of the lines on the first three photographs; this shift is so great that the last line, 310, does not occur on the second and third photographs, and if accurate measurements of lattice parameters are required, CoKα radiation should not be used (6.3.3). Also a superlattice of the Fe_3Al type develops with more than about 7 per cent silicon; this is shown by the appearance of faint extra lines on the second powder photograph.

If the equilibrium diagram of the alloy system being investigated has not previously been satisfactorily determined, the application of X-ray methods is naturally more difficult. The same general principles can be used, but the single-phase alloys must be identified by trial and error. No general principles can be given, but the change in relative intensities of the lines on a powder photograph which may result from a small change in composition often helps in deciding to which pattern certain lines belong. Obviously, as the amount of one phase increases, the intensity of all the lines of its powder pattern will increase together, and all the lines of other phases will decrease in intensity. Difficulties arise, however, when a number of phases have structures which are modifications of the same structure, and so have similar powder patterns; careful work is necessary to distinguish between them.

In a binary system use can also be made of the fact that in a two-phase region the physical properties of the two phases present are independent of the composition of the alloy; the only quantity that can change is the ratio of the quantities of the two phases. Thus the lattice parameters of the two phases will be constant in a two-phase region. The constancy in position of the lines given by different alloys is thus a useful indication that they lie in the same two-phase field.

10.3.3   Minerals

Powder photographs play an important part in the identification of minerals and they are particularly helpful when the polarizing microscope cannot be employed, as, for example, when the material is particularly fine grained. Although reflected polarized light can be used it is not so widely applicable in identification work as transmitted light. Powder methods are, therefore, specially useful in the field of opaque minerals. In the identification of minerals, certain general points should be kept in mind.

(1) The material should always be examined first by means of a microscope, reflected light being used for opaque substances. This is desirable because minerals are often intergrown and frequently contain foreign inclusions. In some cases it may be necessary to use special techniques for separating the mineral constituents in a sample, or to pick out by hand a number of grains of the particular mineral being studied.

(2) Most minerals belong to the systems of lower symmetry, and this means that complex patterns are predominant in powder photographs. The chance of accidental similarities between the patterns of different minerals is consequently very much smaller than in metals or in other substances with simple structures.

(3) Atomic replacements within the same structure are very common in minerals; this results in considerable variation in the positions and relative intensities of corresponding lines in members of mineral series with the same structure. Since the shifting of the lines occurs in complicated patterns, the effect is more confusing than it is with alloys.

Although each specimen should be treated as a problem on its own, certain general points can be taken as a guide. If the substance is transparent, it should first be examined by means of the polarizing microscope and its characteristic optical properties observed and measured. This is the chief method of the mineralogist, and even when it does not give an un-ambiguous result, it will give clues to the nature of the most likely possi-bilities. The powder method can be used, either to check the result of an identification by optical methods or to study the various possibilities suggested by it. It is not advisable to rely solely on the powder diffraction file as the primary method of identifying a completely unknown mineral because of the difficulties outlined above. Instead the index **X-ray Powder Data for Ore Minerals** compiled by Berry and Thompson (1962) is avail-able. The data include visual intensities, observed glancing angles, measured and calculated interplanar spacings and the indices of the reflecting planes for about three hundred minerals. The safest way of using powder photo-graphs in this work is by direct comparison, after the number of possi-bilities has been reduced by means of optical data or by means of simple chemical tests.

X-ray powder photographs are now also widely used to study artificial minerals and examples of their use for this purpose are given in the *Symposium on the Chemistry of Cements* (1938), Bogue (1955), Chesters (1963), Rait, Green and Goldschmidt (1942).

Problems

1. Ignoring fine detail, divide the capital letters of the alphabet into symmetry groups. How many (i) classes, (ii) systems are there?

2. In an oblique two-dimensional lattice with $a = 5$ Å, $b = 3$ Å, $\gamma = 100°$, draw in the planes $(3\bar{2})$. Measure the spacing and compare it with the calculated value.

3. Deduce the unit cells, plane groups and full symmetries of the following patterns.

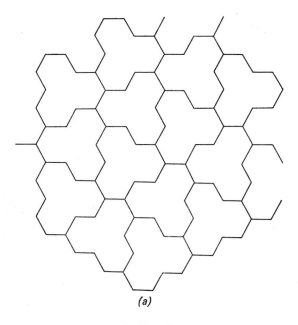

(a)

Problem 3

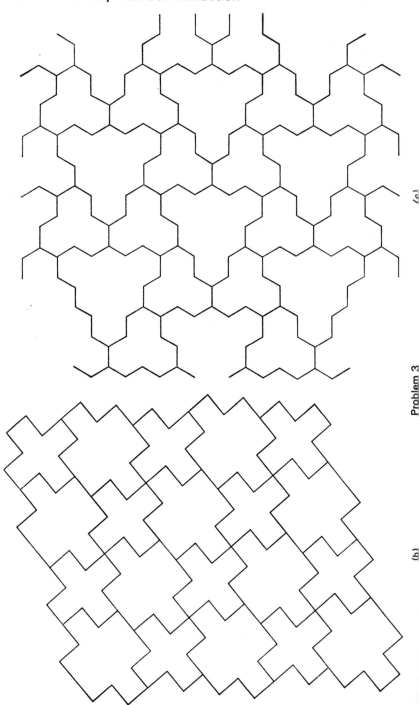

(c)

(b)

Problem 3

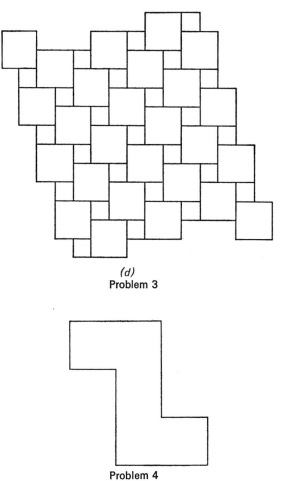

(d)
Problem 3

Problem 4

4. As exercises in 'crystal growth', cut out about thirty pieces of stiff paper in the shape shown. Arrange them in periodically repeating space-filling patterns, and then find their unit cells, symmetry elements and plane groups. Try to make patterns with different symmetries, including those that arise if the unit can be turned over (enantiomorphous forms).

5. (a) The unit cell of a crystal has $a = 8·55$ Å, $b = 8·80$ Å, $c = 16·0$ Å, $\beta = 97·5°$. Calculate the reciprocal-lattice parameters in dimensional units and in dimensionless units with $\lambda = 1·54$ Å.

   (b) The unit cell of a triclinic crystal has $a = 6·64$ Å, $b = 8·31$ Å, $c = 11·18$ Å, $\alpha = 64·0°$, $\beta = 46·3°$, $\gamma = 77·4°$. Calculate the reciprocal-lattice parameters in dimensional units.

   (c) Repeat the calculation for the data in 5(b) except that γ is now $180° - 77·4° = 102·6°$.

6. The unit cell of diphenyl ($C_6H_5 . C_6H_5$) has

$$a = 8.24 \text{ Å}, \, b = 5.73 \text{ Å}, \, c = 9.51 \text{ Å}, \, \beta = 94.5°$$

If the density is 1.16 Mg m^{-3}, find the number of molecules in the unit cell. Given that the systematic absences are $h0l$ when h is odd and $0k0$ when k is odd, find the space group of the crystal. What deductions can be made about the symmetry of the molecule and its position in the unit cell?

7. Derive the relative transparencies of the following windows for CuKα and CrKα radiations
   (a) Aluminium of thickness 0.02 mm
   (b) Beryllium of thickness 0.3 mm
   (c) Lithium of thickness 1.0 mm

8. It is required to take a photograph of an iron specimen with MoKα radiation. Find the thickness of aluminium that will act as a screen to remove the fluorescent FeK radiation—it should reduce the intensity of FeKα radiation to 1 per cent of its original value. Find the reduction of intensity of the MoKα radiation produced by such a screen.

9. Derive the thickness of a β filter for CoK radiation. Assume the filter has to reduce the intensity ratio Kβ/Kα from 1/5 to 1/600. Find by what factor the Kα radiation is reduced in intensity by such a filter.

10. Derive the linear absorption coefficient for CoKα radiation of
   (a) an alloy of 30 per cent Cu, 70 per cent Al by weight (density $= 5.44$ Mg m^{-3})
   (b) the compound $FeCo_2O_4$ (density $= 5.34$ Mg m^{-3})

11. What radiations would you use to examine an alloy of 50 per cent iron and 50 per cent chromium?

12. What radiation would you use to increase the difference between the scattering factors of iron and nickel?

13. An impurity radiation was found to give an alpha doublet with lines at Bragg angles of $73.084°$ and $73.595°$. What is the radiation?

14. What radiation would you use to find an accurate value for the lattice parameter of (a) aluminium, which is face-centred cubic with a lattice parameter of about 4.049 Å, (b) iron, which is body-centred cubic with a lattice parameter of about 2.867 Å?

15. Given that the unit-cell dimensions of gallium, which is orthorhombic, are $a = 4.526$ Å, $b = 4.520$ Å, $c = 7.660$ Å, find which lines occur with Bragg angles above $80°$ for the Kα radiations of iron, nickel and copper.

16. The high-angle lines on a powder photograph of a cubic material, taken with CuKα radiation, have the following values of θ. Find an accurate value for the lattice parameter.

Radiation	$\theta°$
α_1	83·825
α_2	79·318
α_1	78·582
α_2	77·309
α_1	76·685
α_2	72·550
α_1	72·098

17. The positions of the lines on a powder photograph taken by the van Arkel method with CuKα radiation are as follows:

Line	Intensity	Readings (mm) left	right
a	10	124·57	363·39
b	6	139·65	348·28
c	4	146·83	341·33
d	1	157·08	330·99
e	2	162·73	325·38
f	2	171·99	316·13
g	3	177·33	310·78
h	1	186·65	301·57
Kα_1 ⎧ i	2	192·42	295·83
⎪ j	1	202·99	285·34
⎨ k	1	210·38	278·00
⎩ l	1	227·31	260·88
knife edge		108·65	379·38

Given that the angle of the camera is 85·974°, find the indices of the lines and derive an accurate value for the lattice parameter. Use the powder data file to identify the material.

18. Repeat the calculations asked for in problem 17 with the following data, taken with CuKα radiation in the same camera.

Intensity	Readings (mm) left	right	Intensity	Readings (mm) left	right
10	382·99	147·99	1	325·59	205·28
9	381·90	149·13	1	324·35	206·48
4	372·42	158·55	1	318·71	212·20
8	371·64	159·39	1	317·76	213·03
2	363·42	167·58	1	317·27	213·67
2	357·78	173·24	⎧ 1	312·01	218·82
1	356·71	174·18	⎪ 1	310·00	220·93
2	356·10	174·91	⎪ 1	304·33	226·52
2	349·98	181·01	Kα_1 ⎨ 2	303·94	227·00
2	349·31	181·70	⎪ 1	301·96	228·93
2	344·32	186·67	⎪ 1	293·37	237·68
2	336·50	194·47	⎩ 1	282·60	248·29
1	331·73	199·16	1	277·00	253·91
1	330·84	200·09	knife edge	401·03	129·95
1	330·19	200·68			

19. Quartz is hexagonal with $a = 4.913$ Å and $c = 5.405$ Å. A powder photograph taken with CuKα radiation shows lines with the following Bragg angles:

a, 10·44°; b, 13·34°; c, 18·32°; d, 19·78°; e, 20·18°; f, 21·28°; g, 22·90°; h, 25·15°; j, 27·54°; k, 27·76°; l, 30·06°; m, 32·12°
Find the indices of these lines.

20. The following four sets of values of sin² θ were obtained from powder photographs taken with CoKα radiation. Index the lines in each set and find the dimensions of the corresponding unit cell.

a 0·1060, 0·3180, 0·4240, 0·7420, 0·9540
b 0·0393, 0·0597, 0·0786, 0·1184, 0·1572
c 0·0332, 0·1226, 0·1328, 0·1558, 0·2452, 0·2554, 0·2784, 0·4904, 0·5236
d 0·0514, 0·0769, 0·1283, 0·2307, 0·2821, 0·2825, 0·3076, 0·4363, 0·5383

21. A powder photograph taken with CoKα radiation has lines with the following Bragg angles:

	θ		θ		θ		θ
a	13·92	f	31·65	l	47·40	q	61·59
b	19·24	g	35·71	m	50·77	r	68·61
c	24·09	h	41·20	n	52·52	s	74·05
d	27·78	j	42·05	o	54·65	t	77·59
e	28·80	k	44·53	p	59·20		

Find the indices of the lines and the dimensions of the unit cell.

22. Five powder photographs show lines with the following values of sin² θ. Index the photographs and find as much information about the space groups as possible. The values of sin² θ may be assumed to be correct to within 0·0004.

a 0·0318, 0·0373, 0·0476, 0·0569, 0·0850, 0·0917, 0·0999, 0·1094, 0·1169, 0·1227, 0·1258, 0·1286, 0·1322, 0·1368, 0·1475, 0·1594, 0·1824, 0·1920, 0·1953, 0·2004
b 0·0232, 0·0340, 0·0365, 0·0571, 0·0597, 0·1002, 0·1077, 0·1230, 0·1255, 0·1280, 0·1306, 0·1475, 0·1600, 0·1914, 0·2062, 0·2284
c 0·0321, 0·0368, 0·0479, 0·0571, 0·0847, 0·0916, 0·1000, 0·1057, 0·1092, 0·1258, 0·1284, 0·1307, 0·1321, 0·1368, 0·1472, 0·1534, 0·1597, 0·1828, 0·1916, 0·1951
d 0·0250, 0·0321, 0·0342, 0·0368, 0·0479, 0·0571, 0·0597, 0·0618, 0·0847, 0·0916, 0·1000, 0·1057, 0·1078, 0·1092, 0·1166, 0·1229
e 0·0184, 0·0458, 0·0500, 0·0642, 0·0684, 0·0736, 0·0958, 0·1142, 0·1194, 0·1236

23. The structure of aluminium is based upon a face-centred cubic lattice, of parameter 4·05 Å. Given that the density of aluminium is 2·69 Mg m^{-3} calculate the number of atoms in the unit cell and then derive the complete structure. (It is not usual for a complete structure to be evident from the number of atoms in the unit cell.)

24. The unit cell of marcasite, FeS_2, contains 2 Fe and 4 S atoms. The space group is Pmnn, and the intensities of some of the low-order powder reflections are as follows: 200, strong; 400, medium; 020, zero; 040, medium; 002, strong; 004, zero; 110, very strong; 103, very strong. Find as much information as you can about the structure.

25. Four powder photographs showed lines with the following sets of spacings and rough relative intensities. Use the A.S.T.M. Powder Data File to identify the materials present.

(a)		(b)		(c)		(d)	
d in Å	I	d in Å	I	d in Å	I	d in Å	I
11·6	5	3·56	2	4·85	2	10·2	2
6·1	1	3·03	10	3·68	1	7·1	1
5·6	10	2·72	4	3·00	2	6·8	3
5·0	5	2·24	1	2·91	1	5·2	1
4·50	8	2·06	2	2·69	2	4·87	6
4·06	1	1·85	1	2·60	4	4·83	1
3·68	10	1·64	1	2·50	10	4·50	2
3·48	1	1·41	1	2·42	2	4·27	1
3·30	7			2·20	1	3·94	4
3·02	4			2·18	2	3·58	10
2·93	5			2·06	4	3·40	2
				1·94	1	3·35	1
				1·84	1	3·21	1
				1·75	1	2·84	5
				1·69	4	2·61	2
				1·67	2	2·53	2
				1·61	2	2·39	2
				1·59	4	2·19	4
				1·50	8		
				1·48	2		
				1·45	6		

10+

Solutions to Problems

1.

Symmetry	Letters
1	FGJKLPQR
2	NSZ
m (vertical)	AMTUVWY
m (horizontal)	BCDE
mm	HIOX

Four crystal classes if the third and fourth are considered to be the same.
Two crystal systems—1 and 2 oblique; m and mm rectangular.

2.

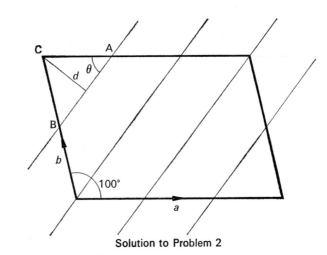

Solution to Problem 2

$CB = \frac{3}{2}$ Å; $CA = \frac{5}{3}$ Å

From the cosine law,
$$(AB)^2 = (\tfrac{5}{3})^2 + (\tfrac{3}{2})^2 - 2.\tfrac{5}{3}.\tfrac{3}{2}\cos 80°$$

i.e.
$$AB = 2 \cdot 040 \text{ Å}$$

Now
$$\frac{AB}{\sin 80} = \frac{3}{2\sin\theta}$$

from which $\sin\theta = \dfrac{3}{2}.\dfrac{0 \cdot 9848}{2 \cdot 040}$

But $\sin\theta = d/\frac{5}{3}$; therefore
$$d = \frac{5}{3}.\frac{3}{2}.\frac{0 \cdot 9848}{2 \cdot 040} = 1 \cdot 21 \text{ Å}$$

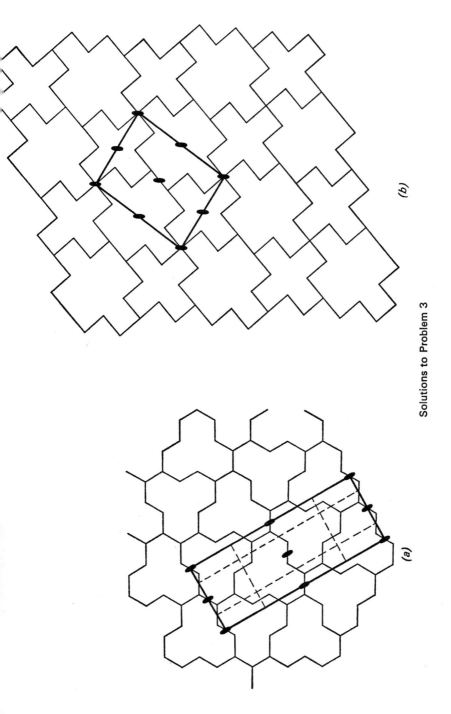

(b)

(a)

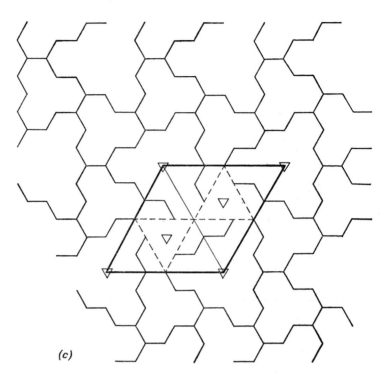

(c)

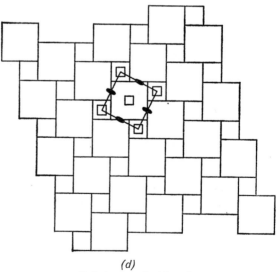

(d)

Solutions to Problem 3

4.

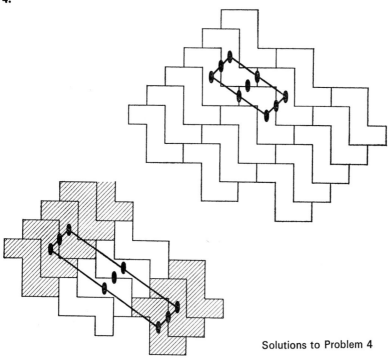

Solutions to Problem 4

Many other possibilities also exist.

5. (a) $\qquad a^* = \dfrac{bc \sin \alpha}{V}; \quad b^* = \dfrac{ca \sin \beta}{V}; \quad c^* = \dfrac{ab \sin \gamma}{V}$

where the volume of the unit cell is given by

$$V = abc(1 - \cos^2 \beta)^{1/2} = abc \sin \beta$$

$$\cos \alpha^* = 0, \quad \cos \beta^* = -\cos \beta, \quad \cos \gamma^* = 0$$

i.e. $\qquad \alpha^* = \gamma^* = 90° \quad \text{and} \quad \beta^* = 82.5°$

In dimensional units,

$$a^* = \frac{bc \sin \alpha}{abc \sin \beta} = \frac{\sin \alpha}{a \sin \beta} = 0.118 \text{ Å}^{-1}$$

$$b^* = 1/b = 0.114 \text{ Å}^{-1}$$

$$c^* = \frac{\sin \gamma}{c \sin \beta} = 0.0632 \text{ Å}^{-1}$$

In dimensionless units,

$$a^* = 0.118.1.54 = 0.182$$
$$b^* = 0.114.1.54 = 0.175$$
$$c^* = 0.0632.1.54 = 0.0975$$

(b) The relevant formulae are:

$$a^* = \frac{bc \sin \alpha}{V}; \qquad b^* = \frac{ca \sin \beta}{V}; \qquad c^* = \frac{ab \sin \gamma}{V}$$

$$\cos \alpha^* = \frac{\cos \beta \cos \gamma - \cos \alpha}{\sin \beta \sin \gamma}$$

$$\cos \beta^* = \frac{\cos \gamma \cos \alpha - \cos \beta}{\sin \gamma \sin \alpha}$$

$$\cos \gamma^* = \frac{\cos \alpha \cos \beta - \cos \gamma}{\sin \alpha \sin \beta}$$

$$V = abc(1 - \cos^2 \alpha - \cos^2 \beta - \cos^2 \gamma + 2 \cos \alpha \cos \beta \cos \gamma)^{1/2}$$

Substitution of numerical values gives

$$\begin{array}{lll} a^* = 0.210 \text{ Å}^{-1}, & \alpha^* = 67° \\ b^* = 0.134 \text{ Å}^{-1}, & \beta^* = 48° \\ c^* = 0.134 \text{ Å}^{-1}, & \gamma^* = 83.5° \end{array} \quad \text{or} \quad \begin{array}{l} \alpha^* = 113° \\ \beta^* = 132° \\ \gamma^* = 83.5° \end{array}$$

(c) Substitution in the formula given in the solution to 5(b) gives

$$\begin{array}{lll} a^* = 0.348 \text{ Å}^{-1}, & \alpha^* = 33.3° \\ b^* = 0.223 \text{ Å}^{-1}, & \beta^* = 26.2° \\ c^* = 0.224 \text{ Å}^{-1}, & \gamma^* = 36.9° \end{array} \quad \text{or} \quad \begin{array}{l} \alpha^* = 146.7° \\ \beta^* = 153.8° \\ \gamma^* = 36.9° \end{array}$$

6. Volume of unit cell $= abc \sin \beta = 8.24.5.73.9.51.0.9969$ Å^3

Mass of unit-cell contents $= 1.16.8.24.5.73.9.51.0.9969$ g $= 520.10^{-24}$ g

Molecular weight of diphenyl $(C_6H_5.C_6H_5) = 154.20$

∴ Mass of one molecule

$$\frac{154.20}{\text{Avogadro's number}} = \frac{154.20}{6.03.10^{23}} = 256.10^{-24} \text{ g}$$

∴ No. of molecules in unit cell $= 520/256 = 2$

$h0l$ absent when h odd means an a glide perpendicular to b; $0k0$ absent when k odd means two-fold screw axis parallel to b. Hence space group is $P2_1/a$. The molecule must have a centre of symmetry. The centre of symmetry is on the 2_1 axis.

7. In the table are given the values of the linear absorption coefficient, μ, the values of the product μt, where t is the thickness of the filter, and the values of I/I_0, the ratio of the transmitted intensity to the incident intensity.

CuKα RADIATION

Element	μ	μt	$I/I_0 = \exp(-\mu t)$
Al	$2.7.10^3.4.86$	0.262	0.77
Be	$1.8.10^3.1.50$	0.081	0.92
Li	$0.53.10^3.0.072$	0.038	0.96

CrKα RADIATION

Element	μ	μt	$I/I_0 = \exp(-\mu t)$
Al	$2 \cdot 7.10^3 . 1 \cdot 52$	0·821	0·44
Be	$1 \cdot 8.10^3 . 0 \cdot 45$	0·243	0·78
Li	$0 \cdot 53.10^3 . 0 \cdot 196$	0·104	0·87

8.
$$I/I_0 = \exp(-\mu t), \quad \text{where } I/I_0 = \tfrac{1}{100}$$
$$\therefore \log_e 0 \cdot 01 = -\mu t = -2 \cdot 7.10^3 . 9 \cdot 39 t$$

i.e. $\qquad t = 0 \cdot 181$ mm

The reduction in intensity of MoKα radiation is given by

$$\log_e I/I_0 = -2 \cdot 7.10^3 . 0 \cdot 516 . 0 \cdot 181 . 10^{-3}$$

from which $I/I_0 = 77 \cdot 7$ per cent.

9. The filter must be made of iron. Let I_0 be the intensity of the Kα radiation. The intensity of the Kβ radiation will be $I_0/5$. If the thickness of the filter is t, then

$$I_\alpha = I_0 \exp(-\mu_\alpha t)$$
$$5 I_\beta = I_0 \exp(-\mu_\beta t)$$

i.e. $\qquad 5 I_\beta / I_\alpha = \exp(\mu_\alpha - \mu_\beta) t$

If the filter reduces the ratio I_β / I_α to $\tfrac{1}{600}$ then

$$\tfrac{5}{600} = \exp(\mu_\alpha - \mu_\beta) t$$

For iron, $\mu_\alpha = 7 \cdot 9.10^3 . 5 \cdot 28$ and $\mu_\beta = 7 \cdot 9.10^3 . 34.9$. Therefore,

$$\tfrac{5}{600} = \exp(7 \cdot 9.10^3 . 29 \cdot 6 t)$$

i.e. $\qquad t = 2 \cdot 05.10^{-2}$ mm

The reduction in intensity of the CoKα radiation produced by this filter is given by

$$I/I_0 = \exp(7 \cdot 9.10^3 . 5 \cdot 28 . 2 \cdot 05 . 10^{-5})$$
$$= 0 \cdot 425$$

i.e. the factor is $42 \cdot 5$ per cent.

10. Linear absorption coefficient, $\mu_1 = \rho \sum p \mu_m$, where ρ is the density of the material and p is the fraction by weight of each element, mass absorption coefficient μ_m, present. Thus for the CuAl alloy

$$\mu = 5 \cdot 44.10^3 (0 \cdot 30.8 \cdot 16 + 0 \cdot 70.7 \cdot 48)$$
$$= 41 \cdot 8 \text{ mm}^{-1}$$

Molecular weight of $FeCo_2O_4$ is

$$55 \cdot 85 + 58 \cdot 94.2 + 16 \cdot 00.4 = 55 \cdot 85 + 117 \cdot 88 + 64 \cdot 00$$
$$= 237 \cdot 73$$

$$\therefore \text{ Fraction by weight of Fe } = \frac{55\cdot85}{237\cdot73}$$

$$\text{,, \quad ,, \quad ,, \quad ,, Co } = \frac{117\cdot88}{237\cdot73}$$

$$\text{,, \quad ,, \quad ,, \quad ,, O } = \frac{64\cdot00}{237\cdot73}$$

$$\therefore \mu \text{ for FeCo}_2\text{O}_4 = \frac{5\cdot34}{237\cdot73}\cdot10^3 \ (55\cdot85.5\cdot28 + 117\cdot88.6\cdot11 + 64\cdot00.1\cdot78)$$

$$= 25\cdot2 \text{ mm}^{-1}$$

11. CrKα or MnKα radiation. The wavelength of either of these radiations is larger than that of the absorption edges of both iron and chromium. Alternatively MoKα radiation. The wavelength of this radiation is very much shorter than those of the iron and chromium absorption edges. ZnKα radiation could be tried with a screen of copper foil (3.4.4).

12. CoKα radiation.

13. From $\lambda = 2d \sin \theta$ we have

$$\frac{\lambda\alpha_2}{\lambda\alpha_1} = \frac{\sin 73\cdot595}{\sin 73\cdot084} = 1\cdot0026$$

But

$$\frac{\text{wavelength of ZnK}\alpha_2}{\text{wavelength of ZnK}\alpha_1} = 1\cdot0026$$

and the impurity radiation is therefore ZnKα.

14. For a cubic crystal,

$$\sin^2 \theta = \frac{\lambda^2}{4a^2} (h^2 + k^2 + l^2)$$

When the unit cell is face-centred, h, k, l must either be all even or all odd. When the unit cell is body-centred, $h + k + l$ must be even.
(a) The approximate value of the lattice parameter of aluminium is 4·049 Å. The unit cell is face centred.
At least one reflection must appear in the high-angle region. That is the Bragg angle for at least one reflection must lie between 80° and 85°, say. The values of $\sin \theta$ corresponding to these angles are 0·9848 and 0·9962 respectively. Thus, from $\lambda = 2a \sin \theta(h^2 + k^2 + l^2)^{-1/2}$ the wavelength must lie between $8\cdot098.0\cdot9848 \ (h^2 + k^2 + l^2)^{-1/2}$ and $8\cdot098.0\cdot9962 \ (h^2 + k^2 + l^2)^{-1/2}$.
The wavelength limits for the possible values of $(h^2 + k^2 + l^2)^{1/2}$ are given in the table.

$(h^2 + k^2 + l^2)^{1/2}$	λ (in Å) between
2·82	2·83 and 2·85
3·32	2·40 ,, 2·42
3·46	2·30 ,, 2·32

$(h^2 + k^2 + l^2)^{1/2}$	λ (in Å) *between*
4·00	1·99 and 2·01
4·36	1·83 ,, 1·85
4·50	1·77 ,, 1·79
4·90	1·62 ,, 1·64
5·20	1·53 ,, 1·55
5·65	1·41 ,, 1·43
5·92	1·34 ,, 1·36
6·00	1·33 ,, 1·34
6·33	1·26 ,, 1·27

Suitable radiations are CoKα (1·79 Å) and CuKα (1·54 Å).
(*b*) Similar calculations for iron give as suitable radiations CoKα and possibly NiKα.

15. For an orthorhombic substance,

$$\sin^2 \theta = \frac{\lambda^2}{4a^2} h^2 + \frac{\lambda^2}{4b^2} k^2 + \frac{\lambda^2}{4c^2} l^2$$

$$= Ah^2 + Bk^2 + Cl^2$$

Values of A, B and C for gallium for the Kα radiations of iron, nickel and copper are given in the table.

	A	B	C
FeKα	0·0458	0·0460	0·0160
NiKα	0·0337	0·0338	0·0117
CuKα	0·0290	0·0292	0·0103

From this table, the *hkl* reflections which have sin θ values greater than 80° are:

With FeKα radiation   422, 242, 404, 044, 325, 235, 333, 207, 027, 306, 036.
With NiKα radiation   109, 019, 245, 425, 250, 520, 251, 521, 153, 513.
With CuKα radiation   047, 407, 129, 219, 345, 435, 350, 530, 442.

16. From the values of sin² θ, the lines can be identified as either 71, 69, 68 and 65 or 72, 70, 69 and 66; they cannot, however, be the former because 71 is a forbidden number (5.3.1). The steps in the calculation are set out below. The values of log a are obtained by adding $\frac{1}{2}$ log N, log $\lambda/2$ and log cosec θ; the results are plotted against the values of $f(\theta) = \frac{1}{2}(\cos^2 \theta/\theta + \cos^2 \theta/\sin \theta)$ shown in the last column. On extrapolation to $f(\theta) = 0$, the value of the lattice parameter, a, is 6·5736₆ Å ± 0·00006 Å. Since the measurements were obtained from a particularly good photograph the smallness of the error cannot be taken as typical.

N	$\frac{1}{2}$ log N	log $(N)^{1/2}.\lambda/2$	log cosec θ	log a	a	$f(\theta)$
72	0·928666	0·815298	0·002527	0·817825	6·57393	0·010
70	0·922549	0·810262	0·007591	853	435	0·030
70	0·922549	0·809181	0·008682	863	450	0·034
69	0·919425	0·807138	0·010742	880	476	0·043
69	0·919425	0·806057	0·011834	891	493	0·047

10*

N	$\frac{1}{2}\log N$	$\log (N)^{1/2}.\lambda/2$	$\log\operatorname{cosec}\theta$	$\log a$	a	$f(\theta)$
66	0·909772	0·797485	0·020461	946	576	0·083
66	0·909772	0·796404	0·021552	0·817956	6·57591	0·087

For CuKα_1, $\lambda/2 = 0.77025$; $\log\lambda/2 = \bar{1}.886632$
For CuKα_2, $\lambda/2 = 0.77217$; $\log\lambda/2 = \bar{1}.887713$

When the calculations are made by different people, differences of one or two units in the fifth decimal place in a, and in corresponding places in other columns in the table, may occur. This illustrates the importance of carrying out the calculations to a higher number of figures than is physically significant.

17. The length, S, of the arc for each pair of reflections is the difference between the left-hand and right-hand readings; the distance, S_k, between the knife edges is 379·38 − 108·65 mm. To correct for film shrinkage, the Bragg angle, θ, is calculated from the relation $\theta° = 90 - 85.974\ S/S_k$ after which $\sin^2\theta$ can be determined for each reflection. These calculations are shown in the table.

Line	Length of arc (S)	θ (degrees)	$\sin^2\theta$
a	238·82	14·2	0·0602
b	208·63	23·7	0·1615
c	194·50	28·2	0·2234
d	173·91	34·8	0·3257
e	162·65	38·4	0·3857
f	146·14	44·3	0·4878
g	133·45	47·6	0·5453
h	114·92	53·4	0·6445
i	103·81	57·1	0·7049
j	82·35	64·0	0·8078
k	67·62	68·6	0·8670
l	33·57	79·4	0·9659
knife edge	270·73 (S_k)		

Inspection of the $\sin^2\theta$ values for the low angles shows that there is a common factor of 0·0202; this suggests that the structure is cubic with $\lambda^2/4a^2 = 0.0202$, where $\sin^2\theta = (\lambda^2/4a^2)(h^2 + k^2 + l^2)$.

As shown in the second table, on this assumption the pattern can be readily indexed. The indices are either all odd or all even, indicating that the unit cell is face centred. From $\lambda^2/4a^2 = 0.0202$, the lattice parameter is approximately 5·43 Å.

$\sin^2\theta_{\exp}$	hkl	$\sin^2\theta_{\text{calc}}$	$N = h^2 + k^2 + l^2$	$d = a/N^{1/2}$ (Å)
0·0602	111	0·0606	3	3·10
0·1615	220	0·1616	8	1·92
0·2234	311	0·2222	11	1·63
0·3257	400	0·3232	16	1·36
0·3857	331	0·3838	19	1·24
0·4878	422	0·4848	24	1·11
0·5453	511, 333	0·5454	27	1·045
0·6445	440	0·6464	32	0·096

$\sin^2 \theta_{exp}$	hkl	$\sin^2 \theta_{calc}$	$N = h^2 + k^2 + l^2$	$d = a/N^{1/2}$ (Å)
0·7049	531	0·7070	35	0·092
0·8078	620	0·8080	40	0·086
0·8670	533	0·8686	43	0·083
0·9659	444	0·9696	48	0·0785

The three strongest lines have spacing 3·10, 1·92 and 1·63 Å. From the relative intensities of these lines, the powder data file indicates that the material is silicon. This is confirmed by comparison with all of the reflections listed on the file card.

To determine the lattice parameter accurately, the procedure described in the solution to problem 16 should be applied to the high-angle lines i, j, k and l. This gives a value for a of 5·4335 ± 0·0005 Å.

18. The values of $\sin^2 \theta_{exp}$ listed in the table have been derived from the experimental observations as described in the solution to problem 17.

	$\sin^2 \theta_{exp}$	hkl	$\sin^2 \theta_{calc}$
	0·0712	200	0·0700
	0·0777	101	0·0767
	0·1421	220	0·1400
	0·1489	211	0·1467
	0·2180	301	0·2167
	0·2725	112	0·2718
	0·2825	400	0·2800
	0·2894	321	0·2869
	0·3523	420	0·3500
	0·3595	411	0·3572
	0·4132	312	0·4118
	0·4950	501	0·4972
	0·5515	103, 332	0·5503, 0·5518
	0·5616	440	0·5600
	0·5685	521	0·5667
	0·6184	213	0·6203
	0·6314	600	0·6314
	0·6909	303	0·6893
	0·6999	620	0·7000
	0·7054	611	0·7072
α_1	0·7566	323	0·7580
α_1	0·7760	541	0·7750
α_1	0·8256	413	0·8275
α_1	0·8284	532	0·8300
α_1	0·8451	631	0·8440
α_1	0·9080	640	0·9080
α_1	0·9411	004	0·9440
α_1	0·9839	721	0·9840

These values do not reveal a common factor and therefore the material is not cubic. However, the first, third, seventh, ninth, fifteenth and seventeenth lines have values of $\sin^2 \theta$ which are in the ratios $1:2:4:5:8:9$. This suggests that the structure may be tetragonal and therefore

$$\sin^2 \theta = \frac{\lambda^2}{4a^2}(h^2 + k^2) + \frac{\lambda^2}{4c^2}l^2$$
$$= A(h^2 + k^2) + Cl^2$$

Assume in the first instance that $A = 0.0712$ and that this is the value of $\sin^2 \theta$ for the 100 reflection; a better value can be obtained by division of 0.6314 by 9 giving $A = 0.0702$. Indices 110 can be assigned to $\sin^2 \theta = 0.1421$, 200 to $\sin^2 \theta = 0.2825$ and so on; this means that $\sin^2 \theta = 0.0777$ could correspond to one of the reflection 001, 002, 003, ... Suppose the indices are 001. Then $\sin^2 \theta = 0.1489$ indexes as 101, $\sin^2 \theta = 0.2180$ indexes as 111, but indices cannot be assigned to $\sin^2 \theta = 0.2725$. By further trial it is found that $\sin^2 \theta = 0.2725$ can be indexed if $A = 0.0702/4 = 0.0175$ and $C = 0.0777 - 0.0175 = 0.0602$. On this basis $\sin^2 \theta = 0.6909$ is the 303 reflection and from this reflection a better value for C of 0.0592 can be obtained. The indices shown in the table were assigned assuming $A = 0.0175$, $C = 0.0602$ for the reflections for which $K\alpha_1$ and $K\alpha_2$ are not resolved, and $A = 0.01745$, $C = 0.0590$ for those reflections for which the α-doublet is resolved. From the table it can be seen that values of $\sin^2 \theta_{\text{calc}}$ are too high for those reflections with high values of l. A better value for C is 0.0588.

From the powder data file the substance can be identified as β-tin. To determine accurately the parameter, a, we write

$$\sin^2 \theta = \frac{\lambda^2}{4a^2}\left(h^2 + k^2 + \frac{a^2}{c^2}l^2\right) = \frac{\lambda^2}{4a^2}\left(h^2 + k^2 + \frac{C}{A}l^2\right)$$

where $A = 0.01745$ and $C = 0.0588$.

From this relation, and using the experimental values of $\sin^2 \theta$ for the 541, 631, 640 and 721 reflections, four values of a can be determined. The final value of a is obtained by extrapolating the graph of a against $\frac{1}{2}(\cos^2 \theta/\theta + \cos^2 \theta/\sin \theta)$ to $\theta = 90°$.

To determine c accurately, $\sin^2 \theta$ is expressed in the form

$$\sin^2 \theta = \frac{\lambda^2}{4c^2}\left(\frac{c^2}{a^2}(h^2 + k^2) + l^2\right) = \frac{\lambda^2}{4c^2}\left(\frac{A}{C}(h^2 + k^2) + l^2\right)$$

From the experimental values of $\sin^2 \theta$ for the reflections 323, 413 and 004, three values of c can be obtained. As before, extrapolation to $\theta = 90°$ will give the accurate value of c. After extrapolation

$$a = 5.8308 \pm 0.0003 \text{ Å}; \qquad c = 3.1770 \pm 0.0005 \text{ Å}$$

19. $A = \lambda^2/3a^2 = 0.0328; \qquad C = \lambda^2/4c^2 = 0.02035$

The indices of the lines are:
a, 100; b, 101; c, 110; d, 102; e, 111; f, 200; g, 201; h, 112; j, 202; k, 103; l, 211; m, 113.

20. a: Cubic with $A = 0.03533$

The indices of the lines are 110, 211, 220, 321, 330 + 411. Dimensions of unit cell given by $0.0530 = \lambda^2/4a^2$. Therefore,

$$a = 0.8954/0.2302 = 3.88 \text{ Å}$$

b: Three of the $\sin^2 \theta$ values are in the ratio $1:2:4$ which suggests that the structure is tetragonal. The pattern will index on the assumption that the structure is, in fact, tetragonal with

$$A = 0.01965 \quad \text{and} \quad C = 0.0204$$

The indices are: 110, 111, 200, 211, 220.
The unit-cell dimensions are:

$$a = 0.8954/0.1402 = 6.387 \text{ Å}; \quad c = 0.8954/0.1428 = 6.270 \text{ Å}$$

c: The second, fifth and ninth lines have $\sin^2 \theta$ values in the ratio $1:2:4$, indicating a tetragonal structure. On this assumption the pattern can be indexed with $A = 0.1226$ and $C = 0.0332$.
The indices are: 001, 100, 002, 101, 110, 102, 111, 200, 201.
From $A = 0.1226 = \lambda^2/4a^2$, $a = 2.559$ Å
From $C = 0.0332 = \lambda^2/4c^2$, $c = 4.917$ Å

d: Four of the $\sin^2 \theta$ values are in the ratio $1:3:4:7$ which suggests that the structure is hexagonal. The pattern can be indexed assuming a hexagonal structure with $A = 0.0769$ and $C = 0.0514$.
The indices are: 001, 100, 101, 110, 111, 102, 200, 112, 210.
From $A = 0.0769 = \lambda^2/3a^2$; $a = 3.731$ Å
From $C = 0.0514 = \lambda^2/4c^2$; $c = 3.954$ Å

21. The values of $\sin^2 \theta$ for the first six lines are a, 0.0579; b, 0.1086; c, 0.1666; d, 0.2173; e, 0.2321; f, 0.2754.
    Since $\sin^2 \theta_d$ is twice $\sin^2 \theta_b$, the structure is probably tetragonal, with b as 100 and d as 110, and hence $A = 0.1086$. From the table of differences given below the several occurrences of 0.0580 ± 0.0001 suggest that $C = 0.0580$, and the first six lines can be indexed on this basis as shown in the last column.

Line	$\sin^2 \theta$	$\sin^2 \theta - 0.1086$	$\sin^2 \theta - 0.2173$	hkl
a	0.0579			001
b	0.1086			100
c	0.1666	0.0580		101
d	0.2173	0.1087		110
e	0.2321	0.1235	0.0148	002
f	0.2754	0.1668	0.0581	111

The remaining lines are: g, 102; h, 200; j, 112; k, 201; l, 210; m, 211; n, 103; p, 202; q, 113; r, 112; s, 220; t, 221, 004; u, 203.
    More accurate values of A and C are 0.1084 and 0.0579 respectively, giving $a = 2.72$ Å and $c = 3.72$ Å. The material is FePt (Lipson, Shoenberg and Stupart, 1941).

22. a: The crystal is not obviously cubic, tetragonal or hexagonal and the next step is to determine whether the structure is orthorhombic. A table of

differences is therefore set up by subtracting the first $\sin^2 \theta$ value successively from the second and following values of $\sin^2 \theta$, then subtracting the second $\sin^2 \theta$ value from those following, and so on. When this is done, it is found that, allowing for the experimental error, the differences 0·0093 and 0·0373 occur with high frequency which suggests that B, say, $= 0·0093$ and $4B = 0·0373$. This makes the second reflection 020.

Subtraction of 0·0093 from the $\sin^2 \theta$ value for the first reflection leaves 0·0225, and since 0·0228 occurs frequently in the differences table we assume that $A = 0·0228$. This makes the first reflection 110. The difference $0·0476 - 0·0093 = 0·0383$ and does not lead anywhere, but $0·0476 - 0·0228 = 0·0248$ which, within the limits of experimental error, occurs frequently. If $C = 0·0248$ then the third reflection is 101. With these values for A, B and C, the low-angle reflections can be indexed but adjustments have to be made when the higher-angle reflections are considered. The final values of A, B and C are respectively 0·0229, 0·0092, 0·0250.

The indices of the lines are: 110, 020, 101, 111, 121, 400, 002, 012, 201, 102, 211, 220, 112, 022, 040, 122, 032, 202, 141, 212.

Space group. There are no systematic absences from the general hkl reflections. This means a primitive lattice. The $0kl$ reflections are present only when l is even, thus indicating the presence of a c-glide plane perpendicular to a. The $hk0$ reflections present have $h + k$ even, indicating an n-glide plane perpendicular to c. From this information the space group is either $Pcn2_1$ or $Pcnm$.

b: The symmetry again does not appear to be higher than orthorhombic. From the table of differences, the difference 0·0232 occurs frequently so that we can try $A = 0·0232$, making the first reflection 100. Differences 0·0365 and 0·0250 also occur frequently. Now $0·0365/4 = 0·0091$ and $0·0250 + 0·0091 = 0·0341$ so that we can try $B = 0·0091$ and $C = 0·0250$. With these values of A, B and C, indexing of the low angles is possible and after adjustment to $A = 0·0229$, $B = 0·0092$, $C = 0·0250$ indexing can be completed.

The indices are: 100, 011, 020, 111, 120, 002, 031, 102, 211, 220, 131, 040, 122, 202, 300, 222.

Space group. The general reflections have $k + l$ even, so that the lattice type is A-face centred. The space group could be A222, Ammm, or Amm2.

c: By comparison with 23(a) and 23(b) this structure is also orthorhombic with $A = 0·0229$, $B = 0·0092$, $C = 0·0250$.

The indices are: 110, 020, 101, 111, 121, 200, 002, 130, 012, 211, 220, 131, 112, 022, 040, 221, 122, 032, 202, 141.

Space group. There are no restrictions on the general reflections and therefore the lattice is primitive. Other reflections present are $0kl$ with l even indicating a c-glide perpendicular to a, $hk0$ with $h + k$ even indicating an n-glide perpendicular to c, and $h0l$ with $h + l$ even indicating an n-glide perpendicular to a. The space group is Pcnn.

d: Again, this structure is orthorhombic with $A = 0·0229$, $B = 0·0092$, $C = 0·0250$. The indices of the lines are: 001, 110, 011, 020, 101, 111, 120, 021, 121, 200, 002, 130, 031, 012, 201, 102.

Space group. For the two general reflections $h + l$ is even which could mean that the lattice is centred on the B faces; however $h + l$ is not always even for the $h0l$ reflections and therefore the lattice is, in fact, not centred. Reflections $h00$ and $0k0$ are present only when h is even and k is even respectively and each gives rise to a 2_1 axis. The space group is P 2_122_1.

e: From a table of differences, 0·0184 and 0·0458 each occur three times and 0·0500 occurs five times. If these quantities are taken as A, B and C respectively, the pattern can be indexed as follows: 100, 010, 001, 110, 101, 200, 011, 111, 210, 201.

Space group. In the one general reflection, h, k and l are all odd but there is no supporting evidence from the $hk0$ and $h0l$ reflections; that is, in these two groups $h + k$ is not always even, neither is $h + l$ always even. The only other apparent condition for reflections to occur is that in $0kl$, $k + l$ should be even. This, in fact, is not a condition either, because neither the $0k0$ nor the $00l$ reflections are all even. The space group is thus P222, Pmmm or Pmm2.

23. Mass of unit-cell contents = vol. of unit cell × density of aluminium = $66·42.10^{-24}.2·69g$.

$$\text{Mass of one atom of aluminium} = \frac{\text{atomic weight of aluminium}}{\text{Avogadro's number}}$$

$$= 26·97/6·03.10^{23}$$

$$\therefore \text{ No. of atoms in unit cell} = \frac{66·42.2·69}{269·7}.6·03 = 4.$$

These atoms are at 000, $0\frac{1}{2}\frac{1}{2}$, $\frac{1}{2}0\frac{1}{2}$, $\frac{1}{2}\frac{1}{2}0$.

24. The International Tables show that the space group Pmnn has eight general positions. Since there are only two formula units in the unit cell of FeS_2, it follows that neither the iron nor the sulphur atoms can occupy general positions. According to their nature, the special positions may be in sets of four and in sets of two; clearly the two iron atoms must occupy a set of two. These are centres of symmetry, and the origin can be chosen such that the iron atoms lie on 0, 0, 0 and $\frac{1}{2}, \frac{1}{2}, \frac{1}{2}$. Of the four-fold positions, on which the sulphur atoms lie, there can be two sets on two-fold axes and one set on a mirror plane; the coordinates of the two sets of atoms on the two-fold axes are:

$$x, 0, \tfrac{1}{2}; \quad x, 0, \tfrac{1}{2}; \quad \tfrac{1}{2} - x, \tfrac{1}{2}, 0; \quad \tfrac{1}{2} + x, \tfrac{1}{2}, 0$$

and

$$x, 0, 0; \quad x, 0, 0; \quad \tfrac{1}{2} - x, \tfrac{1}{2}, \tfrac{1}{2}; \quad \tfrac{1}{2} + x, \tfrac{1}{2}, \tfrac{1}{2}$$

From the structure-factor formula

$$F(hkl) = 2f_{\text{Fe}} + 4f_{\text{S}} \cos 2\pi hx \cos 2\pi ky \cos 2\pi lz$$

it is at once clear that if the sulphur atoms occupy either of these sets of four-fold special positions then neither 020 nor 004 can be of zero intensity. Thus the sulphur atoms must lie on the mirror plane and have coordinates

$$0, y, z; \quad 0, \bar{y}, \bar{z}; \quad \tfrac{1}{2}, \tfrac{1}{2} + y, \tfrac{1}{2} - z; \quad \tfrac{1}{2}, \tfrac{1}{2} - y, \tfrac{1}{2} + z$$

By adjustment of the y coordinate to a value of 0·25 the sulphur atoms will lie on planes midway between the (020) planes on which the iron atoms lie. Radiation diffracted from these successive planes of atoms in the direction [020] will thus be out of phase, and since the diffracting power of four sulphur atoms is roughly equal to that of two iron atoms the resulting intensity will be very weak. Similarly 002 can be strong and 004 zero only if the (004) planes have planes of sulphur atoms midway between them; this will be so if $z = \frac{1}{8}$ or $z = \frac{3}{8}$. The reflection 103 will be strong if $z = \frac{3}{8}$, and since with $y = \frac{1}{4}$ the 110 reflection is also strong it is concluded that the sulphur atoms are roughly on

$$0, \tfrac{1}{4}, \tfrac{3}{8}; \quad 0, \bar{\tfrac{1}{4}}, \bar{\tfrac{3}{8}}; \quad \tfrac{1}{2}, \tfrac{3}{4}, \tfrac{1}{8}; \quad \tfrac{1}{2}, \tfrac{1}{4}, \tfrac{7}{8}$$

It must be noted that 110 and 103 are classified as very strong only by virtue of the fact that on a powder photograph there are two sets of cooperating planes; on a single-crystal photograph they would not be so strong as either the 200 or the 002 reflections.

25. (a) 2,5-dimethyl phenyl, $C_8H_{10}O$
    (b) Potassium amide, KNH_2
    (c) Fe_2CuO_4 with some αFe_2O_3
    (d) $BeSO_4 2H_2O$ with a little $Fe_2(SO_4)_3 8H_2O$.

Appendix 1

For successful plating it is essential that the copper base on which it is required to electrodeposit should first be thoroughly prepared. Old plating must be removed from used targets, and this can be done either chemically or by mechanical abrasion. The copper must then be rinsed in water and afterwards given a preliminary degreasing by the application of a solvent such as trichloroethylene. After rinsing, the copper must be pickled in dilute sulphuric acid to remove oxide and scale, and this is followed by bright dipping to give a lustrous clean surface. Rinse again.

A satisfactory solution for bright dipping is:

> 250 cm^3 of water
> 500 cm^3 of concentrated sulphuric acid
> 185 cm^3 of concentrated nitric acid
> 1·5 g of sodium chloride

The next step is to polish with emery paper, finishing with 000 grade, and after yet another rinse the copper should be cleaned in hot caustic soda solution and rinsed again. If the surface has become tarnished during the alkali cleansing then the tarnish should be removed either by scouring or by electrolytic cleaning. After a final rinse, plating may be started. The plating procedure must avoid deposition by displacement; for example when steel is placed in a copper-plating solution it acquires a coating of copper without the passage of current. Such deposits have very poor adhesion.

In the standard text by Canning (1966) are listed common defects in plated surfaces; defects such as imperfect adhesion, pitting of the deposit, rough deposit, uneven deposit and so on, together with the possible causes and the methods of correction. The book also contains details of plating solutions which are made up from materials listed under trade names.

The plating solutions given below are from Blum and Hogaboom (1949). They are made from materials which carry their usual chemical names.

Chromium

Bath : Chromic acid, CrO$_3$ 50 g
         Sulphuric acid, H$_2$SO$_4$ 0·25–0·5 g
         (instead of sulphuric acid the equivalent of chromic sulphate
            may be used)
         Distilled water   200 cm^3
         Temperature : 45°C–55°C

Chromium (contd.)

Bath : Anode: Lead or tin–lead alloy
Current: 1–2 kA m^{-2}

Note : Increased temperature involves increased current. The chromium should preferably be coated on nickel.

Nickel

Bath : Nickel sulphate, Ni SO$_4$. 7 H$_2$O   21 g
Nickel chloride, Ni Cl$_2$ 6H$_2$0   2 g
Ammonium chloride   3 g
Boric acid, H$_3$BO$_3$ 3 g
Distilled water   200 cm^3
Temperature: 20°C–30°C
Anode: Nickel
Current: 0·05–0·2 kA m^{-2}

Note : If insufficient chloride is present, the anode will not be dissolved efficiently.

Manganese

Bath : Manganese sulphate, Mn SO$_4$ H$_2$O   22 g
or Manganese sulphate anhydride   15 g
Ammonium sulphate (NH$_4$)$_2$ SO$_4$ 26 g
Ammonium sulphite (NH$_4$)$_2$ SO$_3$ 0·2 g
Distilled water   200 cm^3
Anode solution: strong ammonium sulphate solution in porous pot
Temperature: Room temperature
Anode: Lead
Current: 0·5–0·7 kA m^{-2}

Note : After plating, wash quickly and dip in dilute K$_2$Cr$_2$O$_7$ solution; then wash again.

Iron

Bath : Ferrous ammonium sulphate   56 g
Ammonium chloride   1 g
Distilled water   200 cm^3
Temperature: 30°C
Anode: Iron
Current: 0·06–0·1 kA m^{-2}

Bath : Boric acid, 10 g dissolved in 220 cm^3 of boiling water
Sodium chloride   5 g
Cobalt sulphate   144 g
Temperature: 34°C
Anode: Cobalt
Current: 0·25 kA m^{-2}

Zinc

Bath : Zinc sulphate, ZnSO$_4$. 7 H$_2$O   82 g
Sodium sulphate, Na$_2$SO$_4$ 15 g
Aluminium chloride, Al Cl$_3$ 6 H$_2$O   4 g
Distilled water   200 cm^3

Zinc (contd.)

Small quantities of concentrated sulphuric acid should be added at intervals to maintain a low pH.

Temperature: Room temperature
Anode: Zinc
Current: 0·15–0·30 kA m^{-2}

Gold

Bath : Gold cyanide, AuCN   0·45 g
or Gold fulminate, 2Au NH NH$_2$.3H$_2$O   0·52 g
Potassium cyanide   3 g
Sodium phosphate, Na$_2$ HPO$_4$.12H$_2$O   0·8 g
Distilled water   200 cm^3

Temperature: 60°C–80°C
Anode: Gold
Current: 0·015 kA m^{-2}

If desired the gold may be plated on to Nickel.

Silver

Bath : Silver nitrate   7 g
Potassium cyanide   5·2 g
Distilled water   100 cm^3

Method of preparation of bath: Dissolve slightly more than 5·2 g of potassium cyanide in distilled water to make a strong solution, and add slowly to a solution of the silver nitrate until precipitation is just complete. Drain away the supernatant liquors, and dissolve the precipitate in more cyanide solution, with warming if necessary. About 10 per cent of the whole cyanide solution is added to form free cyanide.

Temperature: Room temperature
Anode: Silver
Current: 0·1–0·2 kA m^{-2}

Appendix 2

Cobalt Kα radiation scale:

```
— 0·90
—— centre
— 0·90
— 0·91
— 0·92
— 0·93
— 0·94
— 0·95
— 1·00
— 1·05
— 1·10
— 1·15
— 1·20
— 1·25
— 1·30
— 1·40
— 1·50
— 2·00
— 3·00
— 3·00
— 4·00
— 5·00
— 6·00
— 7·00
—10·00
—— centre
```

Copper Kα radiation scale:

```
— 0·77
—— centre
— 0·77
— 0·78
— 0·79
— 0·80
— 0·82
— 0·84
— 0·86
— 0·88
— 0·90
— 1·00
— 1·10
— 1·20
— 1·30
— 1·40
— 1·50
— 2·00
— 3·00
— 3·50
— 4·00
— 5·00
— 6·00
— 7·00
—10·00
—— centre
```

If approximate values of interplanar spacings are adequate then such values can be obtained rapidly from a scale that has been calibrated directly in Å (4.8.4). The calibration will depend upon the wavelength of the incident X-radiation, the diameter of the camera and the type of mounting employed (4.3.1). With the van Arkel mounting in a camera of diameter 90 mm, the distance on the film between corresponding lines is $(2\pi - 4\theta).45$ mm, where $\theta = \sin^{-1} (\lambda/2d)$. These distances can be determined for various values of the interplanar spacing, d, and hence a scale can be produced which will give d directly from the powder pattern.

The halves of two scales, one for CuKα radiation and one for CoKα radiation are drawn full size in fig. A.2; they are for use with powder patterns obtained with the van Arkel type of mounting in a 90 mm diameter camera.

Appendix 3

As shown in section 5.3.1, the lines on a powder photograph can be numbered according to the sum of the squares of the indices. Certain values for the sum, expressible as $m^2(8n - 1)$, are not possible, and this appendix gives a proof of this result. The proof depends upon two theorems:

(1) The square of an even number is expressible as $4p$.
(2) The square of an odd number is expressible as $8q + 1$.

To prove (2) let the odd number be $2r + 1$;

then
$$(2r + 1)^2 = 4r^2 + 4r + 1$$
$$= 4r(r + 1) + 1$$

If r is odd, $r + 1$ is even and vice versa; theorem (2) then follows.

On the basis of these two theorems we can classify the sum of the squares of three numbers as follows:

Three numbers	Sum of squares
All even	$4n$
Two even, one odd	$4n + 1$
One even, two odd	$4n + 2$
Three odd	$8n + 3$

Two results follow from this list:

No set of numbers can have a sum of squares equal to $8n - 1$.

If the sum of three squares is the same for two sets of integers, those sets must have the same distribution of odd and even integers. For example, 9 is expressible as $2^2 + 2^2 + 1^2$ and $3^2 + 0^2 + 0^2$; both sets contain two even integers and one odd integer.

To prove the theorem that $m^2(8n - 1)$ cannot be expressed as the sum of three squares, we consider two possibilities:

If m is odd, it is evident that the quantity $m^2(8n - 1)$ is expressible as $8N - 1$, from theorem (2).

If m is even, $m^2(8n - 1)$ is divisible by 4 and thus could be the sum of the squares of three even numbers (table). If so, it can be divided by 4 and the values of h, k and l divided by 2. If the result of dividing by 4 is odd, then it must be expressible as $8N - 1$, and no values of h, k and l are possible; if the result is even, the process of dividing by 4 can be continued, and so on until an odd number is reached. This odd number cannot be expressed as the sum of three squares.

Therefore, even if m is even, $m^2(8n - 1)$ cannot be expressed as the sum of three squares.

Tables

Table 1 Data for X-ray Targets and Filters

Element, Atomic number	Line	Wavelength, Å	Excitation Potential, kV	Suitable working peak, kV	Element	Thickness, mm
			Target		β-Filter	
Ag, 47	$K\alpha_1$	0·559 41	25·5	90	Rh	0·079
	$K\alpha_2$	0·563 81			(or Pd)	
	$K\beta_1$	0·497 01				
	A.E.†	0·485 5				
Pd, 46	$K\alpha_1$	0·585 45	24·4	90	Rh	0·073
	$K\alpha_2$	0·589 82			(or Ru)	
	$K\beta_1$	0·520 52				
	A.E.	0·509 0				
Rh, 45	$K\alpha_1$	0·613 26	23·2	90	Ru	0·064
	$K\alpha_2$	0·617 62				
	$K\beta_1$	0·545 59				
	A.E.	0·534 1				
Mo, 42	$K\alpha_1$	0·709 26	20·0	80	Zr	0·108
	$K\alpha_2$	0·713 54				
	$K\beta_1$	0·632 25				
	A.E.	0·619 7				
Zn, 30	$K\alpha_1$	1·435 10	9·7	50	Cu	0·021
	$K\alpha_2$	1·438 94				
	$K\beta_1$	1·295 20				
	A.E.	1·283 1				
Cu, 29	$K\alpha_1$	1·540 50	9·0	50	Ni	0·021
	$K\alpha_2$	1·544 34				
	$K\beta_1$	1·392 17				
	A.E.	1·380 2				
Ni, 28	$K\alpha_1$	1·657 83	8·3	50	Co	0·018
	$K\alpha_2$	1·661 68				
	$K\beta_1$	1·500 08				
	A.E.	1·486 9				
Co, 27	$K\alpha_1$	1·788 90	7·7	45	Fe	0·018
	$K\alpha_2$	1·792 79				
	$K\beta_1$	1·620 73				
	A.E.	1·607 2				
Fe, 26	$K\alpha_1$	1·935 97	7·1	40	Mn	0·016
	$K\alpha_2$	1·939 91				
	$K\beta_1$	1·756 54				
	A.E.	1·742 9				
Mn, 25	$K\alpha_1$	2·101 74	6·5	40	Cr	0·016
	$K\alpha_2$	2·105 70				
	$K\beta_1$	1·910 16				
	A.E.	1·895 4				
Cr, 24	$K\alpha_1$	2·289 62	6·0	35	V	0·016
	$K\alpha_2$	2·293 52				
	$K\beta_1$	2·084 79				
	A.E.	2·070 1				

† A.E. = absorption edge

Table 2 Some Data for the Elements

Element	Symbol	Atomic number	Atomic weight	Density† Mg m^{-3}	Absorption edge, Å‡
Aluminium	Al	13	26·97	2·7	
Antimony	Sb	51	121·76	6·6	*2·9967*
Argon	A	18	39·944	0·0018	
Arsenic	As	33	74·91	5·7	1·0447
Barium	Ba	56	137·36	3·5	0·3314
Beryllium	Be	4	9·02	1·8	
Bismuth	Bi	83	209·00	9·8	*0·9240*
Boron	B	5	10·82	2·0	
Bromine	Br	35	79·916	0·0071	0·9200
Cadmium	Cd	48	112·41	8·6	0·4640
Caesium	Cs	55	132·91	1·9	*2·4724*
Calcium	Ca	20	40·08	1·5	3·0705
Carbon	C	6	12·01	2·3	
Cerium	Ce	58	140·13	6·9	
Chlorine	Cl	17	35·457	0·0032	
Chromium	Cr	24	52·01	7·0	2·0701
Cobalt	Co	27	58·94	8·7	1·6072
Copper	Cu	29	63·57	8·9	1·3802
Fluorine	F	9	19·00	0·0017	
Gallium	Ga	31	69·72	5·9	1·1926
Germanium	Ge	32	72·60	5·5	1·1169
Gold	Au	79	197·2	19·3	*1·0403*
Helium	He	2	4·003	0·00018	
Hydrogen	H	1	1·008	0·00009	
Indium	In	49	114·76	7·3	0·4439
Iodine	I	53	126·92	4·9	0·3742
Iridium	Ir	77	193·1	22·4	*1·1060*
Iron	Fe	26	55·85	7·9	1·7429
Krypton	Kr	36	83·7	0·0037	
Lead	Pb	82	207·21	11·3	*0·9511*
Lithium	Li	3	6·940	0·53	
Magnesium	Mg	12	24·32	1·74	
Manganese	Mn	25	54·93	7·4	1·8954
Mercury	Hg	80	200·61	13·6	1·0095
Molybdenum	Mo	42	95·95	10·1	*0·6197*
Neon	Ne	10	20·183	0·00090	
Nickel	Ni	28	58·69	8·9	1·4869
Niobium	Nb (Cb)	41	92·91	8·5	0·6529
Nitrogen	N	7	14·008	0·0012	
Osmium	Os	76	190·2	22·5	*1·1413*
Oxygen	O	8	16·000	0·0014	
Palladium	Pd	46	106·7	12·0	0·5090
Phosphorus	P	15	30·98	2·0	
Platinum	Pt	78	195·23	21·5	*1·0732*
Potassium	K	19	39·096	0·86	3·4379
Radium	Ra	88	226·05	5	
Rhodium	Rh	45	102·91	12·4	0·5341
Rubidium	Rb	37	85·48	1·53	0·8157
Ruthenium	Ru	44	101·7	12·3	0·5595

Table 2 *(continued)*

Element	Symbol	Atomic number	Atomic weight	Density† Mg m⁻³	Absorption edge, Å‡
Scandium	Sc	21	45·10	2·5	2·7573
Selenium	Se	34	78·96	4·5	0·9797
Silicon	Si	14	28·06	2·3	
Silver	Ag	47	107·880	10·5	0·4855
Sodium	Na	11	22·997	0·97	
Strontium	Sr	38	87·63	2·56	0·7700
Sulphur	S	16	32·06	2·0	
Tantalum	Ta	73	180·88	16·6	*1·2542*
Tellurium	Te	52	127·61	6·2	0·3901
Thallium	Tl	81	204·39	11·9	*0·9798*
Thorium	Th	90	232·12	11·3	*0·7615*
Tin	Sn	50	118·70	7·3	0·4248
Titanium	Ti	22	47·90	4·5	2·4962
Tungsten	W	74	183·92	19·3	*1·2154*
Uranium	U	92	238·07	18·7	*0·7223*
Vanadium	V	23	50·95	5·6	2·2676
Xenon	X	54	131·3	0·0058	
Yttrium	Y	39	88·92	3·8	0·7270
Zinc	Zn	30	65·38	7·1	1·2831
Zirconium	Zr	40	91·22	6·4	0·6888

† Densities are given only for the purpose of evaluating linear absorption coefficients; for more accurate purposes standard tables should be consulted. The values given are those under ordinary laboratory conditions, but for some elements (for example, carbon) density is also dependent upon crystal structure.

‡ Wavelengths of absorption edges are given only when they come within the range of commonly used emission wave-lengths. The K absorption edge is usually given, but where the absorption edge is given the figures are italicized.

Table 3 Mass Absorption Coefficients, μ_m, for Kα Radiations (from International Tables) ($m^2 kg^{-1} . 10$)

Atomic number	Element	Mo λ 0·711 Å	Zn 1·436 Å	Cu 1·542 Å	Ni 1·659 Å	Co 1·790 Å	Fe 1·937 Å	Mn 2·103 Å	Cr 2·291 Å
1	H	0·38	0·42	0·43	0·45	0·46	0·48	0·51	0·54
2	He	0·21	0·35	0·38	0·43	0·49	0·57	0·67	0·81
3	Li	0·22	0·61	0·72	0·85	1·03	1·25	1·56	1·96
4	Be	0·30	1·25	1·50	1·82	2·25	2·80	3·53	4·50
5	B	0·39	1·97	2·39	2·93	3·63	4·55	5·76	7·38
6	C	0·62	3·76	4·60	5·68	7·07	8·90	11·3	14·5
7	N	0·92	6·13	7·52	9·31	11·6	14·6	18·6	23·9
8	O	1·31	9·34	11·5	14·2	17·8	22·4	28·5	36·6
9	F	1·80	13·3	16·4	20·3	25·4	32·1	40·8	52·4
10	Ne	2·47	18·6	22·9	28·4	35·4	44·6	56·7	72·8
11	Na	3·21	24·5	30·1	37·3	46·5	58·6	74·4	95·3
12	Mg	4·11	31·4	38·6	47·7	59·5	74·8	94·8	121
13	Al	5·16	39·6	48·6	60·1	74·8	93·9	119	152
14	Si	6·44	49·4	60·6	74·9	93·3	117	148	189
15	P	7·89	60·5	74·1	91·5	114	142	180	229
16	S	9·6	72·8	89·1	110	136	170	214	272
17	Cl	11·4	86·3	106	130	161	200	252	318
18	A	13·5	101	123	151	187	232	291	366
19	K	15·8	117	143	175	215	266	332	417
20	Ca	18·3	133	162	198	243	299	371	463
21	Sc	21·1	152	184	223	273	336	414	513
22	Ti	24·2	172	208	252	308	377	463	571
23	V	27·5	193	233	282	343	419	513	68·4
24	Cr	31·1	216	260	314	381	463	62·3	79·8
25	Mn	34·7	237	285	343	414	57·2	72·6	93·0

11+

Table 3 (continued)

Atomic number	Element	Mo λ 0·711 Å	Zn 1·436 Å	Cu 1·542 Å	Ni 1·659 Å	Co 1·790 Å	Fe 1·937 Å	Mn 2·103 Å	Cr 2·291 Å
26	Fe	38·5	258	308	370	52·8	66·4	84·3	108
27	Co	42·5	278	313	49·0	61·1	76·8	97·4	125
28	Ni	46·6	297	457	56·5	70·5	88·6	112	144
29	Cu	50·9	43·1	52·9	65·5	81·6	103	130	166
30	Zn	55·4	49·1	60·3	74·6	93·0	117	148	189
31	Ga	60·1	55·3	67·9	83·9	105	131	166	212
32	Ge	64·8	61·6	75·6	93·4	116	146	184	235
33	As	69·7	68·0	83·4	103	128	160	203	258
34	Se	74·7	74·6	91·4	113	140	175	221	281
35	Br	79·8	81·3	99·6	123	152	190	240	305
36	Kr	84·9	88·2	108	133	165	206	258	327
37	Rb	90·0	95·4	117	143	177	221	277	351
38	Sr	95·0	103	125	154	190	236	296	373
39	Y	100	110	134	165	203	252	315	396
40	Zr	15·9	118	143	176	216	268	334	419
41	Nb	17·1	126	153	187	230	284	353	441
42	Mo	18·4	134	162	198	243	300	372	463
44	Ru	21·1	151	183	223	272	334	412	509
45	Rh	22·6	161	194	236	288	352	433	534
46	Pd	24·1	170	206	250	304	371	455	559
47	Ag	25·8	181	218	264	321	391	478	586
48	Cd	27·5	191	231	279	338	412	502	613
49	In	29·3	202	243	294	356	432	525	638
50	Sn	31·1	213	256	309	373	451	547	662

51	Sb	33·1	225	270	324	391	472	570	688
52	Te	35·0	236	282	339	407	490	589	707
53	I	37·1	247	294	352	422	506	606	722
54	Xe	39·2	257	306	366	436	521	620	763
55	Cs	41·3	268	318	378	450	534	632	793
56	Ba	43·5	278	330	391	463	546	675	461
58	Ce	48·2	299	352	414	486	601	418	219
73	Ta	95·4	137	166	201	244	297	363	444
74	W	99·1	143	172	208	253	308	375	458
76	Os	106	155	186	225	272	330	401	487
77	Ir	110	161	193	233	282	341	414	502
78	Pt	113	167	200	241	291	353	427	517
79	Au	115	173	208	250	302	365	441	532
80	Hg	117	180	216	259	312	377	455	547
81	Tl	119	187	224	268	323	389	469	563
82	Pb	120	194	232	278	334	402	483	579
83	Bi	120	201	240	288	346	415	498	596
88	Ra	172	258	304	371	433	509	598	708
90	Th	143	286	327	399	460	536	633	755
92	U	153	310	352	423	488	566	672	805

Table 4 Values of sin² θ

(The decimal point has been omitted)

θ°	·0	·1	·2	·3	·4	·5	·6	·7	·8	·9	Differences ·01	·02	·03	·04	·05
00	0000	0000	0000	0000	0000	0001	0001	0001	0002	0002					
01	0003	0004	0004	0005	0006	0007	0008	0009	0010	0011					
02	0012	0013	0015	0016	0018	0019	0021	0022	0024	0026		Interpolate			
03	0027	0029	0031	0033	0035	0037	0039	0042	0044	0046					
04	0049	0051	0054	0056	0059	0062	0064	0067	0070	0073					
05	0076	0079	0082	0085	0089	0092	0095	0099	0102	0106					
06	0109	0113	0117	0120	0124	0128	0132	0136	0140	0144					
07	0149	0153	0157	0161	0166	0170	0175	0180	0184	0189					
08	0194	0199	0203	0208	0213	0218	0224	0229	0234	0239					
09	0245	0250	0256	0261	0267	0272	0278	0284	0290	0296					
10	0302	0308	0314	0320	0326	0332	0338	0345	0351	0358	1	1	2	2	3
11	0364	0371	0377	0384	0391	0397	0404	0411	0418	0425	1	1	2	3	3
12	0432	0439	0447	0454	0461	0468	0476	0483	0491	0498	1	1	2	3	4
13	0506	0514	0521	0529	0537	0545	0553	0561	0569	0577	1	2	2	3	4
14	0585	0593	0602	0610	0618	0627	0635	0644	0653	0661	1	2	3	3	4
15	0670	0679	0687	0696	0705	0714	0723	0732	0741	0751	1	2	3	4	4
16	0760	0769	0778	0788	0797	0807	0816	0826	0835	0845	1	2	3	4	5
17	0855	0865	0874	0884	0894	0904	0914	0924	0934	0945	1	2	3	4	5
18	0955	0965	0976	0986	0996	1007	1017	1028	1039	1049	1	2	3	4	5
19	1060	1071	1082	1092	1103	1114	1125	1136	1147	1159	1	2	3	4	6
20	1170	1181	1192	1204	1215	1226	1238	1249	1261	1273	1	2	3	5	6
21	1284	1296	1308	1320	1331	1343	1355	1367	1379	1391	1	2	4	5	6
22	1403	1415	1428	1440	1452	1464	1477	1489	1502	1514	1	2	4	5	6
23	1527	1539	1552	1565	1577	1590	1603	1616	1628	1641	1	3	4	5	6
24	1654	1667	1680	1693	1707	1720	1733	1746	1759	1773	1	3	4	5	7

	0	1	2	3	4	5	6	7	8	9	1	2	3	4	5
25	1786	1799	1813	1826	1840	1853	1867	1881	1894	1908	1	3	4	5	7
6	1922	1935	1949	1963	1977	1991	2005	2019	2033	2047	1	3	4	6	7
7	2061	2075	2089	2104	2118	2132	2146	2161	2175	2190	1	3	4	6	7
8	2204	2219	2233	2248	2262	2277	2291	2306	2321	2336	1	3	4	6	7
9	2350	2365	2380	2395	2410	2425	2440	2455	2470	2485	2	3	5	6	8
30	2500	2515	2530	2545	2561	2576	2591	2607	2622	2637	2	3	5	6	8
1	2653	2668	2684	2699	2715	2730	2746	2761	2777	2792	2	3	5	6	8
2	2808	2824	2840	2855	2871	2887	2903	2919	2934	2950	2	3	5	6	8
3	2966	2982	2998	3014	3030	3046	3062	3079	3095	3111	2	3	5	6	8
4	3127	3143	3159	3176	3192	3208	3224	3241	3257	3274	2	3	5	7	8
35	3290	3306	3323	3339	3356	3372	3389	3405	3422	3438	2	3	5	7	8
6	3455	3472	3488	3505	3521	3538	3555	3572	3588	3605	2	3	5	7	8
7	3622	3639	3655	3672	3689	3706	3723	3740	3757	3773	2	3	5	7	8
8	3790	3807	3824	3841	3858	3875	3892	3909	3926	3943	2	3	5	7	8
9	3960	3978	3995	4012	4029	4046	4063	4080	4097	4115	2	3	5	7	9
40	4132	4149	4166	4183	4201	4218	4235	4252	4270	4287	2	3	5	7	9
1	4304	4321	4339	4356	4373	4391	4408	4425	4443	4460	2	3	5	7	9
2	4477	4495	4512	4529	4547	4564	4582	4599	4616	4634	2	3	5	7	9
3	4651	4669	4686	4703	4721	4738	4756	4773	4791	4808	2	3	5	7	9
4	4826	4843	4860	4878	4895	4913	4930	4948	4965	4983	2	3	5	7	9
45	5000	5017	5035	5052	5070	5087	5105	5122	5140	5157	2	3	5	7	9
6	5174	5192	5209	5227	5244	5262	5279	5297	5314	5331	2	3	5	7	9
7	5349	5366	5384	5401	5418	5436	5453	5471	5488	5505	2	3	5	7	9
8	5523	5540	5557	5575	5592	5609	5627	5644	5661	5679	2	3	5	7	9
9	5696	5713	5730	5748	5765	5782	5799	5817	5834	5851	2	3	5	7	9
50	5868	5885	5903	5920	5937	5954	5971	5988	6005	6022	2	3	5	7	9
1	6040	6057	6074	6091	6108	6125	6142	6159	6176	6193	2	3	5	7	9
2	6210	6227	6243	6260	6277	6294	6311	6328	6345	6361	2	3	5	7	8
3	6378	6395	6412	6428	6445	6462	6479	6495	6512	6528	2	3	5	7	8
4	6545	6562	6578	6595	6611	6628	6644	6661	6677	6694	2	3	5	7	8

Table 4 (continued)

θ°	·0	·1	·2	·3	·4	·5	·6	·7	·8	·9	Differences ·01	·02	·03	·04	·05
55	6710	6726	6743	6759	6776	6792	6808	6824	6841	6857	2	3	5	7	8
6	6873	6889	6905	6921	6938	6954	6970	6986	7002	7018	2	3	5	7	8
7	7034	7050	7066	7081	7097	7113	7129	7145	7160	7176	2	3	5	6	8
8	7192	7208	7223	7239	7254	7270	7285	7301	7316	7332	2	3	5	6	8
	7347	7363	7378	7393	7409	7424	7439	7455	7470	7485	2	3	5	6	8
60	7500	7515	7530	7545	7560	7575	7590	7605	7620	7635	2	3	5	6	8
1	7650	7664	7679	7694	7709	7723	7738	7752	7767	7781	2	3	5	6	8
2	7796	7810	7825	7839	7854	7868	7882	7896	7911	7925	1	3	4	6	7
3	7939	7953	7967	7981	7995	8009	8023	8037	8051	8065	1	3	4	6	7
4	8078	8092	8106	8119	8133	8147	8160	8174	8187	8201	1	3	4	6	7
65	8214	8227	8241	8254	8267	8280	8293	8307	8320	8333	1	3	4	5	7
6	8346	8359	8372	8384	8397	8410	8423	8435	8448	8461	1	3	4	5	7
7	8473	8486	8498	8511	8523	8536	8548	8560	8572	8585	1	3	4	5	6
8	8597	8609	8621	8633	8645	8657	8669	8680	8692	8704	1	2	4	5	6
9	8716	8727	8739	8751	8762	8774	8785	8796	8808	8819	1	2	4	5	6
70	8830	8841	8853	8864	8875	8886	8897	8908	8918	8929	1	2	3	5	6
1	8940	8951	8961	8972	8983	8993	9004	9014	9024	9035	1	2	3	4	6
2	9045	9055	9066	9076	9086	9096	9106	9116	9126	9135	1	2	3	4	5
3	9145	9155	9165	9174	9184	9193	9203	9212	9222	9231	1	2	3	4	5
4	9240	9249	9259	9268	9277	9286	9295	9304	9313	9321	1	2	3	4	5
75	9330	9339	9347	9356	9365	9373	9382	9390	9398	9407	1	2	3	4	4
6	9415	9423	9431	9439	9447	9455	9463	9471	9479	9486	1	2	3	3	4
7	9494	9502	9509	9517	9524	9532	9539	9546	9553	9561	1	2	2	3	4
8	9568	9575	9582	9589	9596	9603	9609	9616	9623	9629	1	1	2	3	4
9	9636	9642	9649	9655	9662	9668	9674	9680	9686	9692	1	1	2	3	3

	0	1	2	3	4	5	6	7	8	9
80	9698	9704	9710	9716	9722	9728	9733	9739	9744	9750
1	9755	9761	9766	9771	9776	9782	9797	9782	9797	9801
2	9806	9811	9816	9820	9825	9830	9834	9839	9843	9847
3	9851	9856	9860	9864	9868	9872	9876	9880	9883	9887
4	9891	9894	9898	9901	9905	9908	9911	9915	9918	9921
85	9924	9927	9930	9933	9936	9938	9941	9944	9946	9949
6	9951	9954	9956	9958	9961	9963	9965	9967	9969	9971
7	9973	9974	9976	9978	9979	9981	9982	9984	9985	9987
8	9988	9989	9990	9991	9992	9993	9994	9995	9996	9996
9	9997	9998	9998	9999	9999	9999	1·00	1·00	1·00	1·00

Interpolate

1	1	2	2	3

Table 5 Values of $(h^2 + k^2 + l^2)$, $(h^2 + k^2)$ and $(h^2 + hk + k^2)$ up to 100

N	Cubic $N = h^2 + k^2 + l^2$			Tetragonal $N = h^2 + k^2$	Hexagonal $N = h^2 + hk + k^2$
	P hkl	F hkl	I hkl	hk	hk
1	100			10	10
2	110		110	11	
3	111	111			11
4	200	200	200	20	20
5	210			21	
6	211				
7					21
8	220	220	220	22	
9	300			30	30
9	221				
10	310		310	31	
11	311	311			
12	222	222	222		22
13	320			32	31
14	321		321		
15					
16	400	400	400	40	40
17	410			41	
34	530		530	53	
34	433		433		
35	531	531			
36	600	600	600	60	60
36	442	442	442		
37	610			61	43
38	611		611		
38	532		532		
39					52
40	620	620	620	62	
41	621				
41	540			54	
41	443				
42	541		541		
43	533	533			61
44	622	622	622		
45	630			63	
45	542				

44
70
53

62

71

70

71
55

64
72

73

631	444	710 550 543	640	721 633 552	642	730
	444	711 551 640			642	731 553
631	444 700 632 710 550 543	711 551 640 720 641		721 633 552	642	722 544 730 731 553

46
47

48
49
49
50
50
50

51
51
52
53
53

54
54
54
55
56

57
57
58
59
59

32
41

50

33
42

51

33

42

50
43
51

52

44

411
330

420

332

422

510
431

521

440

331
420

422

511
333

440

322
411
330

331
420
421
332

422
500
430
510
431

511
333

520
432

521

440
522
441

17
18
18

19
20
21
22
23

24
25
25
26
26

27
27
28
29
29

30
31
32
33
33

11*

Table 5 (continued)

N	Cubic $N=h^2+k^2+l^2$ P hkl	Cubic F hkl	Tetragonal $N=h^2+k^2$ hk	Hexagonal $N=h^2+hk+k^2$ hk
60				
61	650		65	54
61	643			
62	732			
62	651			
63				63
64	800	800	80	80
65	810		81	
65	740		74	
65	652			
66	811			
66	741			
66	554			
67	733	733		72
68	820	820	82	
68	644	644		

N	Cubic $N=h^2+k^2+l^2$ P hkl	Cubic F hkl	Cubic I hkl	Tetragonal $N=h^2+k^2$ hk	Hexagonal $N=h^2+hk+k^2$ hk
82	910		910	91	
82	833		833		
83	911	911			
83	753	753			
84	842	842	842		82
85	920			92	
85	760			76	
86	921		921		
86	761		761		
86	655		655		
87					
88	664	664	664		
89	922			85	
89	850				
89	843				
89	762				

69	821							
69	742							
70	653	653						
71								
72	822	822	822	66				
72	660	660	660					
73	830	831		83	81			
73	661							
74	831	750		75			74	65 91
74	750	743						
74	743							
75	751	751			55			
75	555	555						
76	662	662	662		64			
77	832							83
77	654							
78	752	752						
79	840	840	840	84	73		93	
80	900	840	840	90	90			
81	841						94	
81	744						77	
81	663							

90	930	930	930	851	754	93	
90	851	851					
90	754	754					
91	931	931	931				
91							
92	852			852 932 763			
93							
94	932	932					
94	763	763				94	
95							
96	844	844	844	844			
97	940			665			
97	665						
98	941	941		853 770			
98	853	853				77	
98	770	770					
99	933	933		933 771 755			
99	771	771					
99	755	755		10,00 860		10,0 86	
100	10,00	10,00	10,00				10,0
100	860	860	860				

Table 6 Values of $\frac{1}{2}[(\cos^2 \theta/\sin \theta + (\cos^2 \theta/\theta)]$

$\theta°$	·0	·1	·2	·3	·4	·5	·6	·7	·8	·9
10	5·572	5·513	5·456	5·400	5·345	5·291	5·237	5·185	5·134	5·084
11	5·034	4·986	4·939	4·892	4·846	4·800	4·756	4·712	4·669	4·627
12	4·585	4·544	4·504	4·464	4·425	4·386	4·348	4·311	4·274	4·238
13	4·202	4·167	4·133	4·098	4·065	4·032	3·999	3·967	3·935	3·903
14	3·872	3·842	3·812	3·782	3·753	3·724	3·695	3·667	3·639	3·612
15	3·584	3·558	3·531	3·505	3·479	3·454	3·429	3·404	3·379	3·355
16	3·331	3·307	3·284	3·260	3·237	3·215	3·192	3·170	3·148	3·127
17	3·105	3·084	3·063	3·042	3·022	3·001	2·981	2·962	2·942	2·922
18	2·903	2·884	2·865	2·847	2·828	2·810	2·792	2·774	2·756	2·738
19	2·721	2·704	2·687	2·670	2·653	2·636	2·620	2·604	2·588	2·572
20	2·556	2·540	2·525	2·509	2·494	2·479	2·464	2·449	2·434	2·420
21	2·405	2·391	2·376	2·362	2·348	2·335	2·321	2·307	2·294	2·280
22	2·267	2·254	2·241	2·228	2·215	2·202	2·189	2·177	2·164	2·152
23	2·140	2·128	2·116	2·104	2·092	2·080	2·068	2·056	2·045	2·034
24	2·022	2·011	2·000	1·989	1·978	1·967	1·956	1·945	1·934	1·924
25	1·913	1·903	1·892	1·882	1·872	1·861	1·851	1·841	1·831	1·821
26	1·812	1·802	1·792	1·782	1·773	1·763	1·754	1·745	1·735	1·726
27	1·717	1·708	1·699	1·690	1·681	1·672	1·663	1·654	1·645	1·637
28	1·628	1·619	1·611	1·602	1·594	1·586	1·577	1·569	1·561	1·553
29	1·545	1·537	1·529	1·521	1·513	1·505	1·497	1·489	1·482	1·474
30	1·466	1·459	1·451	1·444	1·436	1·429	1·421	1·414	1·407	1·400
31	1·392	1·385	1·378	1·371	1·364	1·357	1·350	1·343	1·336	1·329
32	1·323	1·316	1·309	1·302	1·296	1·289	1·282	1·276	1·269	1·263
33	1·256	1·250	1·244	1·237	1·231	1·225	1·218	1·212	1·206	1·200
34	1·194	1·188	1·182	1·176	1·170	1·164	1·158	1·152	1·146	1·140
35	1·134	1·128	1·123	1·117	1·111	1·106	1·100	1·094	1·088	1·083
36	1·078	1·072	1·067	1·061	1·056	1·050	1·045	1·040	1·034	1·029
37	1·024	1·019	1·013	1·008	1·003	0·998	0·993	0·988	0·982	0·977
38	0·972	0·967	0·962	0·958	0·953	0·948	0·943	0·938	0·933	0·928
39	0·924	0·919	0·914	0·909	0·905	0·900	0·895	0·891	0·886	0·881
40	0·877	0·872	0·868	0·863	0·859	0·854	0·850	0·845	0·841	0·837
41	0·832	0·828	0·823	0·819	0·815	0·810	0·806	0·802	0·798	0·794
42	0·789	0·785	0·781	0·777	0·773	0·769	0·765	0·761	0·757	0·753
43	0·749	0·745	0·741	0·737	0·733	0·729	0·725	0·721	0·717	0·713
44	0·709	0·706	0·702	0·698	0·694	0·690	0·687	0·683	0·679	0·676
45	0·672	0·668	0·665	0·661	0·657	0·654	0·650	0·647	0·643	0·640
46	0·636	0·632	0·629	0·625	0·622	0·619	0·615	0·612	0·608	0·605
47	0·602	0·598	0·595	0·591	0·588	0·585	0·582	0·578	0·575	0·572
48	0·569	0·565	0·562	0·559	0·556	0·553	0·549	0·546	0·543	0·540
49	0·537	0·534	0·531	0·528	0·525	0·522	0·518	0·515	0·512	0·509
50	0·506	0·504	0·501	0·498	0·495	0·492	0·489	0·486	0·483	0·480
51	0·477	0·474	0·472	0·469	0·466	0·463	0·460	0·458	0·455	0·452
52	0·449	0·447	0·444	0·441	0·439	0·436	0·433	0·430	0·428	0·425
53	0·423	0·420	0·417	0·415	0·412	0·410	0·407	0·404	0·402	0·399
54	0·397	0·394	0·392	0·389	0·387	0·384	0·382	0·379	0·377	0·375
55	0·372	0·370	0·367	0·365	0·363	0·360	0·358	0·356	0·353	0·351
56	0·349	0·346	0·344	0·342	0·339	0·337	0·335	0·333	0·330	0·328
57	0·326	0·324	0·322	0·319	0·317	0·315	0·313	0·311	0·309	0·306
58	0·304	0·302	0·300	0·298	0·296	0·294	0·292	0·290	0·288	0·286
59	0·284	0·282	0·280	0·278	0·276	0·274	0·272	0·270	0·268	0·266
60	0·264	0·262	0·260	0·258	0·256	0·254	0·252	0·250	0·249	0·247
61	0·245	0·243	0·241	0·239	0·237	0·236	0·234	0·232	0·230	0·229
62	0·227	0·225	0·223	0·221	0·220	0·218	0·216	0·215	0·213	0·211
63	0·209	0·208	0·206	0·204	0·203	0·201	0·199	0·198	0·196	0·195
64	0·193	0·191	0·190	0·188	0·187	0·185	0·184	0·182	0·180	0·179

Table 6 *(continued)*

$\theta°$	·0	·1	·2	·3	·4	·5	·6	·7	·8	·9
65	0·177	0·176	0·174	0·173	0·171	0·170	0·168	0·167	0·165	0·164
66	0·162	0·161	0·160	0·158	0·157	0·155	0·154	0·152	0·151	0·150
67	0·148	0·147	0·146	0·144	0·143	0·141	0·140	0·139	0·138	0·136
68	0·135	0·134	0·132	0·131	0·130	0·128	0·127	0·126	0·125	0·123
69	0·122	0·121	0·120	0·119	0·117	0·116	0·115	0·114	0·112	0·111
70	0·110	0·109	0·108	0·107	0·106	0·104	0·103	0·102	0·101	0·100
71	0·099	0·098	0·097	0·096	0·095	0·094	0·092	0·091	0·090	0·089
72	0·088	0·087	0·086	0·085	0·084	0·083	0·082	0·081	0·080	0·079
73	0·078	0·077	0·076	0·075	0·075	0·074	0·073	0·072	0·071	0·070
74	0·069	0·068	0·067	0·066	0·065	0·065	0·064	0·063	0·062	0·061
75	0·060	0·059	0·059	0·058	0·057	0·056	0·055	0·055	0·054	0·053
76	0·052	0·052	0·051	0·050	0·049	0·048	0·048	0·047	0·046	0·045
77	0·045	0·044	0·043	0·043	0·042	0·041	0·041	0·040	0·039	0·039
78	0·038	0·037	0·037	0·036	0·035	0·035	0·034	0·034	0·033	0·032
79	0·032	0·031	0·031	0·030	0·029	0·029	0·028	0·028	0·027	0·027
80	0·026	0·026	0·025	0·025	0·024	0·023	0·023	0·023	0·022	0·022
81	0·021	0·021	0·020	0·020	0·019	0·019	0·018	0·018	0·017	0·017
82	0·017	0·016	0·016	0·015	0·015	0·015	0·014	0·014	0·013	0·013
83	0·013	0·012	0·012	0·012	0·011	0·011	0·010	0·010	0·010	0·010
84	0·009	0·009	0·009	0·008	0·008	0·008	0·007	0·007	0·007	0·007
85	0·006	0·006	0·006	0·006	0·005	0·005	0·005	0·005	0·005	0·004
86	0·004	0·004	0·004	0·003	0·003	0·003	0·003	0·003	0·003	0·002
87	0·002	0·002	0·002	0·002	0·002	0·002	0·001	0·001	0·001	0·001
88	0·001	0·001	0·001	0·001	0·001	0·001	0·001	0·000	0·000	0·000

Reprinted by permission of the Physical Society and the authors, Nelson and Riley (1945)

Table 7 Values of $(1 + \cos^2 2\theta)/\sin^2 \theta \cos \theta$

$\theta°$	.0	.1	.2	.3	.4	.5	.6	.7	.8	.9
2	1639	1486	1354	1239	1138	1048	968·9	898·3	835·1	778·4
3	727·2	680·9	638·8	600·5	565·6	533·6	504·3	477·3	452·3	429·3
4	408·0	388·2	369·9	352·7	336·8	321·9	308·0	294·9	282·6	271·1
5	260·3	250·1	240·5	231·4	222·9	214·7	207·1	199·8	192·9	186·3
6	180·1	174·2	168·5	163·1	158·0	153·1	148·4	144·0	139·7	135·6
7	131·7	128·0	124·4	120·9	117·6	114·4	111·4	108·5	105·6	102·9
8	100·3	97·80	95·37	93·03	90·78	88·60	86·51	84·48	82·52	80·63
9	78·79	77·02	75·31	73·66	72·05	70·49	68·99	67·53	66·12	64·74
10	63·41	62·12	60·87	59·65	58·46	57·32	56·20	55·11	54·06	53·03
11	52·04	51·06	50·12	49·19	48·30	47·43	46·58	45·75	44·94	44·16
12	43·39	42·64	41·91	41·20	40·50	39·82	39·16	38·51	37·88	37·27
13	36·67	36·08	35·50	34·94	34·39	33·85	33·33	32·81	32·31	31·82
14	31·34	30·87	30·41	29·96	29·51	29·08	28·66	28·24	27·83	27·44
15	27·05	26·66	26·29	25·92	25·56	25·21	24·86	24·52	24·19	23·86
16	23·54	23·23	22·92	22·61	22·32	22·02	21·74	21·46	21·18	20·91
17	20·64	20·38	20·12	19·87	19·62	19·38	19·14	18·90	18·67	18·44
18	18·22	18·00	17·78	17·57	17·36	17·15	16·95	16·75	16·56	16·36
19	16·17	15·99	15·80	15·62	15·45	15·27	15·10	14·93	14·76	14·60
20	14·44	14·28	14·12	13·97	13·81	13·66	13·52	13·37	13·23	13·09
21	12·95	12·81	12·68	12·54	12·41	12·28	12·15	12·03	11·91	11·78
22	11·66	11·54	11·43	11·31	11·20	11·09	10·98	10·87	10·76	10·65
23	10·55	10·45	10·35	10·24	10·15	10·05	9·951	9·857	9·763	9·671
24	9·579	9·489	9·400	9·313	9·226	9·141	9·057	8·973	8·891	8·810

25	8·054	8·126	8·198	8·271	8·345	8·420	8·496	8·573	8·651	8·730
26	7·389	7·452	7·515	7·580	7·645	7·711	7·778	7·846	7·915	7·984
27	6·800	6·856	6·912	6·969	7·027	7·086	7·145	7·205	7·266	7·327
28	6·279	6·329	6·379	6·429	6·480	6·532	6·584	6·637	6·692	6·745
29	5·817	5·861	5·905	5·950	5·995	6·042	6·088	6·135	6·183	6·230
30	5·406	5·445	5·484	5·524	5·564	5·605	5·647	5·688	5·731	5·774
31	5·040	5·075	5·110	5·145	5·181	5·218	5·254	5·292	5·329	5·367
32	4·715	4·746	4·777	4·809	4·841	4·873	4·906	4·939	4·972	5·006
33	4·426	4·453	4·481	4·509	4·538	4·566	4·595	4·625	4·655	4·685
34	4·169	4·193	4·218	4·243	4·268	4·294	4·320	4·346	4·372	4·399
35	3·941	3·962	3·984	4·006	4·029	4·052	4·074	4·097	4·121	4·145
36	3·739	3·758	3·777	3·797	3·816	3·836	3·857	3·877	3·898	3·919
37	3·561	3·577	3·594	3·612	3·629	3·647	3·665	3·683	3·701	3·720
38	3·404	3·419	3·434	3·449	3·465	3·481	3·497	3·513	3·527	3·544
39	3·268	3·280	3·293	3·306	3·320	3·333	3·347	3·361	3·375	3·389
40	3·149	3·160	3·171	3·183	3·194	3·206	3·218	3·230	3·242	3·255
41	3·048	3·057	3·067	3·076	3·086	3·096	3·106	3·117	3·127	3·138
42	2·962	2·970	2·978	2·986	2·994	3·003	3·012	3·020	3·029	3·038
43	2·891	2·897	2·904	2·911	2·918	2·925	2·932	2·939	2·946	2·954
44	2·833	2·838	2·844	2·849	2·855	2·860	2·866	2·872	2·878	2·884
45	2·789	2·793	2·797	2·801	2·805	2·810	2·814	2·819	2·824	2·828
46	2·757	2·760	2·763	2·766	2·769	2·772	2·775	2·778	2·782	2·785
47	2·737	2·738	2·740	2·742	2·744	2·746	2·748	2·750	2·752	2·755
48	2·728	2·729	2·729	2·730	2·730	2·731	2·732	2·733	2·735	2·736
49	2·730	2·730	2·729	2·729	2·728	2·728	2·728	2·728	2·728	2·728
50	2·743	2·741	2·740	2·738	2·737	2·735	2·734	2·733	2·732	2·731
51	2·766	2·763	2·760	2·758	2·755	2·753	2·751	2·749	2·747	2·745
52	2·799	2·795	2·792	2·788	2·785	2·782	2·778	2·775	2·772	2·769
53	2·843	2·838	2·833	2·828	2·824	2·820	2·815	2·811	2·807	2·803
54	2·896	2·890	2·885	2·879	2·874	2·868	2·863	2·858	2·853	2·848

Table 7 (continued)

$\theta°$	.0	.1	.2	.3	.4	.5	.6	.7	.8	.9
55	2.902	2.908	2.914	2.921	2.927	2.933	2.940	2.946	2.953	2.960
56	2.967	2.974	2.981	2.988	2.996	3.004	3.011	3.019	3.026	3.034
57	3.042	3.050	3.059	3.067	3.075	3.084	3.092	3.101	3.110	3.119
58	3.128	3.137	3.147	3.156	3.166	3.175	3.185	3.195	3.205	3.215
59	3.225	3.235	3.246	3.256	3.267	3.278	3.289	3.300	3.311	3.322
60	3.333	3.345	3.356	3.368	3.380	3.392	3.404	3.416	3.429	3.441
61	3.454	3.466	3.479	3.492	3.505	3.518	3.532	3.545	3.559	3.573
62	3.587	3.601	3.615	3.629	3.643	3.658	3.673	3.688	3.703	3.718
63	3.733	3.749	3.764	3.780	3.796	3.812	3.828	3.844	3.861	3.878
64	3.894	3.911	3.928	3.946	3.963	3.980	3.998	4.016	4.034	4.052
65	4.071	4.090	4.108	4.127	4.147	4.166	4.185	4.205	4.225	4.245
66	4.265	4.285	4.306	4.327	4.348	4.369	4.390	4.412	4.434	4.456
67	4.478	4.500	4.523	4.546	4.569	4.592	4.616	4.640	4.664	4.688
68	4.712	4.737	4.762	4.787	4.812	4.838	4.864	4.890	4.916	4.943
69	4.970	4.997	5.024	5.052	5.080	5.109	5.137	5.166	5.195	5.224
70	5.254	5.284	5.315	5.345	5.376	5.408	5.440	5.471	5.504	5.536
71	5.569	5.602	5.636	5.670	5.705	5.740	5.775	5.810	5.846	5.883
72	5.919	5.956	5.994	6.032	6.071	6.109	6.149	6.189	6.229	6.270
73	6.311	6.352	6.394	6.437	6.480	6.542	6.568	6.613	6.658	6.703
74	6.750	6.797	6.844	6.892	6.941	6.991	7.041	7.091	7.142	7.194
75	7.247	7.300	7.354	7.409	7.465	7.521	7.578	7.636	7.694	7.753
76	7.813	7.874	7.936	7.999	8.063	8.128	8.193	8.259	8.327	8.395
77	8.465	8.536	8.607	8.680	8.754	8.829	8.905	8.982	9.061	9.142
78	9.223	9.305	9.389	9.474	9.561	9.649	9.739	9.831	9.924	10.02
79	10.12	10.21	10.31	10.41	10.52	10.62	10.73	10.84	10.95	11.06

80	11·18	11·30	11·42	11·54	11·67	11·80	11·93	12·06	12·20	12·34
81	12·48	12·63	12·78	12·93	13·08	13·24	13·40	13·57	13·74	13·92
82	14·10	14·28	14·47	14·66	14·86	15·07	15·28	15·49	15·71	15·94
83	16·17	16·41	16·66	16·91	17·17	17·44	17·72	18·01	18·31	18·61
84	18·93	19·25	19·59	19·94	20·30	20·68	21·07	21·47	21·89	22·32
85	22·77	23·24	23·73	24·24	24·78	25·34	25·92	26·52	27·16	27·83
86	28·53	29·27	30·04	30·86	31·73	32·64	33·60	34·63	35·72	36·88
87	38·11	39·43	40·84	42·36	44·00	45·76	47·68	49·76	52·02	54·50

Table 8 General Multiplicity Factors

System	Laue Group	h00	0k0	00l	0kl $(k\neq l)$ $\begin{matrix}0++\\0+-\\0-0\end{matrix}$	h0l $(h\neq l)$ $\begin{matrix}+0+\\+0-\\-0+\\-0-\end{matrix}$	hk0 $(h\neq k)$ $\begin{matrix}++0\\+-0\\-+0\\-+0\end{matrix}$	hkl $(h\neq k\neq l)$ $\begin{matrix}+++\\++-\\+-+\\+--\end{matrix}$	$\begin{matrix}-++\\-+-\\--+\\---\end{matrix}$
Triclinic	$\bar{1}$	2	2	2	2	2	2	2	
Monoclinic	2/m	2	2	2	2 2	2 2	2	2	2
Orthorhombic	mmm	2	2	2	4 4	4	4 4	4	4

	Laue Group	$\left\{\begin{matrix}h00\\0h0\end{matrix}\right\}$	00l	$\left.\begin{matrix}h0l\\0hl\end{matrix}\right\}(h\neq l)$	hk0	hh0 $(h\neq l)$	hhl $(h\neq l)$	hkl $(l\neq h \text{ or } k)$
Tetragonal	4/m	4	2	8	$\begin{matrix}hk0 & \bar{h}k0\\ \bar{h}k0 & hk0\\ kh0 & \overline{kh}0\\ \overline{kh}0 & kh0\end{matrix}$ 4	hh0 → 4	8	$\begin{matrix}hk & \bar{h}\bar{k}\\ \bar{h}k & hk\\ kh & \overline{kh}\\ k\bar{h} & kh\end{matrix}$ 8
	4/mmm	4	2	8	8	4	8	16

		Q h00	Q hh0	Q hk0 $\begin{matrix}h>k\\h<k\end{matrix}$	hhh	Q hhl $\begin{matrix}h>l\\h<l\end{matrix}$	Q hkl $\begin{matrix}h>k>l\\h<k<l\end{matrix}$
Cubic	m3	6	12	12 / 12 → 24	8	24 / 24	24 / 24
	m3m	6	12	24	8	24	48

This table gives, for the trigonal and hexagonal Laue groups, the reflection forms $Q \pm (pqr)$ and their multiplicities.

Top column groups (left to right):

- $Q \pm (pqr)$ — (Any one symbol may be zero and any two may be equal)
- Sign conditions: $q>p$ same sign / $q>2p$ opp. sign (p may be zero); $q<p$; $q\leqq p$ or $\bar p<q<2\bar p$ (r may be zero)
- $Q \pm (pqr)$ — (One symbol is mean of other two; any one symbol may be zero) $q \neq \bar r$
- $Q \pm (pqr)$ — (Any one symbol may be zero; $q \neq \bar r$ but indices are otherwise independent)

Trigonal (with rhombohedral lattice)

Laue Group	ppp	$\bar p q q$	$p0\bar p$	pqr	prq	(mid)	pqr	prq	pqr	prq	$p'q'r'$	$p'r'q'$
$\bar 3$	2	6	6	6	6	6	6	6	6	6	6	6
$\bar 3 m$	2	6	6	12		6	12		12		12	

Trigonal (with hexagonal lattice) ($l \equiv h$ or k or i)

Forms: $000l$ | $h0\bar h0$, $Q\pm(h0\bar h)$ | $h\bar h 2\bar h 0$, $Q\pm(h\bar h2h)$ | $hki0$, $Q\pm(hki)$ | $ikh0$, $Q\pm(ikh)$ | $\pm h0\bar hl$, $Q\pm(h0\bar h)$ ...

Laue Group	$000l$											
$\bar 3$	2	6	6	6	6	6	6	6	6	6	6	6
$\bar 3 m$	2	6	6	12		6	12		12		12	

Hexagonal ($l \neq h$ or k or i)

Forms: $000l$ | $h0\bar h0$, $Q\pm(h0\bar h)$ | $h\bar h2\bar h0$, $Q\pm(h\bar h2h)$ | $hki0$, $Q\pm(hki)$ | $ikh0$, $Q\pm(ikh)$ | $h0\bar hl$, $Q\pm(h0\bar h)\pm l$ | $h\bar h2\bar hl$, $Q\pm(h\bar h2h)\pm l$ | $hkil$, $Q\pm(hki)\pm l$ | $ikhl$, $Q\pm(ikh)\pm l$

Laue Group	$000l$								
$6/m$	2	6	6	6	6	12	12	12	12
$6/mmm$	2	6	6	12		12	12	24	

Notes: 1. The use of different symbols indicates non-equality of the corresponding indices, except where the symbol $=$ shows that the possible equality is without significance. The use of the same symbol in different positions indicates necessary equality. 2. The sign $\pm$ implies a complete change of sign in the symbol or the part thereof to which it refers. When no indication of sign is given, then all possible combinations have to be made. 3. The symbol Q indicates cyclic permutation of the indices within the following bracket. Except where indicated, no interchange of symbols is permissible. 4. Enclosure of two or more numbers in a rectangle indicates that the different forms so enclosed have the same spacing although they are not related by the symmetry of the Laue group.

Subject Index

References and Author Index

A

Alexander, L. E. (1950), *J. Appl. Phys.*, **21**, 126, see also Klug, H. P.   82
Arkel, A. E. van (1926), *Physica*, **6**, 64.   87
Arndt, U. W. and Riley, D. P. (1952), *Proc. Phys. Soc. (A)*, **65**, 74.   196
Arndt, U. W. and Willis, B. T. M. (1966), *Single Crystal Diffractometry* (Cambridge Univ. Press, London).   97
Averbach, B. L., see Warren, B. E.
Azároff, L. V. (1955), *Acta Cryst.*, **8**, 701.   71, 204
Azároff, L. V. (1968), *Elements of X-ray Crystallography* (McGraw-Hill, New York).   116, 185
Azároff, L. V. and Buerger, M. J. (1958). *The Powder Method in X-ray Crystallography* (McGraw-Hill, New York).   135

B

Bacon, G. E. (1962), *Neutron Diffraction* (Oxford Univ. Press, London).   112
Barrett, C. S. (1956), *Acta Cryst.*, **9**, 671.   103
Barrett, C. S. and Massalski, T. B. (1966), *Structure of Metals* (McGraw-Hill, New York).   183

Bearden, J. A. (1931), *Phys. Rev.*, **37**, 1210.   61
Bearden, J. A. (1967), *Rev. Mod. Phys.*, **39**, 78.   61, 62, 163
Bearden, J. A., Henins, A., Marzalf, J. G., Sauder, W. C. and Thomsen, J. S. (1964), *Phys. Rev. (A)*, **13**, 899.   62
Bearden, J. A., see also Spijkerman, J. J.
Berman, R., Brock, J. C. F. and Huntley, D. J. (1964), *Cryogenics*, **4**, 233.   102
Bernal, J. D. (1926), *Proc. R. Soc. (A)*, **113**, 117.   151
Bernal, J. D. and Crowfoot, D. M. (1934), *Annual Reports of the Chemical Society*, **30**, 379.   219
Berry, L. G. and Thompson, R. M. (1962), 'X-ray Powder Data for Ore Minerals,' *The Peacock Atlas* (The Geological Society of America Memoirs), 85.   274
Beu, K. E., Musil, F. J. and Whitney, D. R. (1963), *Acta Cryst.*, **16**, 1241.   173
Beu, K. E. and Scott, D. L. (1962), *Acta Cryst.*, **15**, 1301.   159
Bhuiya, A. K. and Stanley, E. (1963), *Acta Cryst.*, **16**, 981.   242
Bjurström, T. (1931), *Z. Phys.*, **69**, 346.   144, 147
Black, I. A., Bolz, L. H., Brooks, F. O., Mauer, F. A. and Peiser,

H. S. (1958), *J. Res. National Bureau of Standards*, **61**, 367.   103

Blum, W. and Hogaboom, G. B. (1949), *Principles of Electroplating and Electroforming* (McGraw-Hill, New York).   App. I, 297

Bogue, R. H. (1955), *The Chemistry of Portland Cement* (Reinhold, New York).   274

Boldyrev, A. K., Mikheiev, V. I., Kovalev, G. A. and Dubinina, V. N. (1939), *Ann. Inst. Mines. Leningrad*, **13**, 1.   266

Bolz, L. H., see Black, I. A.

Booth, A. D. (1947), *Nature*, **160**, 196.   240, 241

Booth, A. D. (1949), *Proc. R. Soc. (A)*, **197**, 336.   240, 241

Bozorth, R. M. and Haworth, F. E. (1938), *Phys. Rev.*, **53**, 538.   73

Bradley, A. J. (1924), *Phil. Mag.*, **47**, 657.   93

Bradley, A. J. (1935a), *Proc. Phys. Soc.*, **47**, 879.   185, 199

Bradley, A. J. (1935b), *Z. Krist.*, **91**, 302.   233

Bradley, A. J. and Jay, A. H. (1932a), *Proc. Phys. Soc.*, **44**, 563.   86, 159, 160

Bradley, A. J. and Jay, A. H. (1932b), *Proc. Roy. Soc. (A)*, **136**, 210.   159, 160

Bradley, A. J., Lipson, H. and Petch, N. J. (1941), *J. Sci. Inst.*, **18**, 216.   82, 93

Bradley, A. J. and Taylor, A. (1937), *Phil. Mag.* (7), **23**, 1049.   231

Bradley, A. J. and Thewlis, J. (1926), *Proc. R. Soc. (A)*, **112**, 678.   230

Bragg, W. L. (1933), *The Crystalline State, A General Survey* (Bell, London).   261

Bragg, W. L. (1947), *J. Sci. Inst.*, **24**, 27.   61, 183

Bragg, W. L. and Lipson, H. (1938), *Nature*, **141**, 367.   81

Brill, R. von and Tippe, A. (1967), *Acta Cryst.*, **23**, 343.   182

Brindley, G. W. (1940), *Proc. Phys. Soc.*, **52**, 117.   263

Brindley, G. W. (1945), *Phil. Mag.* (7), **36**, 347.   200

Brock, J. C. F., see Berman, R.

Brooks, F. P., see Black, I. A.

Buerger, M. J. (1942), *X-ray Crystallography* (Wiley, New York).   218

Buerger, M. J. (1956), *Elementary Crystallography* (Wiley, New York).   218

Buerger, M. J., see also Azároff, L. V.

Bunn, C. W. (1961), *Chemical Crystallography* (Oxford Univ. Press, London).   149, 229

Burgers, W. G., see Cauchois, Y.

Burke, J., see Cornish, A. J.

C

Cohn, R. W., see Otte, H. M.

Campbell, W. J., Stecura, S. and Grain, C. (1962), *Adv. in X-ray Analysis*, **5** (Plenum Press, New York).   98

Canning Handbook on Electroplating (1966), (W. Canning & Co. Ltd., Birmingham).   App. I, 297

Cauchois, Y., Tiedema, T. J. and Burgers, W. G. (1950), *Acta Cryst.*, **3**, 372.   73

Chesters, J. H. (1963), *Steel plant Refractories* (Percy Lund, Humphries, London).   274

Cisney, E., see Parrish, W.

Clark, G. L. (1955), *Applied X-rays* (McGraw-Hill, New York).   245

Cochran, W., see Lipson, H.

Cohen, E. R., see DuMond, J. W. M.

Cohen, M. U. (1935), *Rev. Sci. Inst.*, **6**, 68.   105, 164, 173, 176

Cohen, M. U. (1936), *Z. Krist.*, **94**, 288.   89, 157

Cornish, A. J. and Burke, J. (1965), *J. Sci. Inst.*, **42**, 212. 98

Cromer, D. T. (1965), *Acta Cryst.*, **18**, 17. 213

Crowfoot, D. M., see Bernal, J. D.

Cullity, B. D. (1956), *Elements of X-ray Diffraction* (Addison-Wesley, Massachusetts). 185

Cunningham, J., see Goldschmidt, H. J.

D

Dauben, C. H. and Templeton, D. H. (1955), *Acta Cryst.*, **8**, 841. 213

Davey, W. P., see Hull, A. W.

Debye, P. (1914), *Ann. Der Physik.*, **43**, 49. 214

Delaunay, B. (1933), *Z. Krist.*, **84**, 132. 135

Delf, B. W. (1963), *Brit. J. Appl. Phys.*, **14**, 345. 181

Douglas, A. M. B. (1950), *Acta Cryst.*, **3**, 19. 154

Dubinina, V. N., see Boldyrev, A. K.

DuMond, J. W. M. (1950), *Rev. Sci. Inst.*, **21**, 188. 73

DuMond, J. W. M. and Cohen, E. R. (1953), *Rev. Mod. Phys.*, **25**, 691. 61

E

Edlen, B. and Svensson, L. A. (1965), *Arkiv Fysik.*, **28**, 427. 61

Edwards, O. S. and Lipson, H. (1942), *Proc. R. Soc. (A)*, **180**, 268. 263

Edwards, O. S. and Lipson, H. (1943), *J. Inst. Metals.*, **69**, 177. 220

Ehrenberg, W. and Spear, W. (1951), *Proc. Phys. Soc. (B)*, **64**, 67. 77

Endter, F. (1960), *Dechema Mono.*, **38**, 21. 181

Ewald, P. P. (1921), *Z. Krist.*, **56**, 148. 48

D'Eye, R. W. M. and Wait, E. (1960), *X-ray Powder Photography in Organic Chemistry* (Butterworth, London). 87, 162

F

Farquhar, M. C. M., Lipson, H. and Weill, A. R. (1945), *J. Iron and Steel Inst.*, **152**, 457. 272

Fischmeister, H. F. (1961), *Acta Cryst.*, **14**, 113. 108

Foote, F. and Jette, E. R. (1940), *Phys. Rev.*, **58**, 81. 183

Francombe, M. H. (1957), *J. Phys. Chem. Solids.*, **3**, 37. 234

Frazer, B. C., see Keeling, R.

Frevel, L. K. (1944), *Journal Indust. and Eng. Chem. Anal. Edit.*, **16**, 209. 270

Frevel, L. K., see also Hanawalt, J. D.

G

Gay, P., Hirsch, P. B. and Kelly, A. (1954), *Acta Cryst.*, **7**, 41. 263

Gayler, M. V. and Preston, G. D. (1929), *J. Inst. Metals.*, **41**, 211. 105

Goldschmidt, H. J. (1964), *Bibliography on High Temperature X-ray Diffraction Techniques* (International Union of Crystallography). 98

Goldschmidt, H. J. and Cunningham, J. (1950), *J. Sci. Inst.*, **27**, 177. 98

Goldschmidt, H. J., see also Rait, J. R.

Goldsztaub, S. (1947), *C.R. Acad. Sci.* (Paris), **224**, 458. 77

Grain, C., see Campbell, W. J.

Green, A. T., see Rait, J. R.

Grigorovici, R., Mănăilă, R. and Vaipolin, A. A. (1968), *Acta Cryst. (B)*, **24**, 535. 233

Guinier, A. (1937), *C.R. Acad. Sci.* (Paris), **204**, 1115. 72

Jones, F. W. and Sykes, C. (1937), *Proc. R. Soc. (A)*, **161**, 440.   213

Jones, F. W. and Sykes, C. (1938), *Proc. R. Soc. (A)*, **166**, 376.   263

Jones, H. (1934), *Proc. R. Soc. (A)*, **144**, 225.   231

K

Kaelble, E. F. (1967), *Handbook of X-rays* (McGraw-Hill, New York).   96, 135, 185, 213

Karlsson, N., see Hägg, G.

Keeling, R., Frazer, B. C. and Pepinsky, R. (1953), *Rev. Sci. Instrum.*, **24**, 1087.   101

Kelly, A., see Gay, P.

Kempter, C. P., see Vogel, R. E.

Kennard, O., Martin, A. J. P. and Woodget, L. (1957), *J. Sci. Instr.*, **36**, 48.   78

Keplin, E. J., see Houska, C. R.

Klug, H. P. and Alexander, L. E. (1954), *X-ray Diffraction Procedures for Polycrystalline and Amorphous Materials* (Wiley, New York).   82, 97

Kovalev, G. A., see Boldyrev, A. K.

Kuentzel, L. E. (1951), *Codes and Instructions for Wyandotte Punched Cards Indexing X-ray Diffraction Data* (Wyandotte Chem. Corp., Wyandotte, Michigan).   269

L

Lang, A. R. (1951), *Nature*, **168**, 907.   196

Lang, A. R. (1956), *J. Sci. Instr.*, **33**, 96.   195

Laue, M. von (1926), *Z. Krist.*, **64**, 115.   246, 247, 249, 261

Laves, F. (1933), *Z. Krist.*, **84**, 256.   233

Lawson, A. W., see Jamieson, J. C.

Lester, K. M. and Lipson, H. (1970), *J. App. Cryst.* (in press).   156

Lipson, H. (1942), *J. Sci. Instrum.*, **19**, 63.   87

Lipson, H. (1949), *Acta Cryst.*, **2**, 43.   130

Lipson, H. and Cochran, W. (1966), *The Crystalline State*, Vol. III, *The Determination of Crystal Structures* (Bell, London).   216, 222, 224, 229, 230

Lipson, H. and Parker, A. M. B. (1944), *J. Iron and Steel Inst.*, **149**, 123.   109

Lipson, H. and Petch, N. J. (1940), *J. Iron and Steel Inst.*, **142**, 95.   232

Lipson, H. and Rogers, L. E. R. (1944), *Phil. Mag.* (7), **35**, 544.   89

Lipson, H., Shoenberg, D. and Stupart, G. V. (1941), *J. Inst. Metals*, **67**, 333.   108

Lipson, H. and Steeple, H. (1951), *Nature*, **167**, 481.   154

Lipson, H. and Taylor, C. A. (1958), *Fourier Transforms and X-ray Diffraction* (Bell, London).   151

Lipson, H. and Wilson, A. J. C. (1941), *J. Sci. Instrum.*, **18**, 144.   160

Lipson, H., see also Bradley, A. J., Bragg, W. L., Edwards, O. S., Farquhar, M. C. M., Henry, N. F. M., Stokes, A. R.

Lonsdale, K. (1936), *Structure Factor Tables* (Bell, London).   215, 216

Lonsdale, K. (1950), *Acta Cryst.*, **3**, 162.   216

Lonsdale, K. and Smith, H. (1941), *J. Sci. Instrum.*, **18**, 133.   99

Lytle, F. W. (1964), *Adv. in X-ray Analysis*, **7** (Plenum Press, New York).   103

M

Mack, M., see Parrish, W., Taylor, J.

Mănăilă, R., see Grigorovici, R.

Martin, A. J. P., see Kennard, O.

Marzolf, J. G., see Bearden, J. A.

Massalski, T. B., see Barrett, C. S.

Matthews, F. W. (1949), *Anal. Chem.*, **21**, 1172.   269

Matthews, F. W. (1962), *Materials, Res. and Standards*, **2**, 643.   269

Mauer, F. A., see Black, I. A.

McWhan, D. B. (1967), *J. Appl. Phys.*, **38**, 347.   104

Meyerhof, W. E., see West, H. I.

Mikheiev, V. I., see Boldyrev, A. K.

Musil, F. J., see Beu, K. E.

N

Nelson, J. B. and Riley, D. P. (1945), *Proc. Phys. Soc.*, **57**, 160.   160

Novak, C. (1954), *Chechoslov. J. Phys.*, **4**, 496.   138, 139

O

Ogilvie, R. E. (1963), *Rev. Sci. Instrum.*, **34**, 1344.   94

Otte, H. M. and Cahn, R. W. (1959), *J. Sci. Instrum.*, **36**, 463.   110

P

Parker, A. M. B., see Lipson, H.

Parrish, W. (1960), *Acta Cryst.*, **13**, 838.   163

Parrish, W. (1962), *Advances in X-ray Diffractometry and X-ray Spectroscopy* (Centrex Publishing Co., Eindhoven).   97, 178

Parrish, W. and Cisney, E. (1948), *Philips Tech. Rev.*, **10**, 157.   93

Parrish, W., Taylor, J. and Mack, M. (1964), *Advances in X-ray Analysis*, **7** (Plenum Press, New York).   180

Parrish, W. and Wilson, A. J. C. (1959), *International Tables for X-ray Crystallography*, **2** (The Kynoch Press, Birmingham).   96

Parrish, W., see also Taylor, J.

Pascoe, K. J., see Stokes, A. R.

Peiser, H. S., Rooksby, H. P. and Wilson, A. J. C. (1955) Eds., *X-ray Diffraction by Polycrystalline Materials* (Institute of Physics, London).   89, 97, 171, 199, 227, 246, 251

Peiser, H. S., see also Black, I. A.

Pepinsky, R., see Keeling, R.

Petch, N. J., see Bradley, A. J., Lipson, H.

Phillips, F. C. (1963), *An introduction to Crystallography* (Longmans, Green, London).   19

Pike, E. R. (1957), *J. Sci. Instrum.*, **34**, 355.   179

Pitts, E. and Willets, F. W. (1961), *Acta Cryst.*, **14**, 1302.   260

Pollock, A. R., see Thewlis, J.

Preston, G. D., see Gayler, M. V.

Q

Qurashi, M. M. (1949), *Acta Cryst.*, **2**, 404.   242

R

Rachinger, W. A. (1948), *J. Sci. Instrum.*, **25**, 254.   256

Rait, J. R., Green, A. T. and Goldschmidt, H. J. (1942), *Second Report on Refractory Minerals* (Iron & Steel Inst., London).   274

Ramachandran, G. N. (1964), *Advanced Methods of Crystallography* (Academic Press, London).   213

Riley, D. P., see Arndt, U. W., Nelson, J. B.

Rinn, H. W., see Hanawalt, J. D.

Robertson, J. H. (1960), *J. Sci. Instrum.*, **37**, 41.   100

Rogers, L. E. R., see Lipson, H.
Rooksby, H. P. and Willis, B. T. M. (1953), *Nature*, **172**, 1054.   234
Rooksby, H. P., see also Peiser, H. S.

S

Sauder, W. C., see Bearden, J. A.
Scherrer, P. (1920), Article in Book. *Kolloidchemie* by Zsigmondy, A. R. (Leipzig).   261
Schieltz, N. C. (1964), *Advances in X-ray Analysis, Vol. 7* (Plenum Press, New York).   151
Schoening, F. R. L. (1965), *Acta Cryst.*, **18**, 975.   259, 263
Scott, D. L., see Beu, K. E.
Seljakow, N. (1925), *Z. Phys.*, **31**, 439.   261
Shoenberg, D., see Lipson, H.
Siegbahn, M. (1925), *The Spectroscopy of X-rays* (Oxford Univ. Press, London).   60
Sinclair, H., see Taylor, A.
Smith, C. S. and Stickley, E. E. (1943), *Phys. Rev.*, **64**, 191.   256, 263
Smith, H., see Lonsdale, K.
Spear, W., see Ehrenberg, W.
Spijkerman, J. J. and Bearden, J. A. (1964), *Phys. Rev. (A)*, **134**, 871.   62
Stanley, E., see Bhuiya, A. K.
Stecura, S., see Campbell, W. J.
Steeple, H. (1952), *Acta Cryst.*, **5**, 247.   226, 230
Steeple, H., see also Hughes, W., Lipson, H. and Thewlis, J.
Stickley, E. E., see Smith, C. S.
Stockdale, D. (1940), *J. Inst. Metals*, **66**, 287.   183
Stokes, A. R. (1948), *Proc. Phys. Soc.*, **61**, 382.   251, 256
Stokes, A. R., Pascoe, K. J. and Lipson, H. (1943), *Nature*, **151**, 137.   263
Stokes, A. R. and Wilson, A. J. C. (1942), *Proc. Camb. Phil. Soc.*, **38**, 313.   261

Stokes, A. R. and Wilson, A. J. C. (1944), *Proc. Phys. Soc.*, **56**, 174.   250
Straumanis, M. E. and Ieviņš, A. (1940), *Springer* (Berlin). In 1959 pp. 1–50 and 97–99 were translated by Beu, K. E. under the title 'The Precision Determination of Lattice Constants by the Asymmetric Method'. (Obtainable from Office of Technical Services, Dept. of Commerce, Washington 25, D.C., U.S.A.)   89
Straumanis, M., see also Ieviņš, A.
Stupart, G. V., see Lipson, H.
Svensson, L. A., see Edlen, B.
Sykes, C., see Jones, F. W.
Symposium on the Chemistry of Cements (1938), (Ingeniörsvetenskapsakademien, Stockholm).   274

T

Taylor, A. (1944a), *Phil. Mag. (7)*, **35**, 215.   200
Taylor, A. (1944b), *Phil. Mag. (7)*, **35**, 640.   200
Taylor, A. (1945), *Introduction to X-ray Metallography* (Chapman & Hall, London).   199
Taylor, A. and Sinclair, H. (1945a), *Proc. Phys. Soc.*, **57**, 108.   91
Taylor, A. and Sinclair, H. (1945b), *Proc. Phys. Soc.*, **57**, 126.   160
Taylor, A., see also Bradley, A. J.
Taylor, C. A., see Hughes, W., Lipson, H.
Taylor, J., Mack, M. and Parrish, W. (1964), *Acta Cryst.*, **17**, 1229.   178, 181
Taylor, J., see also Parrish, W.
Taylor, W. H., see Hargreaves, A.
Templeton, D. H., see Dauben, C. H.